SUCCESS STORIES IN PRODUCTIVITY IMPROVEMENT

D1472943

SUCCESS STORIES IN PRODUCTIVITY IMPROVEMENT

Edited by Jerry L. Hamlin

Published by
Industrial Engineering and Management Press
Institute of Industrial Engineers

Industrial Engineering & Management Press, 25 Technology Park/Atlanta, Norcross, Georgia 30092

Published in 1985
Printed in the United States of America

ISBN 0-89806-071-0

CONTENTS

IV. COMPANY PRODUCTIVITY PROGRAM APPROACHES

PREFACE

Success Stories in Productivity Improvement is a collection of recent papers by a wide variety of authors. The papers included provide a range of approaches to productivity management. The intent of this book is to present various methods for productivity improvement in the hope that readers will be able to apply them to their organizations.

I am fortunate to know some of the writers and to have heard many of their presentations. I have long believed that the keys to improving our country's productivity performance are the individual efforts of organizations and the employees that make up the U.S. economy. If we all pull together, it will be possible to once again attain the productivity improvement levels reached during the 1960s and '70s. This book is an attempt to aid in this effort by sharing some successful productivity improvement efforts.

The book is divided into four sections. Each section is an important component of a well-organized productivity improvement program. The sections, and the papers in them, are arranged in a logical sequence.

Section I, *Productivity Awareness and Education,* includes six papers about productivity and its importance. The first step in a productivity program is to create general awareness about productivity, then to educate employees about the benefits of productivity improvement.

Section II, *Productivity Measurement and Analysis,* deals with several approaches that measure and analyze productivity at various levels within the organization. The twelve papers in Section II illustrate techniques for approaching both direct and indirect labor, as well as the overall department, division, or company. Productivity measurement is an important means of establishing present status, determining weak areas, and determining levels of improvement once a program has been initiated.

Section III, *Productivity Improvement Techniques and Applications,* has eight papers that share some recent techniques that improved productivity. Sharing of this type of non-proprietary information should be done on a larger scale in this country. Thus, some industries could become more competitive in the international marketplace. Hopefully, some of the techniques presented here will benefit readers.

Section IV, *Company Productivity Program Approaches,* consists of thirteen papers illustrating how several major companies implemented formalized productivity improvement programs. Among them are the American Express Company, Xerox, Wang Laboratories, Inc., and Public Service Company of New Mexico. Other companies have productivity improvement programs, and hopefully many others will soon follow.

Jerry L. Hamlin
Tulsa, Oklahoma
1985

I. PRODUCTIVITY AWARENESS AND EDUCATION

Awareness of productivity in an organization can be created by first explaining and defining what is meant by the term "productivity." The next step is to emphasize its importance so that both management and employees can understand and appreciate it.

*Reprinted from Industrial Engineering,
January 1985.*

Results Of IIE's Fourth Annual Productivity Survey:

IEs Evaluate Productivity Improvement Efforts In Own Organizations And Across U.S.

By Lane Gardner Camp
IIE Communications Coordinator

"**P**roductivity" is the buzz word of the eighties. If you want to get people's attention, tell them how you can increase their output while decreasing their input. Tell them how to produce more efficiently. Guarantee them a future of progress and comfortable living.

Ways to improve productivity are numerous, and all are important to the total picture. All workers at all levels can make contributions, but what is the consensus? What are the best productivity improvement strategies?

Some answers to these questions may be found in the results of IIE's fourth Productivity Opinion Survey, a survey designed to determine industrial engineers' views on current productivity thinking.

Industrial engineers are the perfect source for this information because of their day-to-day concern for and involvement with people, material, equipment and energy. IEs draw upon their unique training to design and evaluate systems, and thereby improve productivity.

Productivity campaign

The survey is conducted in conjunction with IIE's annual Productivity Improvement Campaign, a public service effort to raise the public's awareness of productivity.

The month-long campaign, now in its sixth year, will utilize the theme "Productivity and Quality Work Together," along with a red, white and blue logo. IIE members will spread the campaign's message using bulletin board posters, leaflets, proclamations, public presentations, displays and media announcements.

To kick off the January campaign, a U.S. Senate resolution has been obtained naming January 7-13 National Productivity Improvement Week. Governors and local government officials are also expected to support the campaign as they have in the past with proclamations designating Productivity Improvement Week or Month.

Previous campaigns have been held in October. The change to January, which eliminated a 1984 campaign, was made to coordinate activities with the Broadcasting Industry Council to Improve American Productivity.

BICIAP, an arm of the National Association of Broadcasters, was created to implement a five-year radio and television productivity awareness campaign. Television journalist Howard K. Smith is the official spokesperson.

The change in time of IIE's campaign is expected to improve publicity efforts because broadcasters have a greater amount of public service time available in January.

1984 survey

Survey questionnaires were mailed out in early September, 1984, to a random selection of IIE's U.S. non-student members. Recipients were given approximately eight weeks to respond. Of the 2,500 questionnaires mailed, 765 were returned, giving a response rate of 30.6%.

Biographical information revealed that most of the respondents (69%) work in manufacturing, and present principal job function is divided primarily between manager (35%) and staff engineer (40.7%). The largest number of survey participants (32.1%) work for firms with 300-999 employees. This is followed by 21.1% in firms with 1,000-2,999 employees and 12.5% in firms with 100-299 employees.

Attitudes and work ethic

Probably one of the biggest factors in productivity improvement is workers' attitudes. Do workers avoid absenteeism and tardiness? Are they

Table 1: Formal Incentive Programs

	Base	Undertaken	Effective
Wage incentive plans	664	30.3% (201)	74.4%
Gain sharing plans:			
Improshare	422	3.8% (16)	42.9%
Rucker	402	.5% (2)	26.3%
Scanlon	405	1.5% (6)	31.8%
Other	348	16.1% (56)	78.2%

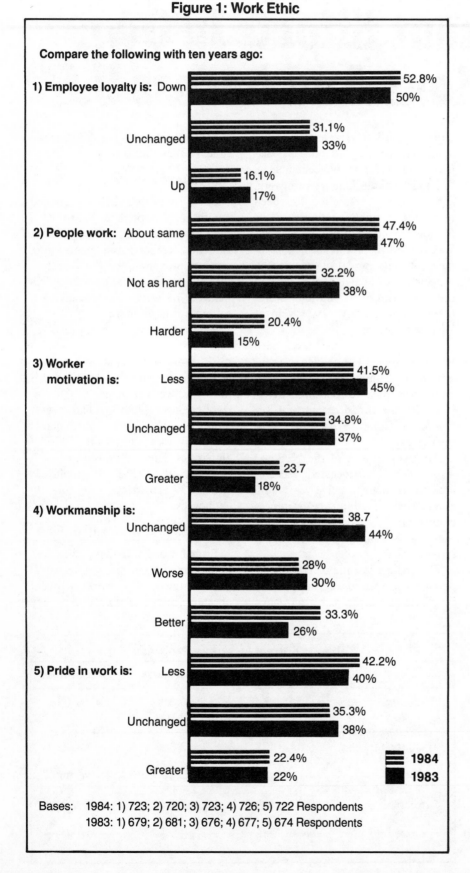

Figure 1: Work Ethic

Compare the following with ten years ago:

1) Employee loyalty is: Down — 52.8% / 50%

Unchanged — 31.1% / 33%

Up — 16.1% / 17%

2) People work: About same — 47.4% / 47%

Not as hard — 32.2% / 38%

Harder — 20.4% / 15%

3) Worker motivation is: Less — 41.5% / 45%

Unchanged — 34.8% / 37%

Greater — 23.7 / 18%

4) Workmanship is: Unchanged — 38.7 / 44%

Worse — 28% / 30%

Better — 33.3% / 26%

5) Pride in work is: Less — 42.2% / 40%

Unchanged — 35.3% / 38%

Greater — 22.4% / 22%

■ 1984
■ 1983

Bases: 1984: 1) 723; 2) 720; 3) 723; 4) 726; 5) 722 Respondents
1983: 1) 679; 2) 681; 3) 676; 4) 677; 5) 674 Respondents

accountable for achievement of goals? Are they responsible? Do they take the initiative to solve problems? Do they openly make suggestions about ways to control costs or improve product quality? Do they trust their superiors, subordinates and peers?

"Yes" answers to these questions would indicate enthusiastic and optimistic worker attitudes. Yet, when survey participants were asked the attitude of most of those who work at their locations, only a little over half (56.8%) said "enthusiastic and optimistic," while 43.2% said "not very enthusiastic and optimistic." However, the split in 1981 was 53% and 47%, respectively, which indicates that enthusiasm and optimism are up in 1984.

Another survey question asked industrial engineers to compare today's work ethic with that of ten years ago. Five areas were pinpointed: effort, pride, loyalty, workmanship and worker motivation. (See Figure 1.)

The most positive response was in the area of workmanship, in which 33.3% said it was better than ten years ago, 28% said it was worse and 38.7% said it was unchanged.

Pride, loyalty and motivation all received unfavorable comparisons: 42.2% said pride in work is less; 52.8% said loyalty is down; and 41.5% said motivation is less. The majority of respondents (47.4%) rated work effort about the same.

Motivation and incentives

As for what employers can do about workers' attitudes and performance, industrial engineers believe personal recognition offers the greatest encouragement. (See Figure 2.) When asked what is the single most effective way to encourage people to generate ideas to improve performance, 55.2% of the respondents said "personal recognition," followed by "money reward" (29.2%), "promo-

Figure 2: Motivating People

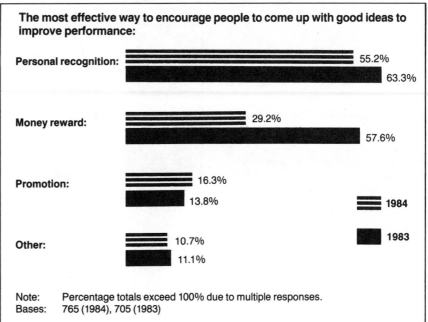

The most effective way to encourage people to come up with good ideas to improve performance:

Personal recognition: 55.2% / 63.3%

Money reward: 29.2% / 57.6%

Promotion: 16.3% / 13.8%

Other: 10.7% / 11.1%

1984
1983

Note: Percentage totals exceed 100% due to multiple responses.
Bases: 765 (1984), 705 (1983)

tion" (16.3%) and "other" (10.7%). This varies somewhat from one year ago when money reward was cited by a much higher 57.6% of the respondents.

Possibly the absence of enthusiasm and optimism and the stagnant work ethic as perceived by industrial engineers can be related to the fact that only 36.7% of respondents' facilities have formal incentive programs.

Most programs fell into the wage incentive category (71.5%) as opposed to the gain sharing category (28.6%). There was also a greater degree of satisfaction with the wage incentive plans. Of those who had implemented such plans, 74.4% felt the expense was justified. (See Table 1.)

Compare this to the small numbers who had even implemented gain sharing plans and the relatively low satisfaction levels. Of those who had implemented Improshare, Rucker and Scanlon, only 42.9%, 26.3% and 31.8%, respectively, felt the expense was justified.

When survey participants were asked what they consider the single most important benefit which results from a formal incentive program, 55.5% said "increased worker efficiency," 18.8% said "improvement of quality of output" and 12.3% indicated "increased company loyalty." "Other" was marked by 13.4%, most of whom defined it as a combination of the other three answers.

Besides being asked to evaluate the attitudes and incentives of workers in general, IEs were asked to compare their current role to when they first became associated with their present employer. Just over half (50.8%) said they were more influential in management decisions. Fourteen percent said they were less influential and 35.2% indicated their influence was about the same. These percentages have not changed significantly over the years since 1981.

Again, this is despite the fact that,

Table 2: Productivity Improvement Activities

		Base	Undertaken Activity? Yes (%)	Effective? Yes (%)
1)	Formal employee involvement in productivity improvement planning and evaluation (quality circles, suggestion programs, etc.)	719	68.7	63.1
2)	Evaluating performance and establishing specific productivity improvement targets	719	63.3	66.7
3)	Introduction or improvement of inventory control methods	708	64.4	70.8
4)	Capital investment for new or automated machinery (not including robotics)	718	75.1	85.7
5)	Introduction or expansion of use of robotics	691	25.9	44.1
6)	Introduction or improvement of quality control methods, etc.	707	66.2	73.0
7)	Systems innovations (integrated factories, advanced material handling techniques, computerized manufacturing methods, etc.)	705	45.1	70.7
8)	Improvement of quality of product through worker training	705	48.7	73.5
9)	Development of indirect labor standards and controls	698	30.9	54.1
10)	Other	46	76.1	94.3

generally speaking, 91.4% of industrial engineers feel they are best qualified for the lead role in developing sound organization productivity programs.

Why? A strong 80.9% said it's due to IEs' broader perspective of workflow production and systems control. Experience and training in management techniques beyond that of other engineering disciplines was noted by 65.4%, and 44.3% cited their knowledge of the relationship of human resources to technology.

One question gave respondents the opportunity to identify particular productivity improvement efforts undertaken at their firms and their degree of effectiveness. (See Table 2.) Capital investment for new or automated machinery (not including robotics) was most often reported (75.1%), followed by employee involvement in productivity improvement planning and evaluation (68.7%), introduction or improvement of quality control/methods (66.2%), introduction or improvement of inventory control methods (64.4%) and evaluating performance/establishing targets (63.3%). Worth noting is that the majority of IEs who saw these programs implemented declared them effective.

Less than half of all respondents have implemented such programs as worker training (48.7%), systems innovations (45.1%), indirect labor standards and controls (30.9%; down from 42% in 1982) and introduction or expansion of use of robotics (25.9%). The majority said these programs were effective with the exception of robotics, which was considered effective by only 44.1%.

Survey respondents were also given the opportunity to provide a written synopsis of activities they felt had had a high level of effectiveness. (See box for some typical answers.)

Why are some of these programs proving ineffective? From a list of six possible obstacles to productivity

improvement, more than half of IEs cited five of the six as "major" obstacles (versus "minor or no obstacle"). (See Figure 3.)

With the exception of "high interest rates," these numbers do not vary much from 1983. The extent to which "high interest rates squeezing investment of sufficient capital" is a major obstacle did tumble from 41.3% to 32%.

Finally, when asked their opinion on what the U.S. productivity ranking would be ten years from now, IEs were more positive than in previous years. (See Figure 4.) Compared to 20% in 1981, 32.6% said the U.S. would rank higher than all other industrial nations in ten years. The number of respondents to the 1984 survey naming other countries that would lead the U.S. dropped from

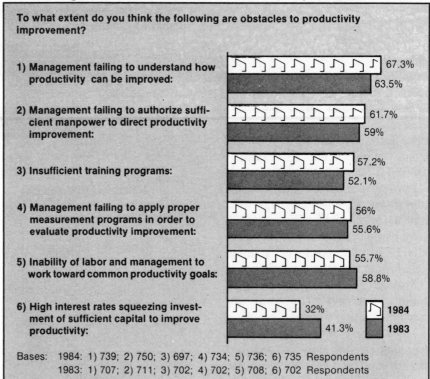

Figure 3: Obstacles to Productivity Improvement

To what extent do you think the following are obstacles to productivity improvement?

1) Management failing to understand how productivity can be improved: 67.3% / 63.5%

2) Management failing to authorize sufficient manpower to direct productivity improvement: 61.7% / 59%

3) Insufficient training programs: 57.2% / 52.1%

4) Management failing to apply proper measurement programs in order to evaluate productivity improvement: 56% / 55.6%

5) Inability of labor and management to work toward common productivity goals: 55.7% / 58.8%

6) High interest rates squeezing investment of sufficient capital to improve productivity: 32% / 41.3%

1984 / 1983

Bases: 1984: 1) 739; 2) 750; 3) 697; 4) 734; 5) 736; 6) 735 Respondents
1983: 1) 707; 2) 711; 3) 702; 4) 702; 5) 708; 6) 702 Respondents

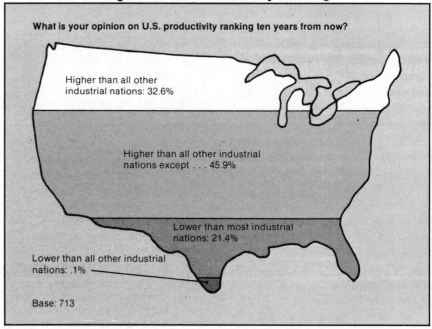

Figure 4: U.S. Productivity Ranking

What is your opinion on U.S. productivity ranking ten years from now?

Higher than all other industrial nations: 32.6%

Higher than all other industrial nations except . . . 45.9%

Lower than most industrial nations: 21.4%

Lower than all other industrial nations: .1%

Base: 713

55% in 1981 to 45.9%.

Among the nations listed as having higher productivity rankings were Japan (92.3%), West Germany (31.6%), Korea (11.6%), Taiwan (6.1%) and China (4.9%). This shows a distinct view by industrial engineers that the center for world productivity strides will be in Asia and not Europe. Countries named in 1981 were Japan (46%), West Germany (29%), France (7.7%) and Sweden (2%).

It's obvious that no survey can accurately pinpoint all the problems and solutions. Each industry and business has its own complex variables. This report is intended only as an overview. The hope is that readers will be inspired to look hard at their own productivity situations and take steps toward improvement. **IE**

An illustrated brochure of the survey findings is available, as is a separate publication containing descriptions of productivity improvement programs reported by IEs surveyed. Single copies of both may be obtained free of charge by writing:
Public Relations/IIE
25 Technology Park/Atlanta
Norcross, GA 30092
or by circling No. 150 on the reader service card.

Productivity Program Results

Following are descriptions of productivity improvement programs as submitted by respondents to the 1984 Productivity Opinion Survey of industrial engineers. Survey participants were asked to describe and document in 100 words or less exceptional productivity activities at the facilities where they were employed.

☐ "We implemented warehouse productivity and inventory control. We used extensive software procedures in the stocking and issuing of electrical and fabricated components. Because of these procedures, it is unnecessary to do the annual wall-to-wall inventory or cycle count. The operation is a 50,000-part warehouse. We saw a 50% increase in inventory accuracy and 20% decrease in labor costs." *(Manufacturing organization, 300-999 employees)*

☐ "Quality circle results are hard to measure, but individual team projects have yielded $5M to $100M in savings. In one case, a machine shop QC team tried to work up a justification for a new boring mill. They had wanted one for a long time, but management had been turning down their request for over five years. After the QC team did their study, they convinced themselves that management was indeed right. The new mill was a 'nice to have,' but not really a 'need to have'." *(Manufacturing organization, 1,000-2,900 employees)*

☐ "Meter repair work management and repair standards have doubled capacity of repair department. Meter reading and collection standards have reduced work force 10% and increased productivity 10%. Meter service work (periodic changing, turn-ons, turn-offs) has reduced job time 40%." *(Service organization, 300-999 employees)*

☐ "During the period 1981-86 approximately $700 million will be spent on new optical character readers and bar code sorters to automate the very costly mail processing operations. The new equipment will be able to sort letters at 10,000 pieces per hour, compared to the present manual and mechanized methods which range from 900 to 1,800 pieces per hour. This new automation, in conjunction with address improvement and the ZIP + 4 code, is projected to save over 19,000 work years and over $900 million per year." *(Governmental organization, 3,000-4,999 employees)*

☐ "We implemented a 'quick die change system.' Two 75-ton punch presses previously took about one hour to change over from one part to the next (die set operations). We installed a die handling system, hydraulic die clamps, standardized tooling and other mechanical improvements that have brought the die change time down to ten minutes on each press. Other presses are being worked on now with hopefully similar results." *(Manufacturing organization, 3,000-4,999 employees)*

☐ "We implemented an 'automated daily time reporting system.' The field personnel (linemen, splicers, etc.) are required to fill out daily time sheets to record their work activities and provide the proper accounting. A new automated system has been developed which identifies indirect labor (i.e. travel time, preparation at headquarters) and delays. This provides a more complete picture of the day's work and identifies obstacles to productivity. The amount of savings estimated in improved work scheduling is approximately $400,000 per year." *(Service organization, 300-999 employees)*

☐ "We started advanced automatic testing of assembled printed circuit boards. The decision was made to invest capital to lower test times on some families of printed circuit boards. Conventional automated test cycle time ran from two to two-and-a-half hours per board. New equipment made possible testing the same boards in about 20 minutes. More defects were detected due to more comprehensive screening of boards. Therefore, quality also increased." *(Manufacturing organization, 1,000-2,999 employees)*

☐ "Five plant operations, two product lines and the general office of this division have been consolidated at a new single location. We are in the start of a learning curve period and struggling with new processes, people, products and plant operation. The efficiency of locating these operations at one location and then segregating processes by type has been unbelievably successful. The work is being done with a third of the supervisory people in half the space and with less time." *(Manufacturing organization, 300-999 employees)*

*Reprinted from **Industrial Engineering,** January 1984.*

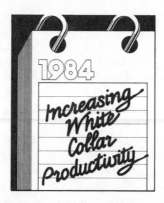

IEs Should Examine Impact Of Management Style On Motivation And Productivity

By Linda J. M. Root, P.E.
U.S. Air Force

Low productivity is merely a symptom of a problem. Management may be a major contributor to the problem itself. As Arnold S. Tannebaum wrote in *Social Psychology of the Work Organization,* "Supervisory style is both a cause and an effect of worker performance."

Typical responses to low productivity are increasing employee training and purchasing more equipment. We routinely expect optimum output from our employees just as we do with machinery. All too often we overlook how management programs and policies are perceived by workers and the negative impact they can have on productivity.

Let's address the problem in terms of decreasing nonproductive work. We as managers may be doing something that is contributing to a motivational problem. How do we eliminate the need to reaccomplish tasks and get a high quality product the first time? How do we get employees interested in job assignments? How do we motivate people to eliminate parochial viewpoints that limit perception of solutions to problems?

Training and equipment are the basic building blocks of any successful operation. However, regardless of how many dollars management spends to provide the latest equipment, if the workers don't know how to use it, wasted resources are the result. Likewise, if you have people trained in the latest equipment and technologies, but don't provide them the opportunity to use their training, what have you gained?

Worker is key

Let us examine the "soul" of the problem: the individual—the worker—the person—who listens to direction, interprets the guidance/policy and tries to provide the boss with the product requested. The individual's perception of the work environment and of his or her role in the company's goals can directly affect productivity. Review your operation and see if you cause, or are victim to, any of the following practices:

☐ Assignments include step-by-step instructions of how they are to be accomplished.
☐ Work assignments are always limited to routine subjects. Potential new areas are never explored.
☐ Choice assignments are given to a select few.
☐ Boss never supports the subordinate if higher management challenges the final product.
☐ Rather than reviewing the product with the worker, the boss redoes it himself or herself every time.
☐ System of recognition and rewards is inequitable.

Negative impacts

Do any of these practices sound familiar? More than likely they sound all too familiar. What impact can these practices have on the productivity of your workers? They can:

☐ Inhibit imagination and approaches to resolution of assignment problems.
☐ Trigger a "Why does he treat me like an idiot?" attitude.
☐ Fail to extend employees beyond their normal working assignments so that potential growth and revitalization of employees' interests are left untapped.
☐ Make employees hesitate to accept responsibility and job assign-

ments, knowing the opportunity to get a fair share of good job assignments is nil.

□ Make workers reluctant to take on controversial assignments, knowing the boss won't support them when the "going gets rough."

□ Lower the quality of the employee's final product by encouraging the attitude that "no matter what I turn in, it's going to be redone, so what's the point in expending the effort the first time?"

□ Cause resentment and job dissatisfaction because of supervisor failure to recognize a job well done or tendency to bestow recognition only upon a select few.

Discouraged employees seldom do quality work or use their own initiative. Therefore, expended efforts are usually wasted. Disharmony breeds discontent in others. The philosophy "lead, follow or get out of the way" is fine in theory, but in reality, people will voice their dissent and may plant the seeds of discontent in others.

At the same time, discouraged workers are likely to be looking for employment elsewhere. Their attention is on the want ads, not their work assignments. Rapid turnover of employees requires frequent orientation of new employees and results in losses of productive time.

In extreme cases, a disgruntled employee may be so upset with higher management that he or she will provide erroneous products or incorrect data to embarrass and possibly eliminate the source of dissatisfaction.

Manage for productivity

What can be done to eliminate management's contribution to nonproductive time?

□ Review management practices and list the various ways policies or practices could be interpreted by employees. Possibly hold a "rap session." If any of the previous symptoms are identified, take steps to eliminate them.

□ Encourage employee participation in decision making when practical. Numerous programs, when implemented under the true conditions for which they were designed—such as management by objectives (MBO) and quality circles—have proven to be enhancers of productivity. As Tannebaum states:

"Hierarchy is divisive; it creates resentment, hostility and opposition. Participation reduces disaffection and increases the identification of members with the organization. Individuals are more likely to feel some sense of commitment and responsibility relative to tasks that are brought before them in their capacity as decision makers."

□ Delegate assignments, responsibility *and credit* for tasks to the worker. Let their expertise be used in solving problems. If it is an area not previously explored, don't discourage—encourage.

□ Be fair to employees. If work is not satisfactory, tell them so and why. What can they do to improve work quality? Ensure that awards and recognition are bestowed *only* on those who deserve it, and on *all* who truly deserve it. A pat on the back is free, but does wonders.

The worker is the key to productivity, and obstacles in his or her path, no matter what the source, must be identified and eliminated. The nontangible factors in the worker's environment may be the key to increased productivity.

For further reading:

Tannenbaum, Arnold S., *Social Psychology of the Work Organization,* Wadsworth Publishing Co. Inc., Belmont, CA, 1966.

Barnard, Chester I., *The Functions of the Executive,* Harvard University Press, Cambridge, MA, 1968.

Linda J. M. Root, P.E., is an industrial engineer with the United States Air Force. She has been the industrial engineer at Maxwell AFB, AL; Osan Air Base, Korea; and Bitburg Air Base, Germany. At present she is assigned to the HQ Military Airlift Command, Engineering and Services DCS, and provides management consultant type services to 16 Air Force bases. She obtained her BS degree in industrial engineering and operations research from Virginia Polytechnic Institute and State University. Her master's degree in business administration is from Auburn University. She is a member of IIE and NSPE and a registered professional engineer in Pennsylvania.

Reprinted from **Industrial Engineering**, *January 1984.*

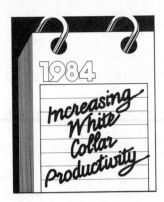

Awareness, Analysis And Improvement Are Keys To White Collar Productivity

By Clarence Smith
North Carolina State University

Creating, processing and distributing information are the primary functions of the white collar worker. This brain intensive sector is becoming increasingly vital as we shift from an industrial to an information society. One estimate by information specialist Marc Porat, in the current bestseller *Megatrends,* indicates that 46% of the United States GNP comes from this sector.

During 1970-78, while the total labor force increased by 18%, the number of managers and administrators rose 58%, health administrators were up 118%, the number of systems analysts increased by 84% and a 76% increase occurred in the number of public officials.

The amount of scientific and technical information is doubling every 5.5 years. Our offices, and our white collar workers, must become more productive to meet these demands.

Even though the above statistics indicate a number of significant problems that the industrial engineering profession will have to resolve in the years ahead, research and the development of new techniques are already providing dramatic productivity gains for numerous organizations. Some of the tools and procedures are similar to those used in blue collar productivity efforts, while others are new and innovative approaches to the problems of low productivity.

The three vital components of successful productivity improvement efforts are awareness, analysis/measurement and improvement. This article will focus on each of these as integral parts of a systematic strategy for increasing white collar productivity.

Awareness

Before the productivity of an office work group can be improved for any long-term sustained period, a basic understanding of the concept of productivity has to be implanted into the day-to-day vocabulary of each worker. Although productivity is considered a fairly well defined term, a number of definitions are necessary for the variety of office activities to clearly appreciate how numerous factors affect the productivity of each group.

Illustration by Jerome Tarpley

Common definitions of productivity include:

☐ It is no more than plain common sense.

☐ Output divided by input.

☐ The function of an organization; it is what sustains jobs.

☐ Doing right things and working right.

☐ How resources are being managed to accomplish organization goals.

IMPROVEMENT

MEASUREMENT

AWARENESS

%

E TARPLEY

and technology.

"You can't measure my work" is a phrase which has been stated over and over in our offices during the past several years. Even though some tasks follow clearly defined procedures, others are creative, developmental and/or judgmental. In some cases, it is almost impossible to quantify the output or measure the quantity and quality of services provided. This facet represents a real challenge to the industrial engineer.

The tools that the white collar worker brings to the work place each day include skill, information and time. Often, formal evaluation of these assets (particularly if quantitative) is actually counterproductive, especially if the worker has never before been measured. Everyone's understanding and commitment are required along with a consistent, operationally sound strategy for productivity enhancement.

A thorough understanding of the benchmarks of performance that are going to be used in any productivity campaign is another important component of an employee's awareness. Key result areas like quantity, quality, time and cost are typically measured and, again, have to be on a personal basis for maximum impact.

Classical techniques often stress efficiency (outputs) and neglect effectiveness (outcomes). It is for this reason that the office area has lagged behind all others in productivity measurement and improvement.

The role of management is absolutely crucial to a successful productivity improvement endeavor. White collar employees' awareness of productivity issues is facilitated when management:
□ Commits to office productivity long term.
□ Understands that education and training are vital to success.
□ Allows those who will be measured to have some say in the perfor-

□ Standard labor hours versus actual labor hours.
□ Reaching the highest level of performance with the lowest possible expenditure of resources.

Certainly, this smorgasbord of definitions suggests that far more is written about productivity than is known about it. However, we must "personalize" the definition such that each employee is totally aware

of the contribution of his or her efforts to organizational productivity.

Productivity mindedness and overall awareness are further enhanced when white collar workers totally understand that nearly all aspects of their work environment affect their performance in one way or another. White collar productivity encompasses people, people management mance benchmarks.

☐ Creates an atmosphere conducive to effective operations through long-term commitments.

☐ Listens to the worker.

☐ Knows that it is sound business practice to let the worker know what is expected of him or her.

☐ Constantly evaluates organizational practices and manages changing organizational styles effectively.

Employee perceptions are crucial in management's role of creating the proper climate or atmosphere for productivity enhancement. It has long been recognized that programs that worked very well in one organization will fall flat in another, simply because of the perceptions of the workers employed at the latter organization. Even when management states its commitment, if employees perceive a lack of commitment, the effect will be the same (or nearly the same) as if management were not committed.

No acid test can truly identify such perceptions. However, open communications, honest and sincere efforts by management and climate/readiness surveys of all those participating can increase the probability of success.

Most successful white collar productivity programs have as their first step a planned, calculated awareness effort. The industrial engineer often plays a strategic part in the formulation of this aspect of the program. In fact, no one is in a better position to do so than the industrial engineer.

Analysis and measurement

Once the white collar worker has become aware of productivity, what it means, how he or she will be measured and the role that management will play in facilitating the effort, the next step in most successful white collar productivity efforts is analysis and measurement. As is often stated in management classes, "before we can control some work

group or functional unit, we must analyze and measure it."

The industrial engineering-based analysis and measurement techniques used most often in white collar productivity efforts include:

☐ *Work unit analysis, process and procedure charting:* Knowing the present methods and procedures is fundamental to the improvement process. Actual work steps and individual responsibilities can be identified using one (or sometimes more than one) of several techniques, including procedure flow charts, work distribution charts, flow process/diagram charts and horizontal procedure charts.

☐ *Activity sampling, group timing and office standard data:* These techniques are especially important for developing time and productivity standards. Work sampling was identified by the U.S. Department of the Navy in 1978 as the most appropriate technique for measuring performance of engineering department functions. Its use in office clerical and support groups has been widely accepted.

Group timing, an extension of activity sampling, has a growing number of applications in stratified support groups, such as in word processing centers. Office standard data systems like Universal Office Controls (H. B. Maynard) and Advanced Office Controls (R. E. Nolan) are but two of several commonly used.

☐ *Multiple linear regression analysis:* Time and productivity standards are often generated by this technique. It is a statistical technique which can establish the relationship among several independent variables (such as number of new orders processed, number of back orders and number of changes for a sales order office) and some dependent variables (often time based, like the number of worker hours required to perform the activities represented by the independent variables). Indirect, expense

and direct labor work have all been measured using this technique.

☐ *Economic measurements:* Most often these are time based staffing or cost comparisons made to instantly compare one work group to another or one work group only over time. Probably the best known of these techniques is the IBM Common Staffing System, which is intended to provide an approximate measurement of groupings of indirect labor activities.

It is certainly a challenge to measure the work of the knowledge/professional/office worker due to the difficulty of defining the output or contribution, the tendency to measure activities rather than results, the matching of inputs and outputs to a time frame and the problem of efficiency versus effectiveness. However, using the above techniques and paying close attention to the reliability and accuracy of the measures, we can measure, and currently are measuring, white collar jobs as a prelude to improving them.

Improvement

Realistically, the improvement process begins with the choice of the measurement/analysis techniques used. The mere analysis often yields immediate short-term improvements. However, continuous ongoing efforts have to be strategically planned, because long-term sustained productivity growth will not just "naturally happen" as a result of the measurement process. Improvement techniques commonly used in office productivity programs include:

☐ *Improvement idea generation techniques:* Brainstorming and nominal group technique are used extensively by white collar workers to identify not only problems or hindrances to productivity, but also solutions to these problems. Both structured processes integrate individual thinking and group interaction, and the benefits of both techniques are well documented. Both are excellent first steps in any improvement program.

☐ *Time management strategies:* Time is one of the three tools that each office worker brings to the work place. How one uses his or her time directly affects that person's productivity. Sound time management approaches like that advocated by Lakein and better work flow scheduling methods such as the short interval scheduling technique recommended by Richardson (see "For further reading" for both references) can significantly improve a white collar worker's performance. These strategies are of particular importance if work (or time) sampling has been used as one of the techniques in the measurement phase of the program.

☐ *Work simplification:* The five-step team approach developed by Mogensen ("For further reading") can lead to significant productivity gains, as in the Texas Instruments and Procter & Gamble efforts. "Working smarter, not harder," the underlying principle of Mogensen's work, sounds like something everyone in the office should be in favor of. Since white collar work output is determined by both worker pace (which is limiting) and worker method (which is not limiting), it makes good business sense to work toward methods, procedures and systems improvements.

☐ *Quality circles and participative management:* Certainly, Mogensen is not the only "people" advocate; the Japanese have perfected the people approach to solving problems, increasing quality, decreasing costs and, most importantly, increasing productivity. So much has been written on the importance of the *worker* in bringing about productivity growth on the theory that no one is better suited to improve a job or work activity than the person doing it. He or she needs only the right education and training.

Today's office environment is filled with change. A sense of participation is essential to maximize cooperation, enhance communications and improve productivity, work output quality and the quality of the work life for the white collar worker.

☐ *Work place/layout changes:* Ergonomics, the science that addresses the compatibility of people and their work places, has tremendous productivity improvement potential in the office environment. Opportunities for productivity enhancement abound in the areas of seating, lighting, noise, glare, worker comfort and keyboard VDT configurations. There is no such thing as an average worker, and no single product can meet the needs of every person, but work place modularity and flexibility can provide maximum comfort *and* productivity.

Other improvement techniques, including layout and procedural changes, office automation and technology improvements and organizational changes, can also lead to enhanced white collar productivity. The five major improvement areas listed above are those most often found in practice; they also offer the most significant gains. No one technique by itself will do. The more varied the work place and the work activities, the more varied the approach needs to be.

Some final thoughts

White collar productivity improvements will not come easy. We have built in many years of inefficiencies in our large staffs, our obsession with job specialties and our lack of initiative to "tackle" this area. We must improve the productivity of this work sector.

A strategy for increasing the productivity of our offices can bring about the desired results if awareness, analysis/measurement and improvement techniques all contribute equally to the process. Technology should be considered a secondary (but important) strategy. Organizations have made great strides and will continue to do so in the future if these three vital components are properly incorporated.

For further reading:

Lakein, Alan, *How to Get Control of Your Time and Your Life,* Peter H. Wyden, Inc., NY, 1973.

Lehrer, Robert N., *White Collar Productivity,* McGraw Hill, NY, 1983.

Mali, Paul, *Improving Total Productivity,* John Wiley & Sons, Inc., NY, 1978.

Mogensen, Allen H., "Training and Support are Key to New Management Success," *Industrial Engineering,* October, 1983.

Naisbitt, John, *Megatrends,* Warner Books, NY, 1982.

Richardson, Wallace J., *Cost Improvement, Work Sampling, and Short Interval Scheduling,* Reston Publishing, VA, 1976. **IE**

Clarence L. Smith, Jr., is an extension specialist in industrial engineering at North Carolina State University. He teaches, consults and conducts research in work analysis and design, productivity improvement strategies and participative management, with particular emphasis on office productivity. A senior member of IIE, Smith is a past Raleigh chapter president and currently serves as Region III membership chairman. He is the founding president of the Central Carolina chapter of the International Association of Quality Circles. A member of Tau Beta Pi and Alpha Pi Mu, Smith received his BSIE and MSIE degrees from North Carolina State University.

Reprinted from 1984 Fall Industrial Engineering Conference Proceedings

DON'T LOSE PRODUCTIVITY AT THE BARGAINING TABLE

W. COLEBROOK COOLING

Consultant

South Orange, New Jersey

Labor agreements contain provisions that result in "feather bedding" and production impediments. Removing this waste will be a major step towards beating the world in productivity. Management must learn to say "no" at the bargaining table. Too many "yes's" have had a disastrous effect upon productivity.

There is an area in our goal of increasing productivity that is often mentioned but frequently ignored or overlooked. Frequently you hear or read the words "take away" or "give backs" which are applied by labor to situations where management is attempting to obtain relief from restrictive provisions of the labor agreement. You hear of quality circles, productivity programs, and the many organizations, profit and not for profit, promoting and selling these approaches. Speaking as an Industrial Engineer, who is presently engaged more in labor negotiations than other areas, it is my opinion that there is more to be gained in less time by the removal of "feather bedding" and work impediments from the labor agreement than by launching these long term programs. It is not meant to imply that these approaches are not valuable. These long term behavioral methods should be done on a continuing basis with sound planning and not attempted with great expectations as a current "buzz word" approach. I have been in the management work force since the second day of 1946 and have seen the rise, fall and resurrection of many behavioral techniques. With regard to labor-management cooperation, the answer to a sure fire increase in productivity is very simple, "get rid of the feather bedding" and work impediments in the labor agreement.

Many times I've heard statements similar to "the union contract won't let us do this". Let's clear up the fact the two parties agreed to the contents of that "umbrella" that covers the employees in the bargaining units. The parties agreed in writing as to how their relationship will function in practice. Management had the opportunity to say "NO" but instead said "YES".

Management always has the opportunity during collective bargaining sessions to amend or eliminate "feather bedding" or work impediment provisions during collective bargaining sessions that come about at the expiration of labor agreements.

I'd like to conclude this introduction by mentioning my approach with regard to union representation. I am not a union "buster". I don't care whether our employees pay dues or not. All I want to do is manage the operation at the lowest possible cost, which means "more good units from each employee on time". When I have union representation, I will "deal above board". I shall train my supervisors to do likewise. I shall avoid an adversary approach, these are my employees first, including the union representatives. Why should I fight with my employees? I believe that many companies waste their time and energies fighting the union at the expense of neglecting supervisory leadership training. I don't mean behavioral bull, but the development of basic leadership skills similar to that outlined in union stewards' manuals and also the improvement of management techniques.

I would like to mention the results of removing "feather bedding" and work impediments from the labor agreement. This occurred in a chemical processing plant with an international union and with a labor agreement having similar restrictions as automotive agreements:

1) Job classifications reduced to 22 from 89 - increasing flexability in work assignments and transfers and eliminating movement in layoff, promotion and recalls.

2) Seniority considerations eliminated from daily job assignments and overtime considerations - provides for placement of the "best man for the job".

3) Agreement reached that production employees would complete all unskilled maintenance work on their own equip-

ment - a big factor in eliminating downtime, job interference and walking in maintenance. Also provided for production employees to function as helpers to maintenance craftsmen.

4) Craft maintenance was converted to General Maintenance except for Electricians and Machinists, with Electricians working as General Maintenance in portions of their job - greatly reduced job manning, walk and wait time, craft interference and simplified the complexities in planning and scheduling maintenance work.

5) The limitation on supervisors working was eliminated.

6) Restrictions on sub-contracting provisions were eliminated.

7) The probationary period was extended from 30 days to up to 45 working days giving more time to determine whether the new hire would be both an acceptable employee and an acceptable member of the bargaining unit.

These changes resulted in a work force of 165 employees completing the work that formerly required 265.

In a plant manufacturing machinery for the paper industry, an agreement was signed, eliminating restrictions similar to the previous examples, that enabled 33 employees to do the work that formerly required 55. At this location the number of union representatives was cut in half and a step in the grievance procedure that was useless was eliminated. Both of which resulted in reduced "hours paid, but not worked".

Maintenance was briefly mentioned as an area of few gains in productivity. I would like to expand my views towards maintenance "feather bedding" provisions that have found their way into our labor agreements and operating methods. Major items of "feather bedding" in maintenance seriously increase our lack of ability to meet foreign competition through the addition of non-productive hours to skilled craftsmen; failing to utilize all the skills of individuals; discouraging individual initiative; causes other employees to be lax in their commitment to time and skills; wastes supervisory time; and increases the cost of management controls such as planning and scheduling.

A good example is the restriction applied to chemical operators or production employees with regard to maintenance work on their own equipment. Why not provide for these employees to complete unskilled maintenance work on their own equipment, saving job interruption and travel costs in maintenance (or assigned maintenance mechanics) and re-

ducing downtime in production? Why should production log downtime because of a simple adjustment? Why should maintenance costs reflect travel, job interruption, and skilled maintenance rates for simple unskilled tasks?

I have negotiated into several labor agreements with locals of international unions, a provision stating that "production employees shall complete unskilled maintenance work on their own equipment". This was not the easiest proposal to present for agreement, but worth the effort. The production employees liked this provision, maintenance employees finally accepted the provision, and from one installation I received a letter stating that production employees "routinely" completed:

° replacing pumps
° replacing hoses
° replacing carriers
° replacing carrier arms
° routine lubrication
° break apart plugged lines and clean
° minor conveyor repair
° replacing valves
° replacing pre-dryer boxes
° replacing belts
° replacing filter
° PVC pipe repair
° open Wildeir pumps and clean

You do not need a great deal of imagination to see the resulting cost improvements.

Some of you may be saddled with agreements that provide for craft trades, and restrictions on working out of the trade. In an automotive agreement, there are fourteen craft trades, and "None shall be interchangeable". There is so much feather bedding in this situation that one can hardly measure the cost effect of manpower waste (both production and maintenance), planning and scheduling waste, de-emphasis on individual pride in skill and accomplishment.

The losses in working along craft lines occur as a result of excess travel time, excess job preparation, inefficient crew balance, and excessive craft interference. The least you can do is to provide for "all maintenance crafts shall complete the unskilled work of other trades" and, "utility employees shall complete unskilled maintenance work using hand and power tools".

The most you can do is to combine the crafts as General Maintenance and assign employees to work within their individual skills and abilities. You may have separations such as General Maintenance, Electricians, Shop Machinists and so forth. But, remember the fewer the classifications in maintenance, the more flexibility you have with the effect of lowering maintenance costs. As long as job classifications have been mentioned, it is necessary to point out that the more job classifications there are, the less flexibility you have and you gain higher costs as a result. The objective should be to hold employee movement down - movement being caused by promotion, layoff, and recall. When job classifications are in excess of your true needs, you get more movement that requires posting, bidding, and so forth. With the recent bottoming out of the economy, I'm sure that some of you ran into situations where there were more classifications than employees.

The purpose of collective bargain is to obtain a labor agreement that establishes wages, benefits, and working agreements over a specified period of time. The objective should be to negotiate to a conclusion on the various issues without any ambiguities in the agreement. In short, reach a loud and clear end point with every provision.

I recently studied an automotive agreement that was so complicated with having to make mutual agreements between the company and the union, the supervisor and the steward, and in addition reports to the union, that it is no wonder that automobiles can be hauled from distant places and be more than competitive with U.S. built cars. Everyone is meeting to mutually agree rather than working. What if "mutual agreement" does not happen after lengthy chats? More cost added to the mutual agreement cost. To paraphrase: "That's a helluva way to run a railroad".

It is appropriate to mention at this point that, when you are operating according to the terms of the current labor agreement, you are establishing proposals that will be submitted across the table at the next negotiations. So, the actions you take that are regarded as unfair or inappropriate will appear at the table. The various restrictive seniority clauses and proposals may be due to actions on the part of supervisors that employees resented.

Originally, the bargaining units' interest in seniority provisions was to protect the senior employee in instances of promotion, recall, and layoff. Some restrictive seniority clauses may have been caused by employees wanting and negotiating more than just protection with regard to promotion, recall, and layoff.

For example:

1) Maintenance men choose jobs by seniority on a daily basis.

2) Overtime by seniority rather than finishing a maintenance job or scheduling the production employee who normally does the work.

3) Temporary transfers by seniority.

4) Qualifications and ability not considered in promotion, layoff, and recall.

Job movement is a large contributor to excess costs as illustrated by the preceding. Other than the examples, you must consider that learning the job, production during a learning period, quality during a learning period are other costs incidents that are affected by seniority movements. The Company's objective must be to reduce job movement, but at the same time, provide a method of upward movement, with training, as a measure of seniority towards long service employees who are able to do the work.

The bargaining unit usually favors plant wide seniority, with all choices of movement based upon seniority considerations. Management favors qualifications and ability considerations with seniority being the governing factor if qualifications and ability are equal. Management also favors departmental seniority in order to achieve an upward line of progression which allows for training through temporary transfers, so the successful bidder very often can do the work and will become 100% effective very rapidly. In the reverse, a layoff situation, employees would then bump down in a department and be 100% effective immediately as the employee had progressed upward through the jobs available for bumping. Some agreements provide movement on a plant wide seniority basis when departmental has been exhausted to protect long term employees who are able to do the work.

Where plant-wide seniority provisions exist, you must build a fence around essential skills such as maintenance, toolroom, and certain key employees. This language could be similar to, "In layoffs and recalls, the Company, in its sole discretion, may designate employees whose services are deemed necessary for starting or maintaining production, for preparing tools, plant or equipment, and such employees shall be retained or recalled as necessary".

Keep in mind that collective bargaining at negotiating sessions means receiving proposals, considering proposals, discussing proposals, and answering proposals. If you have followed those steps and have documentation in your minutes, you don't have to say "Yes" when "No" is the proper response. Keep in mind that the company should present proposals. I have been a chief spokesman on many occasions where this had not been the practice. If you want improvements, the changes must be proposed across the bargain table.

I hope that there is an interest in basic work measurement in this audience. When the IMS ran their annual meetings, they were refreshing to attend as the topics dealt primarily with work methods, work measurement, day work plans, and incentive plans. I still contend that you never know what's going on in the factory without good measurement, tight and accurate control of time and count, and a fast response control method that enables the supervisor to make corrections as the work day progresses. The objective being to feed the supervisor with operating data in time to respond, not to feed the computor. I would like to give my thoughts on several issues with regard to work measurement.

The U.A.W. Education Department publishes a pocket guide titled, "Grievance Handlers and Bargainer's Pocket Guide". This is a worthy edition to one's business library. There are valuable leadership guidelines, which for management application, about the only change required is to substitute the title of supervisor for steward. There is also a section titled, Time Study Warning, well written and for the most part, applicable to good basic industrial engineering practices. Keep in mind two U.A.W. internal policy statements, to which management must say "No" and mean it.

The first, as stated by Thomas Cichocki, Director of Time Study, U.A.W. in the 40th Annual IMS Clinic Proceedings in Chicago, follows:

"The U.A.W. insists in the United States on reserving in its contracts the right to strike on production standards. The heavy cost of a strike act as a balance where to assure that neither the union nor management will try to impose an unreasonable settlement. We believe this is the surest guarantee that collective bargaining will result in mutual acceptance of standards satisfactory to both sides. The proof of that is the fact that for every dispute which ends in strike action, there are thousands settled around the bargaining table." To Tom's statement, one could add "It also proves who has the bigger club."

This type of clause severely weakens the entire labor agreement. Disgruntled employees can use the guise of improper standards to strike on another matter. The strike clause then becomes useless. To avoid the heavy handed punitive economic approach, " No strike " was the reason to provide arbitration as the final settlement for disputes in the first place, standards included. The second, Tom also said in Chicago, "Obviously the easiest way to arrive at mutually acceptable work standards in the case of a dispute, is by the process of collective bargaining". I would say that if this is the practice in the automotive industry, the standards may not be worth a "Tinker's Dam". You don't bargain standards; you determine if they are right or wrong, and if necessary, use a third party, the arbitrator.

Another clause to avoid, to which H. Seroka,

Director of Allied Industrial Workers is an advocate, is:

"Where employees through their own ingenuity, skill and dexterity change methods and/or procedures which tend to make the standard appear inappropriate, the employees shall not have such standards reduced. It is the intent of this provision to preserve for such employees the increase in earnings resulting from the employee's utilization of their own ingenuity, skill, and dexterity unless there is a change in the operation as provided in paragraph 8-02 above".

A proposal for a provision of this nature must be rejected, and, if already in the wage incentive provision of the labor agreement, eliminated. Most unions favor this concept and many companies have agreed, as well as naive Industrial Engineers. This concept should not be judged by emotional considerations. This is a business consideration. The parties must take the same approach in keeping the wage incentive pact in a "pure" state as they do with maintaining a joint apprenticeship program.

From logical considerations, rather than emotional,

1) do you want everyone tinkering with your process, machine speeds, feeds, and so forth - or do you want to maintain a process that will avoid scrap, rework, and insure a quality part?

2) do you want to allow five or six other employees to "ride" the "gravy train developed by another employee?"

3) do you want the earnings of employees working on "loose" standards to be envied by others?

4) are you in a position to let your costs go out of control?

5) do you want to encourage some employee to peg earnings because of "loose" standards?

6) do you want to encourage others to take "short cuts" due to "following the leader?"

7) do you want to increase the cost of tooling or maintenance because of overloads?

8) do you wish to condone safety violations?

All of the previous items are normally regarded as failures of management in wage incentive administration that eventually result in an out of control installation.

The grievance procedure is another area of potential waste. How many union representatives are involved? In an automotive plant the agreement provides for a union representative (Shop Steward) for every 25 employees. In this situation there are more union reps than supervisors with each rep attempting to displace the supervisor's leadership role and not producing automobiles. Obviously, by law, representation is required, but at a practical level. Another consideration is the steps of the grievance procedure. If there are steps where decisions are not made, eliminate this step. Obviously, discussions of suspensions and discharges are not practical at the lower steps. Provide for the movement of such issues into the final step prior to arbitration to avoid the waste of productive time.

Obviously, these are just a few examples that wreck the efficiency of a facility. Every labor agreement has clauses that impede production and these provisions must be removed. So, examine your agreements, estimate the value of each impediment, and plan to make some changes.

Management must choose to say "No" on occasions, and if management can't say "No" when cost, the destruction of work measurement methods and relationships are at stake, the unpleasant results will be experienced by all involved.

In conclusion, I'd like to quote (underlining added) from an arbitration award, reference, Shell Oil Company, 44LA 1219, 1223 (Turkus 1965):

" The basic and primary function of management is to operate on the most efficient basis attainable, to see that unnecessary costs are eliminated and that the labor force is efficiently and productively utilized. <u>Unless restricted by the labor agreement or by oral understandings interpretations and/or mutual commitments of the parties which have grown up over the course of time so as to form an implied term of the contract</u>, management cannot and should not be put in a straight jacket preventing it from exercising its basic responsibility and legitimate function to control methods of operation and to direct the work force."

The portion of Arbitrator Turkus' award that I have underlined has a message to all of us. Say "No" to additional proposals that restrict productivity (feather-bedding) and get rid of those already in the labor agreement and those that have been established through past practice.

BIOGRAPHICAL SKETCH

Cole Cooling, a consultant, has assisted many companies in the areas of labor relations, production labor and maintenance, with the mission of reducing costs.

Many of his articles have appeared in technical journals both here and abroad. He is the author of Front Line Cost Administration, Simplified Low Cost Maintenance Control, and three chapters in Maintenance Engineering Handbook. His presentations at many National Conferences have been subsequently published in volumes of the proceedings.

Cole has taught management courses at Temple and Rutgers Universities and lectured for Drexel University, the University of Pennsylvania, and the University of Wisconsin. He has also presented over 180 seminars in the United States, Canada, Mexico and Mideast. An industrial engineering graduate of Pennsylvania Military College, he received an M.B.A. in industrial management from Temple University. He is a Fellow of the IIE and the Society for the Advancement of Management and has received that organization's Phil Carroll Award, Professional Manager's Citation, and Advancement of Management Award.

Reprinted from *Industrial Engineering*, October 1983.

Rethinking Productivity:

Wanted: A System That Provides 'Box Scores For Productivity'

By Marvin E. Mundel, P.E.

Marvin E. Mundel, P.E., is the principal of M. E. Mundel & Associates. A past president of the Institute, he has served as principal staff officer for industrial engineering, U.S. Bureau of the Budget; director, Army Management Engineering Training Agency; and chairman, industrial engineering, at Purdue University. He is a fellow of IIE and has been awarded the Gilbreth medal by SAM, the APO gold medal by the Asian Productivity Organization and the Frank and Lillian Gilbreth Award by IIE. He was named Engineer of the Year by the Washington Engineering Council in 1978.

It has long been held that the standard of living of a country is a function of its productivity and its natural resources. Other factors come into play, but they do not override these two basic components. It seems reasonable to assume that not much can be done about natural resources; it is productivity which must be altered.

Alteration means change! We are not talking about creeping, crawling change. The pace of change of the past is insufficient. We need dramatic change. Perhaps we need to learn to find exhilaration in change, and adapting to it, rather than trying to preserve the past. The old habits and customs, if unchanged, may well be the death knell of our economy.

Let us look at some small examples. An American invented the industrial robot; our industries were uninterested. He took his idea to Japan. Japan picked up the concept and further developed it, and now our industrialists go to Japan to see robots while America desperately attempts to catch up.

An American invented the food processor, but American factory management was not interested in producing them. France was. France sent them here at $200 each until we finally began to make them in the $40 to $60 range, years later.

Deming and Juran went to Japan to introduce statistical quality control with worker participation and responsibility for quality—something introduced in the U.S. during World War II and then largely neglected. Again, Americans journey to Japan to see the fruits of their own innovations.

Another American engineer, during the early 1960s, reduced the depth of Japanese organizations, which used to be layered as:

1.) Worker.
2.) Hancho.
3.) Kumicho.
4.) Shokucho.
5.) Kakariin.
6.) Kakaricho.
7.) Fuku-katcho.
8.) Katcho.
9.) Butcho-dairi.
10.) Butcho.

At that time the 10th man on the list was the first employee *not* a member of the union. There were more layers above him. The engineer flattened the organizations he worked on to four or five levels; now Americans flock to Japan to see this American innovation.

It's fashionable to look, but will we merely copy QC circles as a fad of a few years, like *job enlargement, work simplification, suggestion systems,* etc.? It is not that I denigrate these approaches as such, but have we ever pushed long and hard and steady enough, at enough of them at the same time, to obtain maximum benefit from them?

However, the types of changes mentioned so far are micro. If auto and steel workers in America are to be paid twice as much as their counter-parts in other countries, they will have to be *more than twice* as productive if we want to reverse recent trends; the same applies to every other occupation and skill.

Such a drastic change cannot come merely from micro changes at the workplace level. We need a restructuring of our entire manufacturing, processing and service plant.

Edward Hood of the board of directors of General Electric, in his keynote address at the 1983 IIE Annual Conference in Louisville, put forth a basic concept that is so central to drastic change:

We need to make the productivity Olympics as visible as baseball box scores or professional football standings.

In other words, we need a macro measurement system that will enable us to evaluate the total impact of *all* relevant factors on our productive establishments year by year, not just at the workplace, not just at the plant level, but between and among plants; not necessarily nationwide, but at the organization level, where concrete actions can be taken and results obtained, rather than national exhortations of pious wishes. We need a means of evaluating the total organization and bringing large changes into being.

Productivity may be defined simply as:

$$\text{Productivity} = \frac{\text{Sum } AO_m / \text{Sum } RI_m}{\text{Sum } AO_b / \text{Sum } RI_b}$$

Where:
AO = Aggregated outputs.
RI = Resource inputs.
m = Measured period.
b = Base period.

Productivity should not be confused with two other terms; effectiveness and performance. *Effectiveness,* in the private sector, means making a profit and preserving future profits. The latter is frequently threatened by:

Perhaps we need to
learn to find exhilaration
in change, and adapting
to it, rather than trying
to preserve the past.
The old habits and
customs if unchanged,
may well be the death
knell of our economy.

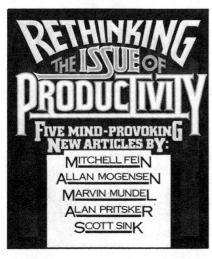

RETHINKING THE ISSUE OF PRODUCTIVITY

FIVE MIND-PROVOKING NEW ARTICLES BY:

MITCHELL FEIN
ALLAN MOGENSEN
MARVIN MUNDEL
ALAN PRITSKER
SCOTT SINK

Table 1: SSA Field Offices: Array of Services Constituting 95% of Total Time

No.	Rank Order	Name	% of total study	% of total study cumulative	% of services	% of services cumulative
14B	1	Title II and XVIII Claims, Auxiliary; Survivor; Uninsured Claims	5.34	5.34	12.45	12.45
14C	2	Title II and XVIII Claims, Disability Claims	5.07	10.41	11.82	24.27
16	3	SSI Blind and Disabled Claims	4.68	15.09	10.91	35.18
14A	4	Title II and SVIII Claims, Retirement Claims	4.36	19.45	10.18	45.36
42	5	Title XVI Redeterminations, No Change, Sched. & Unsched.	2.54	21.99	5.93	51.29
63	6	Post Entitlement RSDHI Inquiries	2.14	24.13	4.98	56.27
51B	7	Overpayment Collection, Title XVI	1.99	26.12	4.65	60.92
01	8	Identify Client Records	1.83	27.95	4.26	65.18
77	9	All Other PI	1.67	29.62	3.89	69.07
93A	10	Misc., Other Title XVI Post Entitlement	1.08	30.70	2.51	71.58
15	11	SSI Aged Claims	.82	31.52	1.92	73.50
38	12	Title XVI Earned/Unearned Income Changes	.81	32.33	1.89	75.39
89B	13	HI/SMI Post Entitlement Activity, SMI	.76	33.09	1.78	77.17
65A	14	RSDHI Reconsiderations	.69	33.78	1.61	78.78
12	15	Pre-claim Client Status/Benefit Inquiries	.63	34.31	1.46	80.24
91	16	Assistance Requests	.58	34.99	1.35	81.59
30A	17	Name/Payee Changes, Title II	.56	35.55	1.31	82.90
51A	18	Overpayment Collection, Title II	.54	36.09	1.26	84.16
29A	19	Change of Address, Title II	.47	36.56	1.09	85.25
34	20	Title SVI Living Arrangement Changes	.46	37.02	1.08	86.33
49A	21	Non-receipt of Checks, Title II	.42	37.44	.98	87.31
65B	22	Reconsiderations, SSI	.38	37.82	.90	88.21
58A	23	Individual Direct Deposit Changes, Title II	.29	38.11	.67	88.88
66	24	RSDHI Hearings	.28	38.39	.65	89.53
18	25	Annual Reports	.27	38.66	.62	90.15
31A	26	Death Notices, Title II	.26	38.92	.61	90.76
29B	27	Changes of Address, Title XVI	.25	39.17	.59	91.85
67	28	SSI Hearings, Title XVI	.24	39.41	.57	91.92
05	29	Resolve Earnings Discrepancies	.21	39.62	.49	92.41
46A	30	DIB Trial Work Enforcement, Title II	.20	39.82	.47	92.88
89A	31	Post Entitlement Activity, HI	.20	40.02	.47	93.35
32A	32	Student Status Change, Title II	.19	40.21	.44	93.79
33	33	Student Enforcements (Title II/XVI)	.19	40.40	.44	94.23
30B	34	Name/Payee Changes, Title XVI	.18	40.58	.43	94.66
50A	35	Check Returned by Individuals or Treasury, Title II	.16	40.89	.36	95.36

□ Pressure for short run dividends.
□ Rotating periods for managers to make a showing without regard for the future.
□ Interest rates that pull money away from productive investments.
□ Taxation that restricts capital formation.
□ Unwillingness to take a chance.

Performance is defined in terms of how close we come to meeting a target of quantity and quality of output.

A simple illustration may adequately separate these concepts. Let us take a group of typists, each of whom manually types 150 pages of letters per day without typos—an excellent performance. However, the letters may be so verbose as to defy understanding, in which case the recipients are not served, and effectiveness is low. The real desired output, a recipient served, is not produced; productivity is abysmal; we are deluded by counting pages!

To measure productivity one must aggregate the outputs which contribute to the achievement of the objective, in an appropriate fashion. Some art is involved at this step. Also, wherever possible, the computation of resource inputs must be related to the totality of the organization.

Further, the maximum comparability between the measurement procedures used in like organizations must be provided to reach Hood's simply stated but vital objective. The measurement should assist in pinpointing places for change.

To this end, I am going to examine two kinds of organizations, a government organization and a commercial bank, giving suitable algorithms for evaluating productivity in each sector.

The government organization to be examined is the Social Security Administration. Field offices of the SSA were the focus of a study completed in 1980. The objec-

tives of the organization were delineated first to provide criteria for identifying inputs. The researchers identified 118 outputs which represented completed services that contributed to the objectives of the SSA.

A sampling-observation study, during which worker performance was rated against a reasonable level of human diligence, was used to develop standard times. In all, 698,000 observations were made in 25 locations over a six-month period.

Of the 118 outputs, 35 consumed 95% of the time. See Table 1 for a listing of these outputs.

Performance against the standard time, with due allowance for overhead and personal categories, was 69.5%. It would appear that there is room for an improvement in performance. The same 698,000 observations referred to above were also classified into 54 activity categories, with 90% of the time being spent on 16 of these categories, as shown in Table 2.

Note that these 16 activities consume about 36,000 staff years per year. Surely drastic change, by means of improvement techniques, seems feasible and desirable.

The following appeared in the *Engineering Times* (NSPE), July 1983, under the headline, "Bank Customers Balk at Losing Human Tellers":

The seemingly unrelenting march into the computer age took a small step backwards recently when the nation's largest bank decided to reintroduce human tellers to do what had previously been handled by automatic teller machines.

Customers at 38 local branches of Citibank in New York who had routine business such as minor withdrawals or deposits had been forced to con-

We need a macro measurement system that will enable us to evaluate the total impact of *all* relevant factors on our productive establishments . . . at the organization level, where concrete actions can be taken and results obtained.

RETHINKING THE ISSUE OF PRODUCTIVITY
FIVE MIND-PROVOKING NEW ARTICLES BY:
MITCHELL FEIN
ALLAN MOGENSEN
MARVIN MUNDEL
ALAN PRITSKER
SCOTT SINK

Table 2: Percentages of Time Spent on Top 16 Activities

Rank Order	Name	% of 100	Cumulative %
1	Read/hard copy	20.36	20.36
2	Discuss/face to face/ non-interview	12.23	32.59
3	Initial FFI-discuss*	10.37	42.96
4	Write	10.30	53.26
5	Telephone interview	10.23	63.49
6	Associate	5.03	68.52
7	Type	4.35	72.87
8	Initial FFI-write	3.94	76.81
9	Code/paper	3.66	80.47
10	Retrieve/storage device	2.06	82.53
11	All other functional activity-unclassifiable	1.75	84.28
12	Store/storage device	1.54	85.82
13	Initial FFI-read/hard copy	1.21	87.03
14	Store/into/jacket	1.14	88.17
15	Subsequent FFI-discuss	1.13	89.30
16	Copier/operate	1.10	90.40

*FFI = Face-to-face interview

Table 3: Condensed Listing of Outputs of Commercial Bank, by Groups*

L ; Commercial loans processed 010101 thru 010309
I ; Import and export assistance provided 020101 thru 020304
S ; Share and stock transactions processed 030101 thru 030601
C ; Credit and visa transactions processed 040101 thru 040202
T ; Traveller's checque transactions processed 050101 thru 050102
F ; Foreign and interbank transactions completed 060101 thru 060202
B ; Stock and bond transactions completed 070101 thru 070202
G ; Gold, silver and other precious metals transactions 080101 thru 080105
E ; Foreign currency transactions completed 090101 thru 090105
O ; Leasing transactions completed 100101 thru 100102
etc.

*The numbers are for identifying the specific work-units in each group.

duct all transactions with robots, according to a New York Times report. The robot-only policy applied only to bank customers dealing with sums less than $5,000.

Citibank corporate strategists concluded earlier this year that most bank customers don't require the services of tellers that walk and talk and breathe. Besides, the automated machines were judged to be much cheaper and, yes, generally more efficient than their human counterparts.

New Yorkers cried foul, however, and Citibank agreed to return the very 20th Century human tellers.

Such is where you wind up if you concentrate on sub-functions, like tellers, rather than on total bank productivity and effectiveness. The following algorithm is offered as an approach to total bank productivity measurement:

☐ Outputs of the bank which contribute to bank's objectives, and symbols for them, are given in Table 3 (the list is compressed into groups, to shorten this illustration).

☐ The algorithms for the computation of resource inputs appear in Tables 4 and 5.

Table 4: Support Resource Inputs of Commercial Bank
RIP1 Partial, Support (RIP = Resource inputs, partial)

Categories	Factors
TH = Teller hours paid	$T = Base year pay per hour, tellers
MH = Manager hours paid	$M = Base year pay per hour, managers
CH = Clerical hours paid	$C = Base year pay per hour, managers
GH = Guard hours paid	$G = Base year pay per hour, guards
RH = Maintenance hours paid	$R = Base year pay per hour, maintenance and repair
OH = Janitorial hours paid	$0 = Base year pay per hour, janitors
SH = Service hours contracted	$S = Cost of service contracts in base year, per hour of operation

$$RIP1_b = TH_b \times \$T + MH_b \times \$M + CH_b \times \$C + GH_b \times \$G + RH_b \times \$R + OH_b \times \$0 + SH_b \times \$S$$
$$RIP1_m = TH_m \times \$T + MH_m \times \$M + CH_m \times \$C + GH_m \times \$G + RH_m \times \$R + OH_m \times \$0 + SH_m \times \$S$$

Where:
b = base year
m = measured year

Table 4 is the partial resource input (RIP1) representing the cost of the support personnel—those who do not directly produce the true, income producing outputs of the bank although they are part of the cost of producing such outputs; b = base year; and m = measured year or period.

It should be noted that all of the equations have been formulated so that productivity may be calculated readily for periods shorter than a year. A wait of a year might only reveal disaster too late to do anything

Table 5: Direct Resource Inputs for Commercial Bank
RIP2 Partial, Direct (RIP = Resource inputs, partial)

Categories	Factors
HL = Commercial loan officer hours paid	$L = Base year pay per hour, commercial loan officers
HI = Import-Export assistance officer hours paid	$I = Same, import-export assistance officers
HS = Share and bond underwriting, advise and port-folio mngt. transactions hours paid	$S = Same, share and bond, etc. officers
HC = Credit and visa transactions officer hours paid	$C = Same, credit and visa officers
HT = Travelers checques transactions hours paid	$T = Same, travelers checques handlers
HF = Foreign and interbank office hours paid	$F = Same, Foreign and interbank officers
HB = Stock and bond transactions officer hours paid	$B = Same, stock and bond transaction officers
HG = Gold, silver and precious metal officer hours paid	$G = Same, gold, silver and precious metal officers
HE = Foreign currency transaction officer hours paid	$E = Same, foreign currency transaction officers
HO = Leasing transaction officer hours paid	$0 = Same, leasing transaction officers
HD = Safe deposit box officer hours paid	$D = Same, safe deposit box officers

$$RIP2_b = HL_b \times \$L + HI_b \times \$I + HS_b \times \$S + HC_b \times \$C + HF_b \times \$F + HB_b \times \$B + HG_b \times \$G + HE_b \times \$E + HO_b \times \$0 + HD_b \times \$D$$

$$RIP2_m = HL_m \times \$L + HI_m \times \$I + HS_m \times \$S + HC_m \times \$C + HF_m \times \$F + HB_m \times \$B + HG_m \times \$G + HE_m \times \$E + HO_m \times \$0 + HD_m \times \$D$$

Table 6: Computing the Aggregated Outputs of a Commercial Bank
Outputs, Direct, Weighting, AOP1 (partial) (AOP = Aggregated outputs, partial)

Categories	Other Symbols
WHL = $ earned per hour of commercial loan officers	b, subscript = base year values
WHI = $ earned per hour of import-export assistance officers;	m, subscript = measured period values
down to	010101 etc. = work-unit identifiers
WHD = $ earned per hour of safe deposit box officers	ST_{xxxxx} = standard time for work-unit of subscript

Work counts
WC_{010101} = work count of work-unit 010101 completed
WC_{020101} = work count of work-unit 020101 completed
etc.

$$AOP1_b = WC_{010101b} \times ST_{010101} \times WHL + WC_{010102b} \times ST_{010102} \times WHL \ldots + WC_{100101b} \times ST_{100101} \times WHO$$

$$AOP1_m = WC_{010101m} \times ST_{010101} \times WHL + WC_{010102m} \times ST_{010102} \times WHL \ldots WC_{100101m} \times ST_{100101} \times WHO$$

Sum of outputs, AO
$$AO_b = AOP1_b + RIP1_b$$
$$AO_m = AOP1_m + (AOP1_m / AOP1_b) \times RIP1_b$$

Note: The subscript b is not an error. Support is earned at base year rates.

about it. The only proviso with respect to the equations is that *all data relating to the measured period be for the same period of time.*

Table 5 is the partial resource input (designated RIP2 to separate it from the support inputs) representing the cost of the income producers of the bank: the loan officers, etc.

☐ The computation of aggregated outputs appears in Table 6. Table 6 shows both the value of the partial outputs (AOP1) and the total aggregated outputs (AO). It should be noted that in AOP1 the standard hours earned by the loan officers and other income producing officials are weighted to reflect the profitability of each standard hour earned.

The weightings are base period weightings to prevent the productivity measure from being distorted by external fiscal conditions. A standard hour earned by a safe deposit box rental officer is probably of far less value than a standard hour earned by a commercial loan officer; the weightings reflect such differences.

Note also that in both of the aggregated output equations, AO_b and AO_m, the support value of the outputs is earned at base year rates, with the measured period showing only base year support costs multiplied by the ratio of measured period direct outputs to direct outputs in the base year. The logic is simple. Increasing support costs does not automatically raise the value of the outputs. If the decision to increase support costs was a wise one, doing so should increase AOP1 sufficiently that the increased support, in the denominator of the upper half of the productivity equation, will be more than offset by the increased direct outputs in the numerator.

A series of algorithms for the iron, steel and manufacturing sectors of industry appeared in the Institute's *1983 Annual Industrial Engineering Conference Proceedings* (M. E. Mundel, "Total Productivity Mea-surement for Manufacturing Organi-zations," pages 333-337).

Similar algorithms are being developed for other industrial and service sectors, but where? The development work is being sponsored by the Asian Productivity Organization; the Secretariat is in Tokyo. Are we going to make pilgrimages there, ten years from now, to view other work originated here in America?

Individual companies are limited by anti-trust legislation as to how they may act in concert or discussion, but the problem of raising our productivity transcends any one company's interest; it is of vital concern to our whole nation.

We need to have our various industry associations working with existing state and university productivity centers and with IIE to develop "total productivity measurement box scores," compare results and assist in the transfer of productivity improvement technology. We need to enlist the unions and the non-unionized worker in the effort; if it fails, their present standard of living will be unsupportable.

The longest journey starts with but a single step. The first step toward improving productivity is measuring it. We must then focus on the measure and on changing it.

The challenge may seem formidable. However, it would seem that all the necessary steps are smaller than the step of crossing the Pacific ten years from now, to see what American ingenuity has produced in someone else's more fertile climate. Let's get started!

How many of our IE curricula have courses on total productivity measurement and improvement? How many of our trade associations are focusing on the same subject? Who is carrying the message to the workers, to the Congress? What are we waiting for? **IE**

> **It would seem that all the necessary steps are smaller than the step of crossing the Pacific ten years from now to see what American ingenuity has produced in someone else's more fertile climate.**

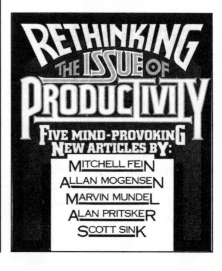

*Reprinted from **Industrial Management**, September-October 1983.*

Five Questions, and a Value, for Productivity

Charles M. Kelly

Abstract: "Five Questions, and a Value, for Producitivity" presents five major issues that influence the productivity of companies in any industry, although it is especially relevant to hightech industries.

It also demonstrates how the value of equality (or lack of it) can affect a company in many subtle ways, from the type of training given to employees, to the degree of commitment that they feel for their jobs.

The two major results of a climate of inequality are that management abstracts itself from both its products and its producers, and workers' jobs lack substance. This article suggests two approaches that have demonstrated success in helping to correct these conditions.

It is generally acknowledged that the quality and productivity of U.S. industry has been severely challenged by other countries in recent years. But possibly of greater importance, there also are challenges to many of our comfortable assumptions about professional management itself.

There is ample evidence that one assumption — that our theories of management are valid because they have worked in the past — may be our most fundamental error. We know that as a nation we have been successful, but what we can never know is how successful we could have been. Other countries, however, have given us some idea by demonstrating what is possible: levels of quality production that we had come to believe were unrealistic, in the entire range of products from electronic equipment to automobiles.

A Cycle of An Industry

The recent polyester business cycle can serve as a capsulized example of what may be true of much of American industry. Within the time span of my own consulting

---

CHARLES M. KELLY, Ph.D., is with Charles M. Kelly and Associates. Author of several articles in professional publications, he was formerly manager of Training and Development, Fibers Industries, Inc., a subsidiary of Celanese Corporation.

experience, it went from booming growth and overcapitalization to market saturation and downsizing.

The following brief description represents a subjective personal viewpoint and is for illustrative purposes only. It does not relate to a particular facility or company; however, I feel that it is a fairly accurate composite summary of what has evolved in some cases. (I can claim a degree of objectivity because I consider my own role as consultant to be a contributing cause of some of the problems cited.)

Explosive Growth

In the initial stages of growth the facilities were small, everyone was on a steep learning curve and participation was the name of the game. Although many decisions may have been made at the top, they were preceded by discussions that included everyone directly involved with a project. Such discussions frequently included several levels of the organization; this was a necessity because no one level had all the relevant information. To a large degree all managers were functioning as technical professionals.

As new equipment was installed and put into operation, new employees were hired, given minimal training (in order to get production as soon as possible), and new projects were initiated. Because of rapid project development, much operator time was devoted to monitoring equipment performance, correcting it when it deviated from specifications, or seeking help if a problem was beyond routine. In other words, manning requirements were high because the process itself was not designed to be "interruption free" in the first place.

Management development was basically on-the-job, and focused more on the improvement of technical knowledge than on managerial skills. Since they were developing new technology as well as learning it, the core start-up team knew the technical aspects of their jobs as well as anyone else in the industry. New employees at all levels, however, were taught only the essentials of what their jobs required at the time of hiring. If they ran into problems beyond their skills, more experienced persons, usually at a higher level, had to intervene.

This was a period of high morale. Everyone had a personal feel for the climate of the organization and top level managers had a working relationship with persons several levels lower. Results were rapid and observable and operations were profitable. Promotions and the movement of people were frequent. Organizational excitement was at a peak.

Maturation

As market demand increased and the industry grew, facilities got large and more complex. Plant managers no longer knew all the employees (even the foremen) by name. The accompanying technical development explosion made it impossible for higher levels of management to make knowledgeable decisions as well as they had in the past.

Younger managers and technical professionals also resented not having the authority to make important decisions without approval; when approval was required (which was frequently) it significantly slowed down the process.

It became obvious that a new approach was needed. Enter: professional management development and systematic management. Both internal and external consultants were extensively used to help management actually manage rather than being "super engineers."

Management was persuaded that it needed to delegate more of its decision making and authority, and that planning and the achievement of objectives had to be more systematic. Proper management development also would require broader experiences for key personnel (which was just about everyone in a supervisory capacity), and personnel at all levels needed to improve their interpersonal and problem solving skills.

It was generally conceded that these new managerial directions were effective. The theories made sense and received wide support, both from company personnel and the then current literature. Supposedly everyone was "managing by objectives," and there were frequent and numerous examples of individuals who became better managers and of natural work groups that quickly become more effective teams.

But somehow within this fast paced and apparently improving environment there was also dissatisfaction. In the opinions of many, especially at lower levels, there wasn't enough participation, promotions were going to the wrong persons, and people were being moved too much. Different groups felt alienated from each other: operators from foremen, foremen from lower management, technical from production. etc.

People seemed to know more about management and interpersonal relationships, but specific improvements in actual conditions seemed short lived. Morale was average, and for many persons organizational excitement was replaced by apathy.

Downsizing

Rapidly deteriorating market conditions for polyester products in 1981-82 was certainly the major cause of cutbacks in capacity and personnel in U.S. companies. For some, however, home-grown quality and productivity deficiencies played a key role in the reduction of customer demand for specific lines.

At such a time morale is obviously at its lowest, with whole plants being permanently shut down. And for the first time with a self-critical orientation, some companies made serious efforts to analyze the effectiveness of the ways they went about managing and solving problems. This led to the exchange of ideas and information with agreeable companies in other countries.

To say that these exchange visits were eye openers to U.S. managers and professionals is an understatement. The attainable levels of quality and the effectiveness of production efforts of, e.g., some Japanese facilities, were in themselves enough to stimulate almost immediate improvements in some aspects of polyester production.

The insights gained from this critical self analysis, as well as recent successful experiences in quality and productivity improvement projects, suggest at least five pertinent questions about the way that management theories are put into practice.

The questions at first glance seem deceptively simple and commonplace; they are not at all. The answers to them are fundamental to any management philosophy and strategy.

- Do people know how to do their jobs?

- Do lower levels of the organization know what upper levels expect?

- Is the production equipment developed to its fullest potential?

- Are people organized in the most effective way?

- Do people want to do their jobs?

These questions have implications well beyond the effectiveness of technical training and the application of motivational theory. They also relate to the style of management delegation, the type and amount of employee participation, the supervisor/subordinate ratio, job rotation practices, management development and promotion policies, MBO practices, and the size and maturity level of the organization.

Most attention will be given to the first question since it is the most fundamental. Many of its implications are also true of the other questions.

Do People Know Their Jobs?

There is a significant difference between being technically competent and knowing the technology.

When new engineers are hired from the top 20% of their classes, and when they demonstrate obvious technical competence in their conversations with others, it is easy to assume that they know how to do their present jobs. Similar assumptions can be made about middle managers who successfully take over a new assignment, about operators who quickly detect and correct an equipment malfunction, etc.

These can be dangerous assumptions because they allow a management to continue to grow and to expand with its current style without facing up to its known deficiences. It is only when compared to what is demonstrated to be possible that the assumptions are seen to be erroneous.

Two major reasons for the lack of knowledge or skill of American workers are both caused by management: (1) rapid movement of personnel at all levels; and (2) the philosophy (at least in practical effect) that a supervisor is primarily a manager or administrator, almost to the exclusion of being a teacher or technical decision maker.

A common dilemma in a large fast growing company is the newly appointed manager who has four subordinates with an average of nine months in their jobs, who are also managing inexperienced people. The attempt to be a "professional manager" by delegating and holding subordinates responsible for their results is disastrous.

The top manager cannot educate or help in technical decisions. Unfortunately, instead of becoming a true team member, some elect to criticize subordinates for not being willing to make decisions or accept responsibility. Failures are blamed on subordinates' incompetence; technically smart but inexperienced persons are replaced by "proven high flyers" and the situation is exacerbated elsewhere.

It is correct that managers need to delegate, leave technical decisions to subordinates, manage objectives vs. details, etc., but only if subordinates have the prerequisite experience and competence. A major mistake in any hightech industry is to make the transition from management-by-technicians to management-by-managers without ever going through the education phase.

Actually neither extreme is acceptable. There are always technical decisions that can be delegated, and yet there also are product or process decisions appropriate for every level of the organization. In the same vein, the education of subordinates shouldn't be a "phase," although it almost has to be, given the fast start-up strategy of some highly competitive technological companies.

The old bromide, that all development is self-development and is the responsibility of the individual, is a concession to the preferences of managers-on-the-move who prefer the sink-or-swim approach. It is true of course, in the sense that no one can develop anyone else. But it also makes it easy for supervisors not to consider the education of their subordinates as their responsibility. (In some cases it is a virtual impossibility, with an operator:foreman ratio of 20:1.)

It is always dangerous to draw conclusions based on comparisons of different cultures, but in at least one successful Japanese company the average person worked in only one functional area, even through a series of three or four promotions, for thirteen years before being broadened by actual job assignment into other areas. The primary responsibility for technical training was the supervisor's, who also taught persons what they needed to know about other functions. The operator:foreman ratio was 5:1.

Problems were solved at the lowest possible level. But if required, several organizational levels could get involved in the discussions, and all had the technical training and experience commensurate with their particular level.

In summary, management must strike a balance between training people in depth within a given functional area, or cross-training persons for greater flexibility and breadth of understanding. (Of course, much personnel movement simply is due to changing market demand and a lack of production stability.)

Crucial to this decision is the rate of growth that is pursued, the role of supervisors as teachers, promotional policies (short term oriented persons currently have a tremendous advantage), and the size and complexity of the organization. The larger and more complex, the more lower level persons must know their jobs and have the authority to do them.

Do Lower Levels Know What Upper Levels Expect?

One of the problems that accompanies size and a fast turnover of personnel is that continuity is hard to maintain and effective communication between levels deteriorates. This is especially true of the communication of values and strategy.

It is quite possible for top management to publicize its genuine commitment to improving quality, engage in comprehensive "awareness" programs, and actually provide funds to support the effort. Yet in OD sessions with lower levels of the organization one frequently will hear that management is only interested in production, not quality. Somehow the message does not get through.

Again, the traditional short term orientation of management at all levels can prevent even a statement of a genuine change in management expectations from being believed. This element of management credibility will be considered later.

Is the Equipment Developed to Its Fullest Potential?

A surprise to Americans who visited other countries was that their basic technology and equipment was not much different from our own. However, they were doing more with it via modifications and innovative alternatives to prohibitively expensive solutions to problems. Whereas we exerted more effort to monitor and correct process variances, they spent more time analyzing and reducing their causes.

During the growth period of the American polyester industry, management strategy was to improve productivity primarily via "the big bang," or the technological breakthrough. Genuine breakthroughs did occur and the strategy was profitable. Unfortunately, this was done too much to the exclusion of exerting technical effort to improve the existing equipment.

More importantly, a genuine effort was never made to train or encourage nonprofessionals or nontechnical professionals to improve existing equipment.

For example, an existing network of supply lines for a given chemical process may create periodic problems because the tube diameters are too small. Since the expense of replacing them may be prohibitive, the quick solution is to train the operator to take corrective actions at certain times.

Alternatives, however, may be to insulate the lines so that the chemical does not cool down so fast, or to insert an automatic drain valve at a certain point to remove sediment, or whatever. The point is that our orientation is to manage the present problem or replace the equipment, vs. taking front-end time to remove the cause.

Are People Organized Effectively?

Some organizational structures currently in use are the result of the perceived need to quickly train persons to do their jobs, or simply not realizing the need to change an original design.

This issue has been treated extensively in the literature and will not be elaborated here except to note that, as with the other questions, this one is closely tied to management philosophy.

The reduction of jobs to the most simple functions that need to be performed results in fast training and a much greater flexibility in the movement of people. It also communicates the value of management that workers are merely part of the equipment and that they aren't expected to use their brains.

On the other hand, workers can be given more complete tasks that require judgement, given the necessary training, and allowed to develop stable work relationships. This communicates a management value that recognizes the potential of workers, not only to handle the product, but to contribute creatively to the improvement of the way the product is produced.

This issue is not limited to production workers. The same is true of professionals and managers, whose jobs sometimes become so specialized that they are little more than "gofers."

Some of our greatest benefits from productivity/quality improvement efforts have been the result of reorganizing persons in a more effective way. And most of the beneficial organizational changes have been suggested by the persons directly involved in the activity.

Do People Want To Do Their Jobs?

I propose that a unifying theme among this and the previous four questions is an issue of value, and it has a lot to do with what we usually think of as motivation.

It is the degree to which there is an environment of equity within the organization. Not equity in a monetary or formal status sense, although these are factors, but in the sense of genuine respect for the human worth of all the members of the organization.

This value, or lack of it, permeates all management action and has profound effects on individuals' desires to truly do their jobs.

In a somewhat superior fashion, American management in many cases has abstracted itself several levels from the production floor, or from wherever the hands-on work is occurring. It identifies with the results of production rather than with the products and producers themselves.

In an almost detached manner, it sees scientific management and behavioral science methodologies as means of getting others to do their work better. This has to be a major cause of so many short lived successes or outright failures of efforts to improve productivity, the interpersonal climate, or team effectiveness.

The American worker too often sees himself as a recipient (victim?) of another program geared to shape him up, rather than as a fully accredited participant in an effort of substance. This understandably causes bewilderment among managers who are operating with the best of intentions and who can't understand why lower levels never really join the team.

A symptom of the problem was described by a Japanese visitor who criticized American workers for not having a sense of inquiry — "not being interested enough to ask why." When he had asked them about aspects of the process, or why certain things were done, they often could not answer him with the level of sophistication of Japanese workers in comparable positions.

It should be no mystery why some workers have a low level of curiosity about their jobs when: they lack technical background, they are frequently transferred from job to job, they are constantly reporting to new supervisors, and there is little tradition of asking for their opinions in the first place.

Again, this is a management problem. The flood of questions and suggestions that are generated under properly designed conditions (job enrichment projects, sociotechnical efforts, quality of worklife programs, etc.) indicates that the curiosity and desire is there; it always surfaces in those groups that are "turned on," even temporarily.

Unfortunately, these conditions are not often a permanent part of management style and exist only when the time and opportunity are designed into the schedule. Management tries to design motivation and participation into the system *via* programs for others, rather than becoming an active part of the system itself.

And the message is not lost on employees.

Some Beginnings

There are two promising procedures that have proved to be

at least partially successful in helping to provide some answers to the questions and problems described above. The first is an almost radical attempt (by traditional standards only) to improve the climate of equity in the organization. The second is an expanded version of the traditional type of training usually given to those who actually operate the equipment.

The Fusion Process

A new type of "diagonal slice" approach for introducing change or a new management direction has been shown to be remarkably effective. It has three elements that make it unusually powerful in breaking down barriers between levels and functions, and in fusing a natural work group into a team.

First, the diagonal slice includes all levels of a department or facility, e.g., from plant manager to operator. The seating arrangements are "random stratified," with all levels and functions mixed in the subgroups throughout the room. All participate equally in whatever discussion topics have been selected.

Second, the procedures are designed in such a way that there are opportunities for genuine confrontation of relevant issues. The experience of raising, discussing and sometimes solving sensitive group-identified problems has a very positive effect on later teamwork.

Third, the stated goal of the process is to improve the climate of equity. After prepatory discussions, the groups can deal with issues such as male/female, black/white, exempt/nonexempt, or maintenance/production relationships, as well as the overall climate of the facility.

In effect, the Fusion Process clarifies the behavioral and performance norms of the facility in such a way that they have total team involvement and commitment.

The benefits go well beyond the actual subject at hand. The process itself allows a high level manager to re-establish contact with the realities of the production floor, and lower levels get a first-hand appreciation of the problems faced by upper management. And all have a much better understanding of the expectations of others. The process also provides the format to discuss subjects that are most important to people, such as the conditions that relate to the five questions cited here.

Certification Workshops

The traditional training of operators is limited to the handling of equipment and to practice in building physical endurance. In effect, workers become an extension of the machine.

In a certification workshop the training deals with the relevant theoretical aspects of the process and equipment. Besides operators, it can include new engineers, technicians, supervisors, and R&D personnel. There is a much greater emphasis on "why" and "how come."

It has been found that such training often is two-way. That is, experienced engineers who were instructors sometimes learned new insights from the experienced operators. Also, insightful operator questions sometimes led the way to the exploration of profitable solutions to problems.

This beneficial exchange, if for no other reason than its practical value, leads to better communication later on the job.

Some have observed that a single workshop such as this, with a mix of personnel as both participants and instructors, does far more for motivation than the typical "interpersonal skills" approach. (It should be strongly emphasized, however, that the facilities with which I am familiar that have used this approach already had had interpersonal-skills kinds of training.) The simple fact of recognition which is real does more for self-esteem than a hundred "attaboys" (which are important also).

It should be recognized that these two suggestions are only first steps, although quite effective ones. The ultimate ideal is that different levels of the organization consider themselves as having equal membership in the team. To the degree that this occurs, the educational exchange of useful information will be a standard part of any interpersonal relationship.

Conclusion

Thomas Friedman and Paul Solman reviewed the current debate about the long vs. short term orientation of American management by concluding:

> . . . they come up against an age-old dilemma of capitalism itself: how to efficiently harness the self-interest on which the system itself runs. [1]

I suggest that this problem will get worse if managers continue to see their profession as a higher level abstraction away from their products and their producers. Any facility that produces a product or service has a technology that encompasses it, and there are technical decisions that are appropriate for every level of the organization.

A company can neglect the technical development of its personnel, and still attempt growth, only at its own peril. The type and degree of this development also communicate what values the company places on its human resources, and what roles it expects them to play. In addition, it has far more to do with the quality of employee interpersonal relationships than do sophisticated interpersonal skills themselves.

When the solution to growth and complexity is seen to be primarily a management control issue, the problems

---

[1] Friedman, Thomas and Paul Solman, "Is American Management Selfish?" *Forbes*, January 17, 1983, p. 77.

get worse. Instead of encouraging persons to work with each other, the application of systems such as MBO separate people and encourage them to work against each other.

Although designed as participative management tools, they actually substitute form and structure for personal involvement. The situation can be especially destructive when the wrong technically inexperienced but "professional manager" is appointed to "turn around" a group that is having serious technical problems.

This is not an indictment of MBO, financial controls, management systems, or currently popular productivity programs, but rather of the reasons they are chosen and of the ways and the conditions in which they sometimes are implemented. In the long run, these solutions relate less to productivity than the far more basic issues discussed here.

Somehow a greater sense of community, with a climate of equity, must be developed in our organizations. I fail to see how this can be accomplished without more meaningful personal contact among organizational members. The amount of contact does not have to be great, but it must be of substance and enough so that persons at all levels can identify with the larger group, as opposed to their own narrowly defined special interests.

II. PRODUCTIVITY MEASUREMENT AND ANALYSIS

Productivity measurement and analysis involve the systematic selection of the productivity measurement approaches that are to be used, and the inputs and outputs for the organization. In measuring productivity several factors must be considered, including the nature of the inputs and outputs, the complexity of the operations, and the level on which productivity is being measured (company, department, individual, etc.).

Reprinted from **Industrial Engineering,** *July 1984.*

Work Environment Survey Generates Ideas On Increasing White Collar Productivity

Jeffrey K. Liker
and Walton M. Hancock
University of Michigan

The full productive potential of the white collar work force in manufacturing organizations is often not realized. This is a serious problem that is likely to become more serious as the number of white collar jobs increases.

Much attention in past research and practical applications has focused on the blue collar work force, while white collar workers in business organizations have been largely insulated from studies of their efficiency and effectiveness. The assumption often seems to be that carefully selecting an array of talented professionals—engineers, draftsmen, designers, etc.—is all that is necessary for the functioning of any medium to large-sized organization and that, left to their own devices, these white collar employees and their managers have the knowledge, resources and organization to be productive.

When white collar employees make up a minority of the work force, some inefficiency may be tolerable in that the additional costs associated with this inefficiency are not significant. However, in world markets where United States manufacturing organizations are disadvantaged, in part because of the number and cost of white collar employees, their productivity becomes cause for major concern.

The purpose of this article is to describe a method for improving the performance of white collar work forces. This survey-based method has been used on blue collar and white collar work forces with impressive results (see Hancock and Karger, "For further reading"). Our intent is to further develop this method so that it can become a standard tool for industrial engineers or other professionals seeking to diagnose and reduce work place deficits.

A management concern

The fact that there has historically been much more interest in the productivity of the blue collar work force than in that of white collar workers has several possible causes.

First, the tasks of blue collar workers are clearly definable, predictable and measurable. Much of classical industrial engineering consists of breaking an operation into smaller component parts and grouping these parts into clearly specified jobs. With the work divided in this way, workers can be taught to perform their particular jobs, and their performance times can be compared to standard data on expected performance times and quality levels. The immediate results of their activity can be measured as "outputs."

The main attraction of Taylor's scientific management was that it provided foremen with specific guidelines on how and what to manage (see "For further reading"). They knew what the workers should be capable of, what they should do, how they should do it and how long it should take them to complete the job. Performance could be evaluated, improvements could be detected, controls could be implemented and production planning became possible with reasonable accuracy.

Compared to the often well defined and clearly identifiable jobs and tasks of their blue collar counterparts, the work of white collar employees is relatively intangible. In fact, it is unclear what "output" is for white collar workers; the results of their efforts might more appropriately be termed "outcomes."

It is seldom clear, for example, what an engineer should do, let alone

33

Productivity improvement methods that have traditionally been used with blue collar workers are difficult to apply to white collar jobs.

exactly what steps will be required and how long each step will take. At best, experience often provides rough guidelines on how long it may take to design a specific part. In such cases an experienced design engineer is probably in the best position to estimate completion time.

Second, the blue collar work force historically has outnumbered the white collar work force and thus has comprised a major component of labor cost. Currently, however, with production schedules cut back and with increased use of automated manufacturing systems, the need for blue collar workers is declining.

At the same time, an increase in the use of high technology equipment has led to an increase in the demand for white collar workers. Thus in many industrial organizations the major component of labor cost is no longer the blue collar, but rather the white collar, work force.

Third, when there has not been enough work for direct labor, blue collar workers have been laid off, whereas layoffs of white collar work-

ers have generally not been considered a viable option. Therefore, there has been a strong financial incentive for increasing the efficiency of blue collar workers in order to reduce labor needs.

Since white collar workers have not been laid off due to variations in demand, no similar incentives have existed for them, at least not until recently. Why measure something that you are not going to change, anyway?

Finally, considerable research and several company programs have focused on the use of performance standards to motivate blue collar employees, but the motivation of white collar employees has received little attention. In part, this is due to the relative financial benefits of making the blue collar employees more productive. In addition, there is the common assumption that white collar workers are already intrinsically motivated and do not require close supervision, whereas the repetitive job of the blue collar worker lacks inherent motivation (see Hack-

man and Oldham, "For further reading").

Approaches to improvement

Whatever their strengths and weaknesses, the productivity methodologies that have been used traditionally for blue collar workers do not apply to most jobs found in the white collar work force. Time study, work sampling, predetermined time systems, etc., work reasonably well for more predictable and repetitive clerical operations, and they are being successfully applied in these areas. However, methods need to be developed for improving the productivity of engineers, technicians, draftspersons, accountants, purchasers, inventory controllers and other white collar personnel.

There are many computer-aided approaches to productivity improvement with tremendous potential (computer conferencing, CAD, spreadsheet programs and MIS, for example); however, computer-aided tools such as these will not solve the problem of "poor organization."

To define "poor organization," one needs a model of good organization. We define a good organization as a working environment that supports, motivates and directs the productivity of work places. The general model of a productive work place described below provides a basis for assessing the functioning of existing work places.

A work place model

Figure 1 is a diagram of a generalized work place for any kind of productive work involving a human-machine system. From this model, a set of axiomatic statements can be developed which define ideal conditions for a productive work place as follows:

☐ *Inputs:* The worker should have the materials and information necessary to do the work before he or she starts the work. Otherwise, he/she

34

Figure 1: Generalized Work Place for Productive Work

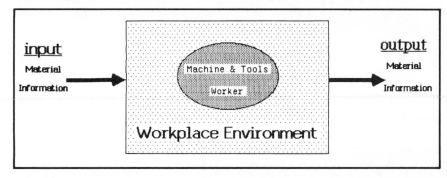

should be able to find and promptly obtain the needed materials and information when required.

☐ *Machines and equipment* should be in good repair.

☐ The appropriate *tooling* should be available in the proper quantity and in good repair.

☐ *The worker* should be capable of doing the work, trained, present and in good health.

☐ *The worker* should be motivated to work hard to accomplish the work consistent with the goals of the organization.

☐ *Task allocation:* Staffing should be adequate (including support staff) and the worker should have enough lead time to accomplish the work.

☐ The *working environment* should not threaten the health of the worker or distract the worker from accomplishing the work.

☐ *Material output* should not impede the continuance of the work. (Generally, this is a problem only where a physical product is produced that is not promptly removed from the work place.)

The model and axiomatic statements will not be new to industrial engineers. In essence, this is a definition of the goals of industrial engineering—to see that none of these aspects of the work place impede the work flow. We argue that this model applies to white collar work as well as to the more repetitive, and clearly definable, tasks on which industrial engineers have traditionally concentrated.

All the factors above, of course, have different degrees of relative importance depending on the work that has to be done. For example, the tools of a design engineer are different from those used by draftspersons. In most white collar work, input will generally be information, and materials input will be minimal. The output will also generally be information (e.g., drawings, a plan of action, etc.).

What is the optimum use of existing personnel? A proper assessment of whether productivity is optimum and where productivity gains are possible depends on three factors:

1.) An estimate of the productive potential of a given organization or unit within an organization.

2.) A quantitative method of estimating discrepancies between the present utilization and the productive potential.

3.) A method of identifying the specific barriers to full productive potential.

Since it is so difficult to predict how long a work assignment will take and exactly what has to be done, the productive potential is not directly predictable for most white collar jobs. However, it is possible to substantially improve productivity without directly measuring its level on any absolute scale. This is done by estimating the differences between ideal work place functioning and actual work place functioning.

The proposed approach draws on employees' knowledge of their jobs. Particularly in the case of white collar work, the specific tasks necessary are highly variable from company to company, from assignment to assignment, and sometimes even from month to month.

It is seldom possible to accurately measure the extent to which tasks are accomplished efficiently. However, if we present employees with axiomatic statements about what their work places should be like and then ask them if they are functioning as the axiomatic statements say they should, common systems deficiencies can be identified and targeted for improvement.

The procedure developed by Han-

cock (see "For further reading") is designed to measure and identify "work place deficits;" i.e., the extent to which specific work place deficiencies are preventing employees from being as productive as they can be. This technique also provides an estimate of the potential cost reduction from eliminating each deficiency.

Work place assessment

Since we have a model of how any work place should function, we now have the problem of assessing each work place in an organization or unit of an organzation. Typically, there is no systematic way of making such an assessment within existing organizations.

Even if the original organizational and work place design functioned well when it was established, there is a general trend toward disorganization unless work places are periodically assessed with an eye toward maintenance. Such periodic assessments are generally prompted by a change of management or occasional employee complaints. However, employee complaints cannot be counted on, since people tend to adapt to conditions as they are, in which case work place improvements are more likely to be reactive than proactive.

There are a number of ways to assess the functioning of existing work places. One possibility is to send an industrial engineer to observe people at work and interview them. However, this approach requires a substantial amount of the time of an experienced professional who must learn much of what the worker already knows. Also, the presence of the industrial engineer may bias responses or change the behavior of the worker under scruti-

Figure 2: Productivity Improvement Cycle

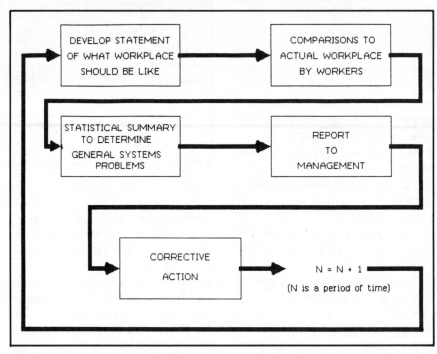

ny, and common problems that span many people cannot be identified without great expense.

It is clearly desirable to obtain such information from the worker in a more efficient and perhaps less threatening manner. Questionnaires can be administered quickly and inexpensively, and anonymity can be offered if desired.

Many employees complain that they fill out many questionnaires that do *not* lead to concrete results. The following guidelines can help ensure that a questionnaire designed to assess work place functioning will not be perceived as "just another attitude survey":

☐ Top management should be committed to productivity improvement.

☐ The questionnaire should be based on an explicit model of productivity improvement. The model above provides a general starting point, but should be refined through discussion with management to fit the particular goals and priorities of the organization.

☐ The questionnaire should focus on aspects of the work place that realistically can be changed. Asking people about problems raises their expectations that the problems will be solved.

☐ Whenever possible, recommendations should be presented in economic terms so that the costs and benefits can be assessed and management has an incentive and justification for action.

☐ The questions should be time constrained to obtain a reasonable time span of recollection and to reduce bias from unusual recent events; for example, how often has information you needed been unavailable over the *last three months?*

☐ Generality versus specificity must be carefully balanced. On the one hand, questions can be so specific they apply to only one person's situation. On the other hand, questions can be so general that the data will

gloss over specific, critical problems faced by some sub-units. This may mean designing different questionnaires for different sub-units; e.g., separate accounting and design engineering questionnaires.

☐ Questionnaires should always be pretested for language clarity and should use terms commonly understood by organization members. A half-dozen workers from a given work group may be sufficient for pretesting the questionnaire.

An improvement cycle

When the first batch of questionnaires has been filled out by workers and statistically analyzed, the work is by no means finished. Work place assessments should become a regular activity, since work places change. Figure 2 shows the sequence of activities, of which data collection and analysis is only a part. The recommended procedure is to:

1.) Develop statements of what the work place should be like based on the general model above as modified by discussions with management. The process of pulling out specific management objectives can be useful in itself, since many managers previously have not had to think through in detail how work places should be functioning.

2.) Elicit comparisons between man-

agement's conception of how the work place should be functioning and the workers' reaction to how it is functioning.

3.) Develop statistical summaries of worker responses to determine general systems problems along with estimates of economic costs.

4.) Summarize these problems and proposed solutions in the form of a report to management.

5.) Take corrective action.

6.) Repeat the whole process to evaluate the corrective actions and plan the next set of corrective actions.

Examples of questionnaires

To test the concept that workers' expertise can be used to improve productivity, a questionnaire was developed for the engineering staff of a large public utility. The engineering unit was involved in the design of substations for electrical switchgear. Questions were asked about each aspect of the worker's environment.

Figure 3 is an example of the information input part of the questionnaire, which was specifically designed for an engineering work place whose main job was the development of design drawings. An axiomatic statement concerning the *ideal* work place environment for that particular aspect of the job preceded each set of questions. Note particu-

Figure 3: A Portion of the Worker Questionnaire

THE INPUTS TO YOUR WORK

IDEALLY, BEFORE STARTING A JOB, YOU SHOULD HAVE ALL THE INFORMATION, DRAWINGS, MATERIALS, AND INSTRUCTIONS YOU NEED TO DO THE JOB. IF YOU BEGIN AN ASSIGNMENT BEFORE YOU HAVE ALL THAT YOU NEED, YOU MAY NOT BE ABLE TO DO AS GOOD A JOB AS YOU CAN. CONSIDER YOUR PRESENT JOB WITH THIS IN MIND.

1. When you are scheduled to start your assignments, do you have all the verbal and written information you need? ..... [1] [2] [3] [4] [5] [6]

2. If you do not have all the information, is it readily available if you contact the appropriate person? (NOTE: Please check "Does Not Apply" if you answered question number 1 with a response of Always or Often) ..... [1] [2] [3] [4] [5] [6]

3. When you do not have all the information needed to start your assignment, how often do you begin anyway? ..... [1] [2] [3] [4] [5] [6]

4. When you do not have all the information needed to start your assignment, does it delay your work? ..... [1] [2] [3] [4] [5] [6]

5. If you were to start your assignment without all the information you need:

   a. Does the quality of work suffer? ..... [1] [2] [3] [4] [5] [6]

   b. What percent (%) of your work suffers? ..... _____ % (NOTE: please fill in the percent - 0 - 100%)

   c. How much does the quality of work suffer?

     [1] To a very great degree
     [2] To a large degree
     [3] To some degree
     [4] To a slight degree
     [5] Not at all
     [6] Does not Apply

Note: Figures 3 and 4 are reproduced from *Assessing Organizational Change*, by S. E. Seashore, E. E. Lawler III, P. H. Mirvis and C. Cammann (eds.), New York: John Wiley & Sons Inc. Reprinted by permission of John Wiley & Sons Inc.

larly questions 4 and 5, the first of a series of questions designed to convert the input problems into economic terms.

Figure 4 is also part of the questionnaire used with the engineering unit. This section is an attempt to get at the economic aspects of the engineering group. The questions concerning rework are not general, but are specifically directed towards the rework issue, which appeared to be a problem during the development of the questionnaire. The results of the questionnaire affirmed the importance of rework and enabled the authors to estimate its economic impact.

The questionnaire, after pretesting, took approximately 20 minutes for each person to complete. It covered information inputs, adequacy of worker training and tools used, and the work place in general. The questionnaire was administered to over 290 engineers and staff personnel (see Hancock, Macy and Peterson, "For further reading").

The results revealed serious work place deficiencies. Lack of sufficient information prior to starting a job resulted in 22.4% of the jobs having to be redone. Rework took an average of 27.0 hours; a normal job (no rework) took 61.0 hours.

Lack of proper environmental impact data prior to starting a job; the necessity of involving higher administrative personnel in decisions that could have been covered by proper guidelines; and poor information flow were the major problem areas.

After attempts were made to correct these situations a second set of data was collected using the same questionnaires. The increase in productivity was 9%, or the equivalent of 26.1 full-time employees. It meant that additional workers did not have to be hired even though the work load was increasing. The annual savings from not having to hire addition-

al employees was estimated at over $600,000.

The same questionnaires were given to a comparable engineering group at a second site during the same period, but at this site the changes suggested by the questionnaire were not made. This engineering organization did not improve in the ways the first site did, which supports the idea that the intervention at the first site was the cause of the observed changes.

More recently, the same approach was tried with the design division of a major automobile company. The entire design engineering staff (100 engineers) was surveyed. Because of insufficient time and poor information flow, engineers were processing incomplete product designs. Other activities that were consuming the

majority of their time included routine clerical and technical work because secretaries and technicians had been dramatically cut back as part of "cost-cutting" efforts.

Figure 5 shows the portion of the questionnaire designed to estimate the percent of engineering time devoted to activities that could be handled by support staff. It begins with an axiomatic statement defining how engineers should *not* be spending their time and then asks them on this basis to describe the portion of a "typical week over the last three months" that: (1) *should* be devoted to specific activities and (2) was *actually* devoted to specific activities.

Through the survey it was determined that design engineers were typically spending 12% of their work

Figure 4: Economic Portion of the Questionnaire

20. Thinking of your work over the past three months, what percentage (%) of your assignments had to be done over or modified? (Please fill in the percent - 0 to 100%)

_____ % of assignments

21. Thinking of these modifications or rework, did the quality of the finished work improve?

[1]	[2]	[3]	[4]	[5]	[6]
To A Very Great Degree	To A Large Degree	To Some Degree	To a Slight Degree	Not At All	Have No Idea

22. Thinking of your work over the last three months, what percentage (%) of your time was spent on rework or modification? (Please fill in the percent - 0 to 100%)

_____ %

23. Thinking of your work over the past three months, what percentage (%) of your time was spent without a work assignment? (Please fill in the percent - 0 to 100%)

_____ % of time without assignments

24. Thinking of your work over the past three months, how many hours did your typical work assignment take to complete? (Please fill in the average number of hours per assignment, 0 to 999 hours; NOTE: If your typical assignemnts lasts longer than three months, report total hours needed for completion.)

_____ average hours per typical assignment

25. Thinking of your work over the past three months, how many hours did you spend redoing or modifying your typical work assignment? (Please fill in the average number of hours per assignment, 0 to 999 hours; NOTE: If your typical assignment lasts longer than three months, report total hours involved in redoing or modifying work.)

_____ average hours per typical assignment

26. On the average, what percentage (%) of time do you spend during an eight hour day being productive (i.e., a work day could be divided up into the following three components: (a) Meaningful work, (b) Less than meaningful work, and (c) No work at all. Please do not consider coffee breaks and lunch as "No work at all").

_____ %
Meaningful Work

_____ %
Less Than Meaningful Work

_____ %
No Work At All

TOTAL = _____ 100% of an eight hour day

---

week on activities they felt could be adequately carried out by secretaries and 21.5% of their work week on activities they felt could be turned over to technicians. Moreover, they estimated that in an average week they spent 19% of their time directly on design work, but should be spending 44% of their time to design component systems for cars right the first time.

The consequences of inadequate design included higher costs for production, prototype design and warranty repairs. In addition, considerable money was spent on processing engineering change requests (ECRs).

This organization did not penalize engineers for processing ECRs, so they were an easy safety valve for fixing hastily designed components. As a result, engineers got caught up in a vicious cycle in which processing ECRs took time away from new product designs, which meant new product designs were not designed correctly, which meant additional ECRs in the future.

Recommendations included additional secretaries and technicians, better information systems and careful monitoring of ECRs to see if their cause was inadequate original design. The potential savings from these changes exceeded the costs by more than 20:1. Since the survey was only recently conducted, it is too early to tell what impact the results will have on the organization.

Conclusions

Traditional industrial engineering tools that have been used to improve the productivity of blue collar workers do not apply to white collar work. However, the underlying model of work used by industrial engineers does apply. The method described here is an efficient way of systematically assessing how actual work place functioning fits this work place design model by drawing on an invaluable resource—workers' knowledge of their jobs.

In both of the applications described above, the questionnaires were easy to write, pretest, administer and analyze. The speed with which employees' assessments relative to axiomatic statements can be developed and analyzed leaves the maximum amount of investigative time for problem solution.

This is in marked contrast to the

Figure 5: Time Use Questions for Design Engineers

<u>TIME USE</u>

KEEP THE FOLLOWING <u>IDEAL</u> CONDITION IN MIND AS YOU FILL IN THE TABLE BELOW: IN ORDER TO USE YOUR TIME MOST <u>EFFECTIVELY</u>, YOU SHOULD NOT BE DOING THINGS THAT (1) DO NOT AID YOU IN YOUR WORK AND (2) COULD BE DONE JUST AS EFFECTIVELY BY SOMEONE AT A LOWER LEVEL WITHIN THE ORGANIZATION.

9. Table I below includes a list of activities an engineer might do. Using column A, estimate the percentage of the time you <u>actually</u> spend on each activity. This percentage should reflect a "typical" week over the last three months. (NOTE: Column A should sum to 100%.) Please read #10 and #11 before answering #9.

10. Again thinking of a typical week, estimate in column B of Table I the percentage of time you think you <u>should be</u> spending on each activity to get the job done right and on time. Please keep in mind the ideal conditions above. (NOTE: Column B should sum to 100%.)

11. For some activities listed in Table I, you may have estimated higher percentages of time spent on the activity (Column A) than the ideal stated in Column B. Some of this additional time might be delegated to someone else. Using the list of personnel below, enter the code in Column C indicating to whom you think some of this activity ought to be delegated. Enter as many or as few codes as necessary for each activity. (EXAMPLE: Enter 1,3 in Column C if you think secretaries and technicians ought to be delegated some of this work.)

1 - Secretary
2 - Draftsman
3 - Technician
4 - Different Engineer in Your Division

5 - Engineer in Another Division
6 - Summer Student
7 - Other Qualified Person

TABLE I: DESIGN ENGINEERING ACTIVITIES	A	B	C
1) Mail handling			
2) Copying (Copy Machine, Blueprinting)			
3) Gathering/searching for files (Microfilm, Vault)			
4) Filling out forms			
5) Telephoning			
6) Scheduling			
7) Proof-reading			
8) Attending scheduled meetings (2 days notice)			
9) Attending unscheduled meetings (< 2 days)			
10) Writing letters			
11) Searching for people/walking in building			
12) Engrg., design work not covered above			
	100%	100%	

usual situation, in which problem identification dominates the investigation. Even the seemingly difficult questions concerning economic impacts were answered by most respondents with little hesitation.

Though white collar work has been the focus here, the method described has also been used to assess the work place of two blue collar work forces with great success. The blue collar workers knew that there were problems that were causing decrements in output, but until the questionnaire, there had not been any systematic outlet for communicating these problems. Complaints of workers tend to be sporadic, and without any statistical evidence of the scope or impact of work place deficits, improvements are not sought.

As production processes become increasingly automated (through computer-aided manufacturing), repetitive blue collar work—the focus of traditional productivity improvement tools like time and motion studies—is apt to become a small part of manufacturing costs. New approaches to productivity improvement will increasingly be needed. We have described one approach.

For further reading:

Hackman, R. and Oldham, G. R. *Work Redesign,* Massachusetts: Addison Wesley, 1980.

Hancock, Walton M., "Quality, Productivity, and Workplace Design: An Engineering Perspective," *Journal of Contemporary Business,* Vol. 11, No. 2, 1982, pp. 107-114.

Hancock, Walton M.; Macy, Barry; and Peterson, Susan, "Assessment of Technologies and Their Utilization," Chapter 9, pp. 257-284 in *Observing and Measuring of Organizational Change, A Guide to Practice,* New York: Wiley Interscience, 1983.

Karger, Delmar and Hancock, Walton, *Advanced Work Measurement,* New York: Industrial Press, 1982.

Packer, M. B., "Measuring the Intangible in Productivity," *Technology Review* Feb/Mar 1983, pp. 48-57.

Taylor, Frederick W., *The Principles of Scientific Management,* New York: W. W. Norton and Co., 1947.

Jeffrey K. Liker is an assistant professor of industrial and operations engineering, University of Michigan. Liker is currently involved in research on white collar productivity and human resource management in computer-integrated manufacturing systems. He is a member of IIE.

Walton M. Hancock is a professor of industrial and operations engineering and professor of hospital administration at the University of Michigan. Hancock is currently involved in productivity improvements as director of the management system division of the Center for Research in Robotics and Integrated Manufacturing. He is an IIE fellow.

Reprinted from 1984 Fall Industrial
Engineering Conference Proceedings

A REVIEW OF SOME APPROACHES TO THE MEASUREMENT OF TOTAL
PRODUCTIVITY IN A COMPANY/ORGANIZATION

Dr. David J. Sumanth, Ph.D.
Department of Industrial Engineering, University of Miami

Kitty Tang, B. S. (I.E.)
Florida Power & Light Co.

ABSTRACT

Since the 1973 oil embargo and energy crisis in the
United States, the downtrend of the U.S. economy is
demonstrated by high unemployment rate, high
inflation rate, increased foreign competition and
numerous other indicators. These indicators have
alerted many U.S. companies who have a growing
concern in knowing how to measure and evaluate
their performance. In the recent years, there has
been a growing number of research papers dedicated
to address the issue of productivity measurement.
The main focus of this paper is to bring together
the central focus on total productivity concepts
of measuring productivity at the company/organiza-
tion level.

INTRODUCTION

The objective of this paper is to catalog and describe six different Total Productivity Measurement
Models which are published in the open literature.
The format for the discussion on each model is
divided into three sections. The first part of the
discussion will provide some background information
about each model. The second part will describe the
model in general terms, and the definition for each
variable required by the model. The third part will
provide a numerical example of the model in a summary format. The six models for measuring the total productivity at the company/organization level
are listed as follows:

1. Kendrick-Creamer Model 1965
2. Craig-Harris Model 1972-1973
3. Hines Model 1979
4. Sumanth Model (TPM) 1979
5. American Productivity Center Model (APC) 1979
6. Swain-Sink Oklahoma Productivity Center
   (OPC) Model 1982

KENDRICK AND CREAMER MODEL (1965)

Background

In 1965, Kendrick and Creamer wrote a book titled
"Measuring Company Productivity"(1965). This book
was written exclusively to address the productivity
measurement issue at the company level. The
Kendrick and Creamer model includes three approaches to productivity measurement: total productivity,
total factor productivity, and partial productivity.
Since the focus of this paper is on total productivity measurement, only the total productivity
measurement approach is discussed here. The concept
of productivity index was introduced in this book.
They stated that productivity indexes can furnish
another reference point for analyzing the profit in
a company. An unfavorable productivity change can
provide signals for early corrective action.

Model Description

They defined total productivity as the ratio of
output to all associated inputs in physical volume.
Kendrick and Creamer explained that it is difficult
to describe the method to measure productivity in
an instructional manual format. There are many
problems when dealing with the measurement of
multiple inputs and outputs. One of the basic
problems is to define the value of physical volume
when dealing with two or more different types of
outputs or inputs in different time periods. To
deal with this issue, they introduced the concept
of total productivity index where each of the physical volumes are weighted by their relative unit
values in a "base" period. This weighting process
transforms the physical unit measure into a "constant price" measure.

In using the constant price concept to measure
productivity, the first step is to weight the
physical units of output - each type of commodity
or service produced, by the base period price. The
next step is to analyze all inputs and categorize
them into four major categories: Labor, Material
and Services, Capital, Indirect Business Taxes.
The real labor cost can be obtained by weighting
the manhours worked in each significant classification by base-period average hourly labor cost. The
real cost of material and contractual services can
also be obtained by using deflator indexes. In
obtaining the real capital cost which includes
depreciation allowances, and income or profit on

equity capital before corporate income tax, the appropriate capital index is suggested to estimate the constant dollar value of net plant, land, equipment and inventories. To estimate the indirect business taxes is to measure the real value of output exclusive of indirect business tax by deducting in each time period the base period ratio of indirect business taxes to value of output. After all the output and input factors have been weighted by the base period value, the total productivity ratio for the company is simply the real value of output divided by the total real cost of input. Therefore, Kendrick and Creamer's model can simply be expressed as follows:

Total Productivity Index = Measured period output in
for the measured period ... base-period price
... Measured period input in
... base-period price

and

Productivity gain (or loss) = Sum of outputs in base-
in the measured period ... period price minus in-
... puts in base period
... price.

They further explained that a base period should be the year when the profit rate is close to the average over one or more business cycles.

Numerical Example

Table 1 is a numerical example to calculate the net gain and index of total productivity of a firm using Kendrick and Creamer's model.

Table 1. Total Productivity of the Firm: Sample Calculation

	1962 A	1965 Current Prices B	1962 Prices C
OUTPUT : Sales value of production (1)	$20,000	$23,000	$22,000
INPUT :			
a. Labor	5,000	5,900	5,500
b. Investor input	4,300	5,000	4,500
c. Raw materials and supplies	6,300	7,200	6,200
d. Outside services	3,000	3,000	2,200
e. Capital equipment depreciation	360	360	417
f. Other	1,200	1,588	1,378
Total Input (2)	20,160	23,048	20,195
TOTAL PRODUCTIVITY :			
Net gain (1C minus 2C)			1,805
Index (1C ÷ 2C)	100.0		108.9

In this example, using Kendrick and Creamer's model, the total productivity of the company in 1965 has increased by 8.9% and resulted in a net gain of $1,805 when compared with 1962-the base year.

CRAIG AND HARRIS MODEL (1973)

Background

In 1973, Craig and Harris developed a model to measure productivity at the firm level. Their model is called a Service Flow Model because physical inputs are converted to dollar values which represent payment for services provided by all inputs. They defined productivity as a measure of the efficiency of this conversion process. Craig and Harris discussed the two types of productivity measurements: Total Productivity and Partial Productivity. They also pointed out that using partial productivity measures (for example, labor productivity) is misleading and the total productivity measurement is a better approach.

Model Description

Craig and Harris defined total productivity as follows:

$$\text{Total Productivity} = \frac{\text{Total Output}}{\text{Total Input}}$$

OR

$$P_t = \frac{O_t}{L + C + R + Q}$$

Where:

P_t = Total productivity
L = Labor input factor
C = Capital input factor
R = Raw material and purchased parts input factor
Q = Other miscellaneous goods and services input factor
O_t = Total output

Output is the summation of all units produced (not units sold) times their base-year selling price. Notice that Craig and Harris also utilized the concept of constant dollar value by using the base-year selling price. The model also includes revenues received from sources other than production such as dividends from securities and interest from bonds and any other such sources.

Input is categorized by Labor, Raw Material and Purchased Parts, Miscellaneous Goods and Services, and Capital. The input factors are defined as follows:

Goods and Services, and Capital. The input factors are defined as follows:

Labor input for current year in base year dollars

= manhours worked in ... base-year wage rate and/
each classification X or salary scale for that
... job classification

Raw material and Purchased Parts for current year in base year dollars

= purchased units adjusted X Base year material
for inventory changes ... price

41

When base-year prices are not available, it is necessary to adjust the current year price by an appropriate commodity price index.

Miscellaneous Other Goods and Services: this group of input consisted of all the other inputs except labor, capital and raw material/purchased parts. All of the input factors in this category must be deflated to the base-year value.

Capital: the capital input factor is defined as the sum of the annuity values calculated for each asset on the basis of its base year cost, productive life, and cost of capital in the base year. The cost of capital can be referred to as the "minimum rate of return" in engineering economic terms. Accounts receivable, securities, inventory and other liquid assets are also part of the capital input factor. The service cost of fixed assets is calculated on the basis of a finite productive life and the liquid asset is based on an infinite productive life. Input cost for these assets is calculated by the formula K x C where K is the value of the asset in base year terms and C is the base year cost of capital for the firm.

Craig and Harris define output strictly as a weighted physical output product of goods and services. That is, interest earned on investments is not included as an output because the firm does not produce interest as capital investment. Likewise, capital input is related to the physical consumption of capital assets. Federal, state and local taxes would not be inputs on the assumption that they are payment for services rendered by the government.

Craig and Harris measure the return to capital as the residual left when all other input factors have been paid for their services. Therefore, profit is the return on capital. In other words, if total productivity is 1.0, the firm breaks even. A value greater than one means profit for the firm and that the firm has made the necessary payment and has some earnings left after making the payment.

Numerical Example

Table 2 shows the total productivity calculations for a company using the Craig and Harris model:

Table 2. Craig and Harris Model Numerical Example
Total Productivity Calculation Summary

	1968	1969	1970	1971
Output (1)	377	415	290	378
Input				
Capital	36	40	41	44
Material	138	141	103	172
Labor	99	101	81	102
Other	99	103	79	130
Total input (2)	372	385	304	448
Total Productivity Index [(1) ÷ (2)] x 100	101.3	107.8	95.3	106.7
Total Productivity in relation to base year	100.0	106.4	94.1	105.3

In comparison, the Craig-Harris model is similar to the Kendrick-Creamer model. The major difference between the two models is in the definitions of the input and output factors. Furthermore, the major improvement in the Craig-Harris model is the handling of capital input. Kendrick and Creamer use the "book value" whereas Craig and Harris express the cost of capital through the concept of "lease value".

HINES' MODEL (1976)

Background

In 1976, W. W. Hines developed a model to measure total productivity at the company/organization level from an Industrial Engineer's point of view. The objective of his model was to be able to track the company performance over time and also to be able to focus attention upon the effects of changes in the mix of input and outputs.

Model Description

Hines' model is expressed as follows

$$\text{Total Productivity} = \frac{\text{Total Output}}{\text{Total Input}}$$

$$TP_i = \frac{O_i}{L_i + C_i + R_i + Q_i}$$

Where:

TP_i = Total productivity for the ith period.
L_i = Labor input factor for ith period.
C_i = Capital input factor for the ith period.
R_i = Raw material and purchased parts input factor for the ith period.
Q_i = Other miscellaneous goods and services input factor for the ith period.
O_i = Total output for the ith period.

Model Description

The following is an outline of Hines' model:

1. Determine the time interval and frequency for the measurement of input and output values. This is necessary because a unit started in one period and finished in another should only account for the fraction of input or output occured during that period.

2. Select the base period for measurement and determine the unit price and cost in that base period. When necessary, estimation of the base period value for the inputs and outputs should be determined using statistical techniques.

3. Define the productivity measurement model in terms of output, and input categories. Included are labor, capital, material input and miscellaneous other inputs in this model.

4. Express output measurement in terms of a sum of products, where each product has two terms:

$$O_i = \sum_j P_j \cdot U_{ij}$$

P_j = price/unit for item j in the base period
U_{ij} = number of production units of item type j produced in period i.

Partially finished output unit is expressed in fractional values credited to the U_{ij} terms for the respective periods.

5. The labor input measure is:

$$L_i = N_{ik} \cdot W_k$$

Where L_i is the labor input measure in period i, N_{ik} is the number of employees in category k in period i, and W_k is the base period wage and salary for category k.

6. Express the capital input as the "Uniform Annual Cost" in engineering economy terms. The value of a piece of capital equipment should reflect the cost of an asset, the desired "minimum attractive cost of return", number of years of service life, and estimated salvage value. To implement the productivity measurement system, it is necessary to obtain a complete inventory of all capital assets. The initial cost must be converted to base year dollars. The productive life and salvage value are estimated for each item and the base year should be used to establish a minimum attractive rate of return; the capital input measure is expressed as:

$$C_i = \sum_j c_{ij}$$

where

c_{ij} = Uniform annual cost for item j in period i.

7. The material input measure is expressed as:

$$R_i = \sum_j V_{ij} \cdot M_j$$

where R_i is the material input for the ith period, V_{ij} is the volume of material type j used in period i and M_j is the base period cost for material j.

8. All other miscellaneous input factors should be expressed in terms of base period cost.

Numerical Example

Table 3 shows a numerical example for the Hines model

Table 3. Numerical Example for Hines Model
Total Productivity Calculation Summary

	1968 Product 1	1968 Product 2	1970 Product 1	1970 Product 2	1968 Total	1970 Total
Output (1)	377	415	290	378	792	668
Input						
Capital	36	40	41	44	76	85
Material	138	141	103	172	279	275
Labor	99	101	81	102	200	183
Other	99	103	79	130	202	209
Total input (2)	372	385	304	448	757	752
Total Productivity (1) ÷ (2)	101.3	107.8	95.3	106.7	104.6	88.8
Total Productivity in relation to base year	100.0	100.0	94.1	98.9	100	84.9

His approach is basically similar to Craig and Harris model. One enhancement is that he expressed the model in mathematical form to deal with productivity measurement in a Company for more than one product in more than one time period.

SUMANTH'S MODEL (1979)

Background

Sumanth (1979) in his doctorial thesis titled "Productivity Measurement and Evaluation Models for Manufacturing Companies" developed a model which he called: The Total Productivity Model (TPM). This model takes into account all the tangible outputs and inputs corresponding to each "product". The model further enables the summation of measurement for each "product" up to the company/organization level where management can evaluate the total productivity of the firm.

Model Description

To explain the model, the following notation is defined:

TPF = Total productivity of the firm
TP_i = Total productivity of the ith "product" ("product" refers to any hard product or operational unit)
j = (H,M,C,E,X)
H = Human input (includes all employees)
M = Material and Purchased parts input (includes raw materials and purchased parts used)
C = Capital input (includes the uniform annual cost of both fixed and working capital)
E = Energy input (includes oil, gas, coal, electricity, etc.)
X = Other Expense input (includes taxes, professional fees, information processing expense, office supplies expense, travel expense, etc.)
i = 1,2,.....,N
N = Total number of "products" in the period under consideration (current period)
O_i = Current-period output of product i in value terms (expressed in constant dollars, or any other monetary unit in base period terms using selling price as the weight)
OF = Total current-period output of the firm in value terms (expressed in constant dollars, or any other monetary unit, in base period terms using selling price as the weight)
= $\sum_i O_i$
I_i = Current-period total input for product i in value terms (expressed in constant dollars, or any other monetary unit in base period terms).
= $\sum_j I_{ij} = I_{iH} + I_{iM} + I_{iC} + I_{iE} + I_{iX}$
I_{ij} = Current-period input of type j for product i in value terms (expressed in constant dollars, or any other monetary unit in base period terms)
IF = Total current-period input used by the firm in value terms (expressed in constant dollars, or any other monetary unit, in base period terms)

$$= \sum_i I_i = \sum_i \sum_j I_{ij}$$

The total productivity of "product" i is the ratio of the total output value of "product" i to the total input cost that is incurred in producing this output.

$$\text{TPF} = \frac{\sum_i O_i}{\sum_i I_i} = \frac{\sum_i O_i}{\sum_i \sum_j I_{ij}} = \text{Total Productivity of a Firm.}$$

or

$$\text{TPF} = \sum_{i=1}^{N} W_i \, TP_i \quad \text{where}$$

W_i represents the fraction of total input for product i with respect to the total of all such inputs combined for the N products manufactured in the firm. Therefore, the equation means that the total productivity of the firm is the weighted sum of total productivities for each of the individual products.

Sumanth further defines the basic model by denoting the period to which the total productivity value refers. He denotes o and t to represent subscripts corresponding to base period, and current period, respectively, so that:

$$\text{TPF}_t = \frac{OF_t}{IF_t} = \frac{\sum_i O_{it}}{\sum_i I_{it}} = \frac{\sum_i O_{it}}{\sum_i \sum_j I_{ijt}} \quad \text{is total productivity of the firm for current period}$$

$$\text{TPF}_o = \frac{OF_o}{IF_o} = \frac{\sum_i O_{io}}{\sum_i I_{io}} = \frac{\sum_i O_{io}}{\sum_i \sum_j I_{ijo}} \quad \text{is total productivity of the firm for the base period}$$

Then, the Total Productivity Index for the entire firm in period t is given by TPIF_t where:

$$\text{TPIF}_t = \frac{TPF_t}{TPF_o}$$

and the Total Productivity Index for "product" i in period t is TPI_{it} where : $\text{TPI}_t = \dfrac{TPF_t}{TPF_t}$

To summarize the Sumanth model, the following equations are appropriate:

Total Productivity for "product" i in period t is TP_{it} where:

$$\text{TP}_{it} = \frac{O_{it}}{I_{it}} = \frac{O_{it}}{\sum_j I_{ijt}} = \frac{O_{it}}{I_{1Ht} + I_{1Mt} + I_{1Ct} + I_{1Et} + I_{1Xt}}$$

Total Productivity for product i in base period is given by TP_{io}, where

$$\text{TP}_{io} = \frac{O_{io}}{I_{io}} = \frac{O_{io}}{\sum_j I_{ijo}} = \frac{O_{io}}{I_{iHo} + I_{iMo} + I_{iCo} + I_{iEo} + I_{iXo}}$$

Total Productivity for the firm in period t is given by TPF_t where:

$$\text{TPF}_t = \sum_i W_{it} \, TP_{it}$$

Total Productivity for the Firm in base period is given by TPF_o where:

$$\text{TPF}_o = \sum_i W_{io} \, TP_{io}$$

and W_{it} and W_{io} are the weights for the current period and base period, respectively. Finally, the Total Productivity Index for the Firm is given by TPIF_t where

$$\text{TPIF}_t = \frac{\sum_i W_{it} \, TP_{it}}{\sum_i W_{io} \, TP_{io}}$$

For more details about the model, see reference [7], chapter 8.

Numerical Example

Tables 4 and 5 provide the numerical example for the Sumanth model. Table 4 is the description of the input factors and the values used to deflate each input factor back to the base period as a reference point. Table 5 is a summary calculation for total productivity of the company having two products in three time periods. As can be noticed, this example can be quite complex when dealing with more than two products and numerous time periods. There is a computer program available for this model on IBM PC, Apple IIe, Univac 1181, and Prime 400 computers, to make the computations easy and routine.

Table 4. Numerical Example for Sumanth Model: Input Variables

	Product 1			Product 2		
	Period 1	period 2	period 3	period 1	period 2	period 3
INPUT:						
1. Human Input						
Total Hours Worked	500	400	350	400	450	500
Ave. Wage rate ($/hr)	6	6.5	6.7	6.2	6.5	6.8
2. Material Input						
Tons of raw Material	3	5	4	6	4	4.5
Price/ton	1	1.3	1.6	1.1	1.2	1.3
Purchased Parts:	100	120	105	160	120	150
Price/Part	1.5	2.0	1.8	1.4	1.5	1.6
3. Capital Input*						
Total Money Value	10000	12000	15000	20000	24000	28000
Deflator	1.0	1.10	1.18	·1.0	1.10	.1.8
4. Energy Input						
Gallon of oil	25	30	28	31	40	42
Price/gal	1.0	1.15	1.18	1.0	1.15	1.18
Tons of coal	3	5	7	4	6	3
Price/ton	5.0	5.5	6.0	5.0	5.5	6.0
KWH of electri.	1000	1500	1600	600	650	700
Price/KWH	0.5	0.55	0.6	0.5	0.55	0.6
5. Other Expense Input						
Consulting Fee	1000	4000	1500	2000	4000	8000
Information Expense	3000	350	400	300	400	450
Marketing Expense	200	100	150	180	700	300
Deflator	1.00	1.15	1.18	1.00	1.15	1.18

* Fixed Capital input only; working capital is assumed zero just for this example. Usually W-Capital input is a finite value so that the break-even point will be less than 1.0.

Table 5. Numerical Example for Sumanth Model:Total Productivity Calculation Summary

	Product 1			Product 2			Total Firm		
	Period 1	period 2	period 3	period 1	period 2	period 3	period 1	period 2	period 3
Output :									
1. Finished units	10000	12500	15000	19000	20000	12000	29000	32500	31000
2. Partial units	2500	2600	5500	2100	2850	3200	4600	5450	8700
3. Dividends (sec)	1000	1818	1304	1200	1802	2252	2200	3620	3556
4. Interest (Bond)									
5. Others Income									
Total Output	13500	16918	21804	22300	24652	17454	35800	41570	39258
Input :									
1. Human	3000	2400	2100	2480	2790	3100	5480	5190	5200
2. Material	153	195	170	231	172	215	384	367	385
3. Capital	10000	10909	12712	20000	21818	23729	30000	32727	36441
4. Energy	540	805	863	351	395	407	891	1200	1270
5. Other Expense	1500	3869	1737	2480	4435	7415	3980	8304	9152
Total Input	15193	18178	17582	25542	29610	34866	40735	47788	52448
Total Productivity	0.89	0.93	1.24	0.87	0.83	0.5	0.88	0.87	0.75
Total Prody. Index	1.0	1.05	1.39	1	0.95	0.57	1.0	0.99	0.85
Break - even Point of Total Prody.	1	1	1	1	1	1	1	1	1

The TPM is a systematic approach to measure and monitor the total productivity of any operational unit over a period of time. It has both the diagnostic and prescriptive capabilities to steer the management in the right direction to productivity improvement. The TPM uses the breakeven concept to relate total productivity of the company to profitability. Associated with TPM is the concept of Productivity Evaluation Tree (PET) as a tool to systematically evaluate the changes in total productivity of the product in the firm. The Sumanth model represents one of the first Total Productivity Models that has been described in a systematic, instructional manual format.

AMERICAN PRODUCTIVITY CENTER MODEL (APC)

Background

In 1979, the American Productivity Center (a privately-funded, non-profit organization, founded in 1977) developed a Total Productivity model called the APCOMP Performance Measurement system to evaluate the productivity of a company/organization.

Model Description

The center defines productivity as output over input. Specifically, productivity is an organization's output divided by its labor, capital, energy and material resources. The data required for the APCOMP Performance Measurement System are value, quantity and price for each time period and for each output and input of the entity being analyzed. The system can treat two time periods, two plants, actual vs. budgeted or any other two element comparison. The Center describes the model more as a general approach whereas the application and implementation are more specific and tailor-made to each organization.

The value, quantity and price of the various outputs and inputs required by the model can be derived from most basic accounting systems. Depending on the circumstances, general and specific price indexes are sometimes used in lieu of actual price. Quantities of unlikes are combined using base period price weighting and price data are combined with current-period quantity weighting.

Numerical Example

Tables 6 and 7 show a numerical example of the American Productivity Center Model. Columns 1,2,4, and 5 are the primary input data required for this model. The model will calculate the input and output values as shown in columns 3 and 6. Columns 7,8,9 are the weighted change ratios for quantity, price, and value. They represent the percentage increase (or decrease) of an item from the base period to the current period. Columns 10,11,12 represent the dollar variance of an item from the base period to the current period for value, quantity, and price, respectively. Columns 13 and 14 are the values for each period, displayed again as in columns 3 and 6. Columns 15, 16, 17 show, respectively, the values for the weighted performance indexes of productivity, price recovery and profitability. Column 15, change in productivity, is calculated by dividing the total output quantity change ratio in column 7 by the corresponding input quantity change ratio in column 7. For example, the 7.6% increase in labor productivity was the result of a 21.1% increase in total output quantity and a 12.5% increase in total labor used; that is, 1.211 / 1.125 = 1.076. The value in column 16, change in price recovery, is the result of dividing the total output price change ratio in column 8 by the corresponding input price change ratio in column 8 (see Table 6). Column 17, change in profitability, is the result of dividing the total revenue value in column 9 by the corresponding input cost value in column 9. The model then calculates the dollar impact of the change in the productivity, and the price recovery of each input; These values are shown in columns 18 and 19. The total dollar effect of both productivity and price recovery are combined for all inputs to explain the total change in profits from one period to the next, and this is displayed in column 20. The model further calculates the output effect on margin as well as the total margin.

The APC model is an economic accounting-based measurement system. The model includes the analysis of the relationship between productivity, price recovery, and profitability. Changes in productivity consist of controllable and uncontrollable components. The controllable components are changes in utilization and efficiency while uncontrollable components may be the change in price recovery which is largely a function of a market place. The APC model analyzes the profit of a company from the standpoint of productivity measurement as well as changes in price recovery.

Table 6. Numerical Example for American Productivity Model

	Period 1			Period 2			WEIGHTED CHANGE RATIO			DOLLAR VARIANCE		
	quantity	price	value	quantity	price	value	quantity	price	value	quanity	price	value
	Q_{i1}	P_{i1}	(1x2)	Q_{i2}	P_{i2}	(4x5)	$\frac{Q_{i2}P_{i1}}{Q_{i1}P_{i1}}$	$\frac{Q_{i2}P_{i2}}{Q_{i2}P_{i1}}$	$\frac{Q_{i2}P_{i2}}{Q_{i1}P_{i1}}$	(4-1)x2	(12-10)	(6-3)
	(1)	(2)	(3)	(4)	(5)	(6)	(7)	(8)	(9)	(10)	(11)	(12)
Product A Product B Product C	16505	12.88	212584	19495	13.40	261233	1.181	1.048	1.229	34511	10137	48648
TOTAL OUTPUT			608362			748535	1.211	1.016	1.23	128475	11694	140173
Inputs :												
Helper Class 1 Leadman Mangement	11232	4.62	51892	7512	5.08	38161	0.669	1.100	0.735	-17185	3454	-13731
Total Labor Input			254840			317956	1.125	1.109	1.248	31939	31177	63116
Oil Electricity Scrap Wood	5825	2.47	14400	5883	2.56	15049	1.01	1.035	1.045	143	506	649
Total Energy Input			17500			18899	1.044	1.034	1.080	769	630	1899
Material Wood Chemical 1 Chemical 2	1310	22.0	28820	1310	23.0	30130	1.00	1.045	1.045	0	1310	1310
Total Materials			37856			38788	0.985	1.040	1.025	-564	1496	932
Inventory Product A Inventory Steel Building Land Equipment Depreciation	11109	0.20	2222	11109	0.21	2311	1.0	1.040	1.040	0	89	89
Total Capital			187921			200485	1.012	1.055	1.067	2200	10364	12564

46

Table 7. Numerical Example for American Productivity Center Model

	Period 1 value	Period 2 value	WEIGHTED PERFORMANCE RATIOS			EFFECT ON PROFITS		
			produc tivity	price recovery	profit ability	produc tivity	price recovery	profit ability
	(13)	(14)	(15)	(16)	(17)	(18)	(19)	(20)
subtotal Helpler								
Subtotal Leadman								
Subtotal Management								
Total Labor	254840	317956	1.076	0.916	0.986	21879	-26277	-4399
Total Energy	17500	18899	1.160	0.982	1.139	2926	-293	2633
Total Material	37856	38788	1.229	0.977	1.201	8559	-768	7790
Total Capital	187921	200485	1.197	0.963	1.153	37486	-6751	30735
Total Input	498116	576128	1.133	0.939	1.064	70849	-34090	36759
Output Effect on Margin						23282	2120	25402
Total Margin						94131	-31970	62161

OKLAHOMA PRODUCTIVITY CENTER MODEL
MULTI-FACTOR PRODUCTIVITY MEASUREMENT MODEL
(MFPMM-OSU/OPC Version)

Background

The Oklahoma Productivity Center (OPC) was establi-shed in 1976. The center is located within the School of Industrial Engineering and Management of the Oklahoma State University. The Objective of the center is to offer the knowledge and assistance to improve productivity and quality of work-life in the U.S. industry. OPC defines productivity as a relationship between quantities of outputs from a given organizational system, and quantities of inputs used by that organizational system to create those outputs.

Model Description

The Multi-Factor Productivity Measurement model is a dynamic aggregated computerized model, particularly applicable at the firm level. The required information for the model is periodic data for output quantities and price, and input quantities and cost. Output and inputs are categorized in a hierarchy structure of Class-Type-Level. Two time periods (base period and current period) of output and input data are required by the Model. Again, like the other models, the selection of the base period is a critical decision. As expected, the base period should be a normal period.

Numerical Example

Tables 8 and 9 show a numerical example of the Oklahoma Productivity Center Model. Like the numerical example in the APC model, columns 1,2,4 and 5 are the primary input data required for this model. The model will calculate the input and output values. Columns 7,8,9 are the weighted change ratios for quantity, price and value. They represent the percentage increase (or decrease) of an item from the base period to the current period. Columns 10 and 11 show the revenue ratios for each period. These ratios represent the percentage of reported revenue consumed by a particular input in a given period. These two columns show where cost reduction will pay the biggest dividends. Columns 12, 13, and 14 hold the value for the weighted performance indexes of productivity, price recovery, and profitability. Columns 15 and 16 hold the value of the dollar impact of changes in productivity, price recovery, and profitability of each input. Column 17 shows the value for the total dollar effect of both productivity and price recovery. This value displays the total change in profits from one period to the next.

The unique feature of this model is that it allows the manangement to conduct sensitivity analysis of the project data. The user has the opportunity to make projections of the desired value of "total inputs effect on changes in profitabiltiy" by trying any combination of output and input quantities and prices. The model utilizes the Monte Carlo simulation method which generates 100 random outcomes which represent 100 simulated values for total inputs' effect on change in profitability based upon projections. The simulation routine also gives a probability of achieving the desired level of profitability based on the input projections. This model gives the user an opportunity to create different scenarios (make different projections) and see the sensitivity of the projected data on the dollar effects as profits.

Conclusion

This paper has summarized six methodologies in measuring the total productivity at the company/organization level. Although, this paper has provided background, description, and numerical example for each of these methods here, the intent of this paper is to merely provide an overview rather than a critical analysis. Such an analysis will require a more thorough investigation and understanding of each of these approaches.

Seven years ago, it was not a common practice to talk about firm-level productivity measurement using the total productivity perspective. This paper demonstrates that, even though the six methodologies have many common features, the total productivity concept is more "acceptable" now. In fact, the 1984 World Productivity Congress in Oslo also emphasized the total productivity perspective.

This research effort also demonstrates that there is a greater interest now in company-level productivity measurement and this is a good sign of knowledge-enhancement in a rather important area of productivity research.

Table 8. Numerical Example for Oklahoma Productivity Center Model

	Period 1			Period 2			WEIGHTED CHANGE RATIO			COST/REVENUE RATIOS	
	quantity	price	value	quantity	price	value	quantity	price	value	period 1	Period 2
	Q_{i1}	P_{i1}	(1x2)	Q_{i2}	P_{i2}	(4x5)	$\frac{Q_{i2}P_{i1}}{Q_{i1}P_{i1}}$	$\frac{Q_{i2}P_{i2}}{Q_{i2}P_{i1}}$	$\frac{Q_{i2}P_{i2}}{Q_{i1}P_{i1}}$	$\frac{I_{ij1}}{Q_{i1}P_{i1}}$	$\frac{I_{ij2}}{Q_{i2}P_{i2}}$
	(1)	(2)	(3)	(4)	(5)	(6)	(7)	(8)	(9)	(10)	(11)
Product A Product B Product C	16505	12.88	212584	19495	13.40	261233	1.181	1.048	1.229		
TOTAL OUTPUT			608362			748535	1.211	1.016	1.23		
Inputs :											
Helper Class 1 Leadman Mangement	11232	4.62	51892	7512	5.08	38161	0.669	1.100	0.735	0.09	0.05
Total Labor Input			254840			317956	1.125	1.109	1.248	0.42	0.42
Oil Electricity Scrap Wood	5825	2.47	14400	5883	2.56	15049	1.01	1.035	1.045	0.02	0.02
Total Energy Input			17500			18899	1.044	1.034	1.080	0.03	0.03
Material Wood Chemical 1 Chemical 2	1310	22.0	28820	1310	23.0	30130	1.00	1.045	1.045	0.05	0.04
Total Materials			37856			38788	0.985	1.040	1.025	0.06	0.05
Inventory Product A Inventory Steel Building Land Equipment Depreciation	11109	0.20	2222	11109	0.21	2311	1.0	1.040	1.040	0	0
Total Capital			187921			200485	1.012	1.055	1.067	0.31	0.27

Table 9. Numerical Example for Oklahoma Productivity Center Model

	WEIGHTED PERFORMANCE RATIOS			EFFECT ON PROFITS		
	productivity	price recovery	profit ability	productivity	price recovery	profit ability
	(12)	(13)	(14)	(15)	(16)	(17)
subtotal Helpler						
Subtotal Leadman						
Subtotal Management						
Total Labor	1.076	0.916	0.986	21879	-26277	-4399
Total Energy	1.160	0.982	1.139	2926	-293	2633
Total Material	1.229	0.977	1.201	8559	-768	7790
Total Capital	1.197	0.963	1.153	37486	-6751	30735
Total Input	1.133	0.939	1.064	70849	-34090	36759

REFERENCES

1. American Productivity Center (APC) 1979; J. W. Kendrick 1981-1982 APCOMP Performance Measurement System, Sales Brochure from APC.

2. Craig, C. E. and Harris, C. R. 1973. Total Productivity Measurement at the Firm Level. Sloan Management Review, 14(3): 13-29.

3. Hines, W. W. 1976. Guidelines for Implementing Productivity Measurement. Industrial Engineering 8(6): 40-43.

4. Kendrick, J. W., and Creamer, D. 1965. Measuring Company Productivity: Handbook with Case Studies. Studies in Business Economics, No. 89. National Industrial Conference Board, New York, NY.

5. Sink, D. S. and Swain, J. C. 1983. Current Development in Firm or Corporate Level Productivity Measurement and Evaluation. Sales Brochure from OPC.

6. Sumanth, D. J. 1979. Productivity Measurement and Evaluation Models for Manufacturing Companies. Published Doctoral Thesis, Illinois Institute of Technology, Chicago.

7. Sumanth, D. J. 1984. Productivity Engineering and Management. McGraw-Hill, Inc., New York, N.Y. (567pp).

BIOGRAPHICAL SKETCHES

Dr. David J. Sumanth is an Associate Professor and Director of the Productivity Research Group in the Department of Industrial Engineering at the University of Miami. Dr. Sumanth has lectured on "Productivity Engineering" and Productivity Management in many countries of the world during the last five years. He has contributed several articles and conference papers to IIE. He is a Senior Member of IIE, and past president of the Miami Chapter of IIE during 1982-1983. He is a recipient of numerous awards, including a gold medal. His book just published by McGraw-Hill, titled PRODUCTIVITY ENGINEERING AND MANAGEMENT (567pp.), is the first book on the subject, suitable as professional reference material and as a text book.

Kitty Tang is a graduate student at the University of Miami working towards the Masters Degree in Industrial Engineering. She is a Senior Analyst in the Energy Management Planning Department at Florida Power & Light Company in Miami, Florida.

Reprinted from 1984 Fall Industrial
Engineering Conference Proceedings

DECISION SUPPORT SYSTEM DEVELOPMENT
FOR THE MULTI-FACTOR PRODUCTIVITY
MEASUREMENT MODEL

by

Jeff Swaim
Associate Industrial Engineer
Frito-Lay, Inc.

D. Scott Sink, Ph.D., P.E.
Director, Virginia Productivity Center
and Associate Professor
Department of Industrial Engineering
and Operations Research
VPI and State University

BACKGROUND

At the 1983 Annual Industrial Engineering Conference in Louisville, Kentucky, Deane Cruze, then Director of Operations for Boeing's 707/727/737 and 757 Divisions, addressed conference attendees on the topic of "Management's Impact on Productivity (and the Role of the Industrial Engineer)." Cruze challenged Industrial Engineers to "develop optimum computer applications for business data and to package that information in a simple language for management decision making." He commented, "someday we will be able to start off our work day by sitting down to a control panel and within a few minutes know the economic health of the factory. . . We will be able to simulate changes of those events that we can control and know what impact they will have on our productivity. . ." (1983).

The term "Decision Support System" has frequently been used to denote the type of system described by Cruze (computerized systems specifically designed to help managers make decisions). Recently, Decision Support Systems (DSS) have received a great deal of attention; similarly, productivity measurement has been the focus of numerous books, articles, seminars, etc. (e.g., APC, 1978; Bain, 1982; Sadler and Grossman, 1982; Siegel, 1980; and Sink, 1983a, 1983b, 1984a, 1984b). However, very little has been written about the potential contribution that productivity measurement models could make as Decision Support Systems. A study was recently completed in which the Multi-Factor Productivity Measurement Model (MFPMM) was modified and enhanced in hopes that it would become a useful DSS for managers concerned with productivity management. This paper will briefly describe the MFPMM, then summarize the study by covering the methodology, the results, and recommendations for further study and development.

MFPMM

The MFPMM is a dynamic, aggregated, indexed, computerized approach to measuring productivity based upon absolute definitions and formulas for productivity ratios and indexes. The data required for the model is classified as either output information or input information. Output information refers to data concerned with the finished product made (or services offered) during an accounting period. Input information refers to data concerned with the resources (e.g., labor, energy, materials, etc.) used in the production of the output. Specifically, quantity and price for each output and input of the entity being analyzed are required to run the model. This information is straightforward and can be derived from most basic accounting systems. The data must be entered for two time periods: a base period and a current period. The basic structure of the output generated by the MFPMM is shown in Figure 1. Since the features of the MFPMM have been well documented (Sink, 1984a; Sink, 1984b; Sink, Tuttle, DeVries, and Swaim, 1984), they will not explicitly be presented in this paper; however, many of them will become evident as the study is described.

METHODOLOGY

The study was divided into two phases: evaluation and development. The evaluation phase of the study consisted of a field study with a local motel and a review of DSS literature. The information acquired during the review of literature on DSS was used in conjunction with the insights gained from the field study to evaluate the MFPMM as it existed prior to the study. The purpose of the evaluation was to provide answers to the following questions:

1. Is the MFPMM a DSS or does it have the potential to become a DSS?

2. What will it take for the MFPMM to become a DSS or how can the MFPMM be improved as a DSS?

The purpose of the development was dependent upon the answers to the two evaluation questions. Figure 2 depicts the evaluation and development phases of the study as a series of questions and activities. As shown in Figure 2, if it was determined through evaluation that the MFPMM was not a DSS and/or did not have the potential to become a DSS, then it would be improved as a management tool. However, if it was determined that the MFPMM was a DSS or had the potential to become a DSS, then it would be developed and/or improved as a DSS. The next section reviews the results of the study in terms of both the evaluation of the MFPMM and its development.

DATA INPUT COL's 1-6						COL's 7-9			COL's 10-11		COL's 12-13		COL's 14-16			COL's 17-19		
QNTY & PRICE						WEIGHTED CHANGE RATIOS			COST/REVENUE RATIOS		PRODUCTIVITY RATIOS		PERFORMANCE INDEXES			EFFECTS ON PROFITS		
P_1			$P_{2(3)}$													$P_1 \rightarrow P_2$ or $P_1 \rightarrow P_3$		
Q_1	p_1	V_1	Q_2	p_2	V_2	Q	p	V	P_1	P_2	P_1	P_2						

OUTPUTS — GOODS, SERVICES, INFORMATION

Weighted change ratios (outputs):
$$Q:\ \frac{Q_2^0 p_1^0}{Q_1^0 p_1^0} \qquad p:\ \frac{Q_2^0 p_2^0}{Q_2^0 p_1^0} \qquad V:\ \frac{Q_2^0 p_2^0}{Q_1^0 p_1^0}$$

INPUTS — DATA, CAPITAL, LABOR, ENERGY, MATERIALS

Weighted change ratios (inputs):
$$Q:\ \frac{Q_2^1 p_1^1}{Q_1^1 p_1^1} \qquad p:\ \frac{Q_2^1 p_2^1}{Q_2^1 p_1^1} \qquad V:\ \frac{Q_2^1 p_2^1}{Q_1^1 p_1^1}$$

COST/REVENUE RATIOS: INDIVIDUAL COST/REVENUE RATIOS FOR EACH PERIOD

PERFORMANCE INDEXES: PRODUCTIVITY INDEXES, PRICE RECOVERY INDEXES, PROFITABILITY INDEXES

EFFECTS ON PROFITS: CHANGE IN PROFITS DUE TO PRODUCTIVITY, CHANGE IN PROFITS DUE TO PRICE RECOVERY, TOTAL CHANGE IN PROFITS

Figure 1.   MFPMM Basic Structure

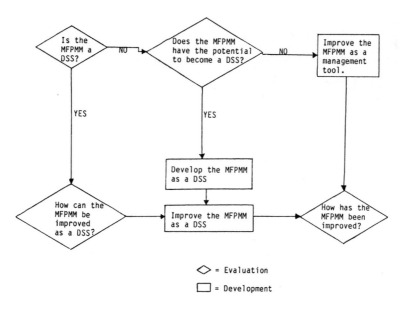

Figure 2.   Evaluation and Development Flow Chart

◇ = Evaluation

▢ = Development

51

RESULTS

Evaluation

During the field study with the local motel, data was collected on a monthly basis and the motel manager was provided with monthly written briefings summarizing the results of the MFPMM run. He was supplied with forms and verbal instructions to assist him in the data collection phase of the study. Interviews with the motel manager were conducted in order to:

1. subjectively evaluate his reaction to the MFPMM;
2. determine whether or not the MFPMM did, or could serve as a DSS; and
3. elicit suggestions on how the MFPMM could be improved as a DSS

The motel manager's reaction to the MFPMM itself was positive. He did express some concern over the absence of written instructions on the data collection forms, and he regretted not having a computer on-site with which to run the MFPMM. He did, however, say that he would purchase the MFPMM software package if he had a computer and he suspected that he would use it almost daily to support decision making. He commented that he would have liked to be able to test (at his convenience) the impacts on performance of such things as new equipment purchases, and changes in labor hours, advertising expenses, and room rates.

The only suggestion that the motel manager offered concerning the improvement of the MFPMM as a DSS was the addition of a year-to-date analysis. His rationale was that it could be used to gain a broader picture of the motel's performance by comparing the year-to-date performance of two years rather than comparing only one particular month of two years. He noted that this capability would especially be useful for budgeting purposes; that is, it could show whether the motel was over-budget or under-budget and identify the areas where surpluses or deficiencies were occurring. He added that the year-to-date information provided by the MFPMM could also be helpful in the preparation of future budgets.

With respect to the field study process, the motel manager agreed that an on-site computer (ideally a terminal in his office) would have improved the process significantly. It would have given him the opportunity to conveniently test as many projected scenarios as he wished, and he would have been able to examine the results of the base-to-current period analysis before entering his projections. Direct access to a computer would have also eliminated the burden of having to submit the data to be run and waiting for the results.

The review of literature on Decision Support Systems disclosed various views on the definitions and characteristics of DSS. The definitions of DSS were composed of words and phrases such as: "interactive," "computer-based," "aimed at semi-structured and unstructured problems," "used on an on-going basis," "provides direct support," "mesh with existing activities," "extend capabilities," "serve decision makers," "query capability," "analytical tools," and "in users' offices." (Sprague and Carlson, 1982; Smartt, 1983; Keen and Scott-Morton, 1978; Welsch, 1981; and Thiel, 1983) It became apparent that any definition of DSS included some combination of those words and/or phrases and described a system that supported decision-making.

The definitions were enhances by identifying characteristics possessed by DSS and which distinguished them from Electronic Data Processing (EDP) Systems, Management Information Systems (MIS), and Operations Research/Management Science (OR/MS) tools and techniques. In an attempt to make the evaluation of the MFPMM a little more objective, a checklist was devised utilizing those characteristics and is shown in Table 1.

It should be clear by now that the MFPMM (as it existed prior to the study) could accurately be referred to as a DSS (based upon the field study and the DSS literature review). The motel manager claimed that the MFPMM supported him in decisions concerning labor hours and salaries, vendor selection, and room rates, even though he did not have the opportunity to use it on an on-going basis. Reinforcing the manager's subjective evaluation, the MFPMM possessed 8 of the 11 characteristics listed in Table 1 and somewhat possessed 1 of the remaining 3. Perhaps the most significant characteristic possessed by the MFPMM (not explicitly stated in Table 1) is the ability to simulate the impact of projected changes in the data which could be vital to productivity-related decisions.

However, some supplementary data (external to the study being described) does exist that should be communicated at this point. Before the field study with the local motel began, a similar field study was proposed to a local manufacturing plant. A management briefing was held in order to present the MFPMM to the plant manager, the controller, and their respective assistants. Essentially, their response to the MFPMM was that it did not offer them any information that they did not already know, and they rejected the proposed field study. Dr. Sink has received similar responses from managers he addresses in seminars and short courses across the country. Over 1000 managers have been exposed to the MFPMM during the last

TABLE 1. DSS CHECKLIST

Characteristics	Does the MFPMM possess them?		
	Yes	No	Somewhat
1. Ability to access, transform, and utilize data external and internal to the DSS.	✓		
2. Ability to examine and validate data easily and efficiently.	✓		
3. Ability to provide the user with a complete array of reporting and display alternatives.		✓	
4. Ability to extend the "What If?" question to the "What's Best?" question.		✓	
5. Provides a friendly, easy to understand dialogue between the user and the DSS.	✓		
6. Aimed at less well structured and underspecified problems.	✓		
7. Focuses on features that make it easy to use by noncomputer people in an interactive mode.	✓		
8. Ability to support all phases (intelligence, design, choice, and implementation) of decision-making.	✓		
9. Emphasize flexibility, adaptability, quick response, user initiation and control, and support for the personal decision-making style of individual managers.			✓
10. Extends the range and capability of a manager's decision processes.	✓		
11. Provides good documentation.	✓		

two years and only six have purchased the software package.

It appears that even though the MFPMM was theoretically a DSS, it was still not viewed by managers (upon initial response) as a system which they could not do without. The challenge to the developers of the MFPMM was, and still is, to continue to search for and implement modifications and enhancements which would/will increase the appeal of the MFPMM to a wide range of managers. In other words, referring to the flowchart depicted in Figure 2, the MFPMM was to be improved as a DSS.

Development

The MFPMM as it existed prior to the study was evaluated to be a DSS. Based upon that determination, several enhancements were made to the MFPMM in order to improve it as a DSS. Those enhancements were made in response to the field study and the DSS literature review and can be categorized as follows:

1. Increased data capacity
2. Improved interactive capabilities
3. Added graphics capabilities
4. Increased flexibility
5. Improved documentation

The following sections will be devoted to each of the five enhancements, respectively. The addition of a year-to-date analysis capability will be addressed in the section titled "Increased Flexibility." (All of these enhancements were made on an IBM-PC.)

Increased Data Capacity

Prior to the study, the IBM-PC version of the MFPMM could handle up to 50 output and input items; that is, 50 categories (including subtotals and totals) could be entered and run by the MFPMM. Now, the program will handle up to 120 categories representing an increase in data capacity of 140%. That increase should set aside enough memory for any organization to include a sufficient amount of their outputs and inputs in a particular MFPMM analysis.

Increased Interactive Capabilities

The interactive capabilities of the MFPMM were improved considerably by the addition of several questions strategically placed in the program. They will be described in the order that they would appear during an MFPMM run.

The first question is:

WOULD YOU LIKE TO SEE A LIST OF THE FILES ON DRIVE B?   Y/N?

When running the MFPMM on the IBM-PC, the disk upon which the MFPMM and the necessary data files are stored is placed in drive B. This question gives the user an opportunity to see which data files he has to choose from before entering the data into the model. An example of this question as it appears during the MFPMM run is shown in Figure 3.

```
WOULD YOU LIKE TO SEE A LIST OF THE FILES ON DRIVE B? Y/N? Y

THE FOLLOWING FILES ARE ON THE CURRENT DIRECTORY OF DRIVE B:
B:\
MFPMMDEV.BAS      PER2   .BAK      PER1   .BAK      ELSOL6C
PER2             PER1             ELSOL5C .BAK      ELSOL5C
ELSOL5B .BAK     MFPMMNEW.BAS     ELSOL5B          ELSOL6B .BAK
ELSOL6C .BAK     ELSOL6B          MFPMM17 .BAS     OE

6
 240640 Bytes free
```

Figure 3.   Listing of the Files on Drive B

The second improvement in the MFPMM's interactive capabilities is the question:

WOULD YOU LIKE TO GIVE MORE DESCRIPTIVE NAMES TO PERIODS 1, 2, AND 3?   (E.G., JUL 1984)   Y/N?

This question gives the user the opportunity to assign more descriptive names to the three periods being analyzed. This question and its subsequent question (assuming the user responds "Y" to the initial question) are displayed in Figure 4.

```
DO YOU WANT TO READ PERIOD 1 DATA FROM A FILE? Y/N? Y
ENTER NAME OF FILE WHERE PERIOD 1 DATA IS STORED? ELSOL6B

DO YOU WANT TO READ PERIOD 2 DATA FROM A FILE? Y/N? Y
ENTER NAME OF FILE WHERE PERIOD 2 DATA IS STORED? ELSOL6C

WOULD YOU LIKE TO GIVE MORE DESCRIPTIVE NAMES
    TO PERIODS 1, 2, & 3 (E.G., JUL 1984) Y/N? Y

ENTER AN 8 CHARACTER NAME FOR PERIOD 1? JUN 1980
ENTER AN 8 CHARACTER NAME FOR PERIOD 2? JUN 1983
ENTER AN 8 CHARACTER NAME FOR PERIOD 3? JUN 1984
```

Figure 4.   Descriptive Names for Periods 1, 2, and 3

The use of these names occurs when the tableaus are displayed. The descriptive names appear where "Period 1," "Period 2," and "Period 3" once appeared as headings above columns 1 through 6 and subheadings under columns 10 through 13. These names are especially beneficial if the user has several MFPMM printouts representing several different periods. A specific printout can easily be identified by the descriptive headings. An example of the use of these names is shown in Table 2.

A third new question appears after the data has been entered and validated. The question is:

DO YOU WANT TO SEE THE PERIOD 1 & 2 TABLEAUS?   Y/N?

Prior to the study, the period 1 and 2 tableaus were automatically displayed after the data was entered and validated. Now, the user has a choice as to whether or not the period 1 & 2 tableaus are displayed. If the user chooses not to see the tableaus, then the program immediately branches to the sensitivity analysis. This option is helpful when the user is well aware of the base period and current period (periods 1 & 2, respectively) results and is concerned only with the sensitivity analysis of projected data.

TABLE 2.   EXAMPLE OF THE USE OF DESCRIPTIVE NAMES

COST/REVENUE RATIOS		PRODUCTIVITY RATIOS	
(10) JUN 1980	(11) JUN 1983	(12) JUN 1980	(13) JUN 1983
0.1856	0.1445	5.39	7.98
0.0866	0.0891	11.55	12.32
0.0510	0.0585	19.60	19.69
0.0359	0.0401	27.88	26.94
0.0186	0.0228	53.89	49.23
0.3776	0.3550	2.65	3.18
0.0247	0.0246	40.42	39.88
0.0062	0.0059	161.68	159.51
0.0263	0.0230	38.04	37.53
0.0572	0.0535	17.48	17.24
0.0402	0.0530	24.86	26.28
0.0494	0.0715	20.23	20.42
0.0145	0.0178	68.95	69.26
0.1042	0.1423	9.60	9.86
0.4020	0.3913	2.49	2.45
0.0294	0.0320	34.04	33.58
0.0124	0.0216	80.84	79.76
0.0309	0.0467	32.34	31.90
0.0155	0.0241	64.67	63.80
0.4902	0.5157	2.04	2.01
1.0291	1.0664	0.97	1.03

A fourth improvement in the interactive capabilities of the MFPMM consists of the following two questions:

WOULD YOU LIKE A HARD COPY OF YOUR QUANTITY ESTIMATES?   Y/N?

WOULD YOU LIKE A HARD COPY OF YOUR PRICE ESTIMATES?   Y/N?

Before the addition of these questions, the only record a user would have of his estimates would be on his data collection forms. Now, he can include a computer printout of his quantity and/or price projections along with the rest of his MFPMM results.

The fifth and final improvement to the MFPMM's interactive capabilities is shown in Figure 5. This option does not offer the user anything new but it only requires one response whereas the old version (Figure 6) potentially required two responses. The new menu-format is much more efficient and more clear to the user.

Added Graphics Capabilities

Prior to the study, the only graphical representation of data offered was a histogram plot and it consisted of simulated values which did not appear anywhere in the MFPMM output. Recognizing that some managers prefer pictures to numbers, columns 10, 11, 14, 15, and 16 of the MPFMM can now be displayed as bar graphs. This option appears in a menu-format (Figures 7 and 8) and is offered

```
YOU HAVE THE FOLLOWING OPTIONS:

    0 - TERMINATE THE PROGRAM
    1 - COMPARE PERIODS 1 & 3
    2 - COMPARE PERIODS 2 & 3

INDICATE YOUR CHOICE BY ENTERING A 0, 1, OR 2?
```

Figure 5.   Selection of Periods to Compare (New Version)

```
DO YOU WANT TO COMPARE PERIODS 2 & 3? Y/N? N

DO YOU WANT TO COMPARE PERIODS 1 & 3? Y/N? Y
```

Figure 6.   Selection of Periods to Compare (Old Version)

```
YOU MAY NOW SELECT FROM THE FOLLOWING GRAPHICS OPTIONS:

    1 - COST/REVENUE RATIOS (BASE PERIOD AND CURRENT PERIOD)
    2 - PRODUCTIVITY INDEXES (BASE PERIOD TO CURRENT PERIOD)
    3 - PRICE RECOVERY INDEXES (BASE PERIOD TO CURRENT PERIOD)
    4 - PROFITABILITY INDEXES (BASE PERIOD TO CURRENT PERIOD)

INDICATE YOUR CHOICE BY ENTERING A NUMBER (ENTER 0 IF
    YOU DO NOT WISH TO SEE ANY GRAPHICS)?
```

Figure 7.   First Graphics Offering

```
YOU MAY NOW SELECT FROM THE FOLLOWING GRAPHICS OPTIONS:

    1 - COST/REVENUE RATIOS (BASE PERIOD AND PROJECTED PERIOD)
    2 - PRODUCTIVITY INDEXES (BASE PERIOD TO PROJECTED PERIOD)
    3 - PRICE RECOVERY INDEXES (BASE PERIOD TO PROJECTED PERIOD)
    4 - PROFITABILITY INDEXES (BASE PERIOD TO PROJECTED PERIOD)

INDICATE YOUR CHOICE BY ENTERING A NUMBER (ENTER 0 IF
    YOU DO NOT WISH TO SEE ANY GRAPHICS)?
```

Figure 8.   Second Graphics Offering

at two different times during the MFPMM run. The first offering (Figure 7) is after the option to see the period 1 & 2 tableaus. The second offering (Figure 8) is after the option to see the histogram plot. Choice 1 displays cost/revenue graphs of the two periods being analyzed. Choices 2 through 4 of Figure 7 display the base-to-current period productivity indexes, price recovery indexes, and profitability indexes, respectively. Choices 2 through 4 of figure 8 display the base or current (depending on which the user chose to compare) to projected period productivity indexes, price recovery indexes, and profitability indexes, respectively.

An example of Choice 1 is shown in Figure 9. Notice that both periods of the analysis are displayed and descriptive names (if entered) are used. Also notice that each "*" represents 0.01 so, for example, the 40 *'s after Note Premium in Figure 9 indicate a cost/revenue ratio of approximately 0.40. In other words, 40% of the motel's revenues were consumed by the cost of the note. (The actual column 10 and 11 values are also printed to the left of the graph.)

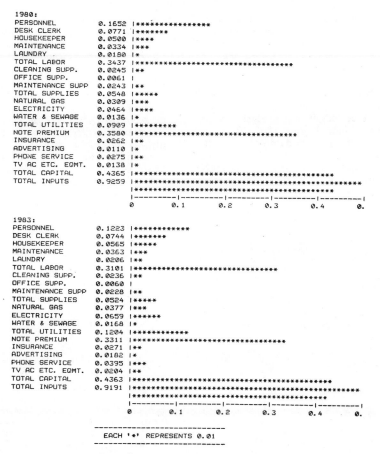

```
1980:
PERSONNEL            0.1652  |***************
DESK CLERK          0.0771  |*******
HOUSEKEEPER         0.0500  |****
MAINTENANCE         0.0334  |***
LAUNDRY             0.0190  |*
TOTAL LABOR         0.3437  |********************************
CLEANING SUPP.      0.0245  |**
OFFICE SUPP.        0.0061  |
MAINTENANCE SUPP    0.0243  |**
TOTAL SUPPLIES      0.0548  |*****
NATURAL GAS         0.0309  |***
ELECTRICITY         0.0464  |****
WATER & SEWAGE      0.0136  |*
TOTAL UTILITIES     0.0909  |*********
NOTE PREMIUM        0.3580  |*********************************
INSURANCE           0.0262  |**
ADVERTISING         0.0110  |*
PHONE SERVICE       0.0275  |**
TV AC ETC. EQMT.    0.0138  |*
TOTAL CAPITAL       0.4365  |*******************************************
TOTAL INPUTS        0.9259  |*******************************************
                            |*******************************************
                            |---------|---------|---------|---------|---------|
                            0        0.1       0.2       0.3       0.4       0.

1983:
PERSONNEL           0.1223  |************
DESK CLERK          0.0744  |*******
HOUSEKEEPER         0.0565  |*****
MAINTENANCE         0.0363  |***
LAUNDRY             0.0206  |**
TOTAL LABOR         0.3101  |*****************************
CLEANING SUPP.      0.0236  |**
OFFICE SUPP.        0.0060  |
MAINTENANCE SUPP    0.0228  |**
TOTAL SUPPLIES      0.0524  |*****
NATURAL GAS         0.0377  |***
ELECTRICITY         0.0659  |******
WATER & SEWAGE      0.0168  |*
TOTAL UTILITIES     0.1204  |***********
NOTE PREMIUM        0.3311  |********************************
INSURANCE           0.0271  |**
ADVERTISING         0.0182  |*
PHONE SERVICE       0.0395  |***
TV AC ETC. EQMT.    0.0204  |**
TOTAL CAPITAL       0.4363  |*******************************************
TOTAL INPUTS        0.9191  |*******************************************
                            |*******************************************
                            |---------|---------|---------|---------|---------|
                            0        0.1       0.2       0.3       0.4       0.

    ----------------------------
    EACH '*' REPRESENTS 0.01
    ----------------------------
```

Figure 9.   Example Cost/Revenue Bar Graphs

An example of Choice 2 is shown in Figure 10. Notice that a "-" indicates below 1.00 (a decrease in productivity) and a "+" indicates above 1.00 (an increase in productivity). Choices 3 and 4 are similar to Choice 2. The user has the option of seeing as many of these bar charts as he wishes before generating new tableaus.

Increased Flexibility

It is sometimes difficult to differentiate between enhancements that improve interactive capabilities and enhancements that increase flexibility. Similarly, increased data capacity and added graphics capabilities

Figure 10. Example Productivity Index Bar Graph

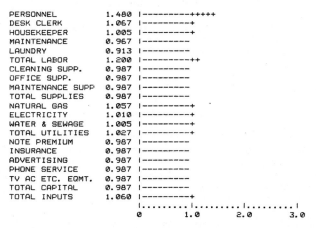

```
            PRODUCTIVITY INDEXES
        (BASE PERIOD TO CURRENT PERIOD)

PERSONNEL          1.480  |----------+++++
DESK CLERK         1.067  |----------+
HOUSEKEEPER        1.005  |----------+
MAINTENANCE        0.967  |---------
LAUNDRY            0.913  |---------
TOTAL LABOR        1.200  |----------++
CLEANING SUPP.     0.987  |---------
OFFICE SUPP.       0.987  |---------
MAINTENANCE SUPP   0.987  |---------
TOTAL SUPPLIES     0.987  |---------
NATURAL GAS        1.057  |----------+
ELECTRICITY        1.010  |----------+
WATER & SEWAGE     1.005  |----------+
TOTAL UTILITIES    1.027  |----------+
NOTE PREMIUM       0.987  |---------
INSURANCE          0.987  |---------
ADVERTISING        0.987  |---------
PHONE SERVICE      0.987  |---------
TV AC ETC. EQMT.   0.987  |---------
TOTAL CAPITAL      0.987  |---------
TOTAL INPUTS       1.060  |----------+
                          |.........|.........|.........|
                          0        1.0       2.0       3.0

    ----------------------------------------------------------
    EACH '-' (BELOW 1.0) OR '+' (ABOVE 1.0) REPRESENTS 0.1
    ----------------------------------------------------------
```

also contribute to increased flexibility. Since the other enhancements have been described in previous sections, only the year-to-date analysis capability will be described in this section.

The question that gives the user the opportunity to run a year-to-date analysis appears early in the MFPMM run. If the user responds "Y" to this question, he is then required to specify the number of periods in the analysis and the names of the data files. (See Figure 11.) Then, the program runs as usual with the year-to-date data compiled into two periods (base and current) consisting of between 2 and 52 sub-periods for each year. Each year can be broken down semi-annually, quarterly, monthly, or weekly, depending on the needs of the user and on how the data is stored.

```
DO YOU WISH TO RUN A YEAR-TO-DATE ANALYSIS? Y/N? Y
HOW MANY PERIODS DO YOU WISH TO INCLUDE? (LIMIT 52)? 4
ENTER NAME OF FILE WHERE PERIOD 1 DATA IS STORED? ELSOL1B
ENTER NAME OF FILE WHERE PERIOD 1 DATA IS STORED? ELSOL2B
ENTER NAME OF FILE WHERE PERIOD 1 DATA IS STORED? ELSOL3B
ENTER NAME OF FILE WHERE PERIOD 1 DATA IS STORED? ELSOL4B
ENTER NAME OF FILE WHERE PERIOD 2 DATA IS STORED? ELSOL1C
ENTER NAME OF FILE WHERE PERIOD 2 DATA IS STORED? ELSOL2C
ENTER NAME OF FILE WHERE PERIOD 2 DATA IS STORED? ELSOL3C
ENTER NAME OF FILE WHERE PERIOD 2 DATA IS STORED? ELSOL4C
```

Figure 11. Year-to-Date Analysis Specifications

Improved Documentation

The only enhancement made during the study that could be considered improved documentation was the design of new data collection forms. The new forms contained written instructions and were much simpler to understand and use.

CONCLUSION AND RECOMMENDATIONS

As mentioned earlier, the potential contribution that productivity measurement models can make as Decision Support Systems has not been sufficiently explored. The study described in this paper not only explored that potential but it took action towards realizing that potential. Prior to the study, the MFPMM was considered to be at least a powerful computerized productivity measurement tool, and through evaluation, it was determined to be, in fact, a DSS. During the study, the MFPMM was significantly improved as a DSS through development based upon input from the motel field study and the DSS literature review. A great deal of progress was made toward meeting Deane Cruze's challenge cited in the opening paragraph of this paper; however, there still remains much to be done and much to be studied.

As with most research and development efforts, the study that took place revealed numerous areas needing further attention. One obvious need is the continued development of the MFPMM. Graphics capabilities could be expanded and improved, interactive capabilities could be enhanced, and the documentation of the MFPMM could be improved and updated. With respect to further study, field studies could be conducted with organizations that have direct access to computers and are able to run the MFPMM on-site. Such field studies could provide much more insight

into how the MFPMM can actually be used as a DSS. Feedback from actual users of the MFPMM would provide invaluable input to further developments.

Due to the everchanging needs of managers and the rapid advancement of computer technology, the possibilities for further study and development utilizing the MFPMM are endless. The field of computerized productivity measurement is in its infancy and the MFPMM provides an excellent mechanism with which to experiment and ultimately advance the field. Perhaps the most important recommendation that can be made is that the MFPMM be at least maintained and hopefully enhanced so that its usefulness as a DSS and as an educational tool does not become obsolete.

REFERENCES

American Productivity Center. (1978). How to Measure Productivity at the Firm Level. Short Course Notebook and Reference Manual. Houston, Texas.

Bain, D. F. (1982). The Productivity Prescription. New York: McGraw-Hill.

Cruze, D. (1983). Management's Impact on Productivity (and the Role of the Industrial Engineer). Institute of Industrial Engineers Spring Conference Proceedings. Norcross, Georgia.

Keen, P. G. W., and Scott-Morton, M.S. (1978). Decision Support Systems: An Organizational Perspective. Reading, Massachusetts: Addison-Wesley.

Sadler, G. E., and Grossman, E. S. (1982). Measuring Your Productivity. Plastics Technology, 28 (1).

Siegel, I. H. (1980). Company Productivity: Measurement for Improvement. Kalamazoo, Michigan: W. E. Upjohn Institute for Employment Research.

Sink, D. S. (1983a). Organizational System Performance: Is Productivity a Critical Component? Institute of Industrial Engineers Spring Conference Proceedings. Norcross Georgia.

Sink, D. S. (1983b, October). Much Ado About Productivity: Where Do We Go From Here? Industrial Engineering, pp. 36-48.

Sink, D. S. (1984a). The Essentials of Productivity Management: Strategic Planning, Measurement, Evaluation, Control, and Improvement. Short Course Notebook. UNPRIM, Inc., Stillwater, Oklahoma.

Sink, D. S. (1984b). Productivity Management: Planning, Measurement, Evaluation, Control, and Improvement. To be published by John Wiley & Sons, New York, New York.

Sink, D. S., Tuttle, T. C., DeVries, S. J., and Swaim, J. C. (1984). Development of a Taxonomy of Productivity Measurement Theories and Techniques. AFBRMC Contract Number F33615-83-C-5071. Wright-Patterson Air Force Base, Ohio.

Sink, D.S., Tuttle, T.C. and DeVries, S.J. "Productivity Measurement and Evaluation: What's Available?" National Productivity Review. Vol. 3, No. 3, Summer 1984.

Smartt, P. C. (1983, January). "Ingredients" for a Successful Decision Support System. Data Management, pp. 26-33.

Sprague, R. H., and Carlson, E. D. (1982). Building Effective Decision Support Systems. Englewood Cliffs, New Jersey: Prentice-Hall, Inc.

Swaim, J. C., and Sink, D. S. (1983). Current Developments in Firm or Corporate Level Productivity Measurement and Evaluation. Institute of Industrial Engineers Fall Conference Proceedings. Norcross, Georgia.

Swaim, J. C., and Sink, D. S. (1984). Productivity Measurement in the Service Sector: A Hotel/Motel Application of the Multi-Factor Productivity Measurement Model. Institute of Industrial Engineers Spring Conference Proceedings. Norcross, Georgia.

Thiel, C. T. (1983, March). DSS Means Computer-Aided Management. Infosystems, pp. 39-44.

Welsch, G. M. (1981). Successful Implementation of Decision Support Systems: The Role of the Information Transfer Specialist. American Institute for Decision Sciences Annual Proceedings. Boston, Massachusetts.

BIOGRAPHICAL SKETCHES

JEFF SWAIM is an Associate Industrial Engineer for Frito-Lay, Inc. at their Irving, Texas Manufacturing Plant. He was recently a Graduate Associate for the Oklahoma Productivity Center at Oklahoma State University. Jeff received his B.S. and M.S. degrees in Industrial Engineering and Management from Oklahoma State University and is a member of IIE and NSPE.

D. SCOTT SINK, P.E., is an associate professor in the Department of Industrial Engineering and Operations Research at VPI and State University and director of the Virginia Productivity Center at VPI. He received his BSIE, MSIE and PhD degrees from Ohio State University. His areas of interest include productivity management and measurement, work measurement and improvement and organizational behavior. Sink is senior member of IIE and is currently director for the management division. Sink is a recipient of the Haliburton Award of Excellence, the Dow Outstanding Young Faculty Award, and the Oklahoma Society of Professional Engineers Outstanding Young Engineer of the Year Award. He is a registered professional engineer in the state of Oklahoma. His book titled, "Productivity Management: Planning, Measurement, Evaluation, Control, and Improvement," will be released by John Wiley and Sons in early 1985.

Reprinted from 1984 Annual International Industrial Engineering Conference Proceedings.

MULTI-CRITERIA PERFORMANCE/ PRODUCTIVITY MEASUREMENT TECHNIQUE

by D. Scott Sink, Ph.D., P.E.
Director, Oklahoma Productivity Center
School of Industrial Engineering & Management
Oklahoma State University
Stillwater, OK 74078

ABSTRACT

An innovative, widely-applicable, and reasonably simple approach to measuring performance or productivity is being developed at the Oregon Productivity Center and at the Oklahoma Productivity Center. The technique can be nicely integrated with the Normative Productivity/Performance Measurement Methodology (Participative approach utilizing the Nominal Group Technique—see issue number 2 of *Productivity Management*) in order to facilitate more effective use of the tremendous number of measures and ratios of productivity or performance that can be attained from groups in organizations. The technique will be briefly presented in this paper.

BACKGROUND

Studies beginning at The Ohio State University in 1975 developed a participative, yet highly structured methodology for identifying consensus productivity/ performance measures for a given organizational system. The value of a participative approach lies in the creation of an "ownership" for the resulting measures. Successful implementation of ensuring measurement, evaluation, and control systems is more assured with effective participative approaches.

However, difficulties in operationalizing measurement systems that have origins in a participative process hindered early efforts. The question of how to evaluate performance against a list of measures that is often highly heterogeneous became a critical one to continued development. William Stewart addressed the issue of how to aggregate and hence evaluate performance against many measures or criteria in his dissertation effort at Ohio State. His approach was to develop a prioritized set of productivity/performance measures utilizing the NGT. (This approach evolved from the NSF sponsored Ohio State Studies of 1975-1977). He then developed a "utility" curve for each of the priority (top eight to ten) measures. A ranking and rating process was executed so as to weight the relative importance of each productivity/performance measure. The utility curve was utilized to transform actual performance against each specific measure or criteria to a common 0 to 1.0 performance score. This performance score, 0 to 1.0, was then multiplied by the relative weight for each measure to obtain a performance value. The various performance values for each of the top priority measures are then added together to obtain a productivity/performance index. Stewart based the procedure upon the works of Morris (1975, 1977), The Ohio State Productivity Research Group (1977), and Keeney and Raiffa (1976).

MANAGEMENT OF MEASUREMENT AND IMPROVEMENT

PRODUCTIVITY MANAGEMENT
Oklahoma State University
Oklahoma Productivity Center
Industrial Engineering and Management

Since those early developments in 1976-1978 at Ohio State, several other efforts have been made in this general area. In 1980, Stewart applied this approach to the common carrier industry. In 1981, William Viana, while a graduate associate in The Oklahoma Productivity Center at Oklahoma State University applied a hybrid design of this procedure in a fairly large, diversified manufacturing firm (gate valves, ball valves, etc.) in Brazil. More recently, Riggs and Felix have developed and published an analogous approach called the "objectives matrix". (1983)

MCPMT PROCEDURE

Assume you have just generated a consensus and prioritized list of productivity/performance measures for a given organizational system utilizing the NGT (See *Productivity Management* issue number 2 and Sink, 1982). You have a list of heterogeneous measures (i.e., apples, oranges, peaches, etc.) You are interested in aggregating or evaluating performance against these criteria in an integrated fashion.

A common performance scale or utility scale needs to be developed that converts all the uncommon measures into some common denominator. The performance scale commonly is rather arbitrarily allowed to range from either 0 to 1.0, 0 to 10.0, or 0 to 100.

10.0 ⊤ "Excellence"

5.0 ┼ "Acceptable"

0 ⊥ "Lowest Possible"

Common Performance Scale for each Productivity/ Performance Measure

Level 0 represents the lowest level of performance possible for a given measure. Level 5 represents a minimally acceptable performance level (MAPL). And, level 10 represents the perception of best performance or excellence. Levels 0, 5, and 10 should be clearly defined and accepted benchmarks.

Each productivity/performance measure or criterion has at least one "natural" scale that it can be or is measured with. Often this "natural" scale is simply an industry consensus or norm. For example, measuring liquids in gallons or liters, measuring coal in tons, measuring profitability performance in terms of ROI, etc.

no complaints   100 complaints
                                 per month

Customer Satisfaction

The objective in the MCPMT is to develop a valid set of natural scales used to measure performance against a given criteria and to match levels of performance on that scale to levels of performance on the common utility scale.

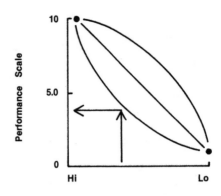

Customer Satisfaction
"Natural" Scale(s) (x-axis)

A utility curve, (such as curves a, b, or c) which can, and often will be, subjective, is developed and used to transform performance on one scale to a common scale. There exist techniques for developing these curves in a valid manner.

The M in the MCPMT stands for multi, which signifies that there are many criteria or measures of productivity/performance against which we are attempting to perform. Therefore, there will often be many of the utility function graphs as depicted above. The question becomes one of how to aggregate performance scores. Since these scores are all from a common scale we might be tempted to just add them up. However, the question of relative impact of performance against various criteria on overall performance becomes critical.

The NGT will have given you a ranked list of productivity/performance measures. The next step is to rate this list. Figure 1 depicts this process. The first step, once the criteria have been ranked, is to arbitrarily assign 100 points to the top priority measures or criteria. Next, the relative importance of the second most important measure is assessed. In the example depicted in Figure 1, which is for a computer center, customer satisfaction is seen as being equally important to projects completed/constant value budget $'s. So, customer satisfaction is also assigned 100 points. This paired comparison relative assignment of points is done for each successive criterion. (i.e., the most important (1st) relative to the second most important (2nd); the 2nd to the 3rd; 6th to 7th, etc.) The total points allocated for all measures or criteria is summed. Relative weights are then determined by dividing the individual points assigned by the total points (i.e., 100/730, . . .80/730). One then has a sense for the relative importance or contribution of each measure or criterion to overall performance/productivity. There are some critical nuances to this procedure that are not described herein, however the approach is basically as straightforward as it seems. These weights can be determined unilaterally by the manager of the group or by an analyst or participatively by the same persons who identified the criteria and their rankings.

#	Criterion	Rank/Priority	Rating	Weight
1	Reports/Projects completed and accepted / Constant value Budget $	1	100	$\frac{100}{730}=.137$
2	Customer Satisfaction	2	100	$\frac{100}{730}=.137$
3	Quality of Decision Support From Systems Developed	3	100	$\frac{100}{730}=.137$
4	Meeting User Flexibility Requirements	4	90	$\frac{90}{730}=.123$
5	Existence of and use of work scheduling/project management	5	90	$\frac{90}{730}=.123$
6	Projects completed on time / Total projects completed	6	85	$\frac{85}{730}=.116$
7	# of requests for rework/ redoing a project	7	85	$\frac{85}{730}=.116$
8	Existence of and Quality of strategic planning for facilities, equipment, staffing, management processes, and operational systems	8	80	$\frac{80}{730}=.111$
			730	1.000

Figure 1.   Ranking and Rating Procedure

The next step in the MCPMT is to integrate the performance (utility) graphs (scales and curves) with the criteria weightings. This will allow the development of one performance/productivity indicator which will indicate the overall performance of the organizational system. Figure 2 conceptually depicts what is happening in this step. Actual performance as measured against the scales represented on the x-axis is transformed into a performance score (0 to 10) on the y-axis. Those performance scores are then multiplied by the criteria weighting factors to obtain weighted scores. Note that these weighted scores all have common units while the x-axis reflects a variety of units. Note also from the computer center example that only one of the eight measures is a pure productivity measure (ratio) and that is criterion #1.

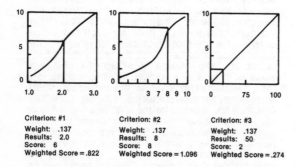

Criterion: #1
Weight: .137
Results: 2.0
Score: 6
Weighted Score = .822

Criterion: #2
Weight: .137
Results: 8
Score: 8
Weighted Score = 1.096

Criterion: #3
Weight: .137
Results: 50
Score: 2
Weighted Score = .274

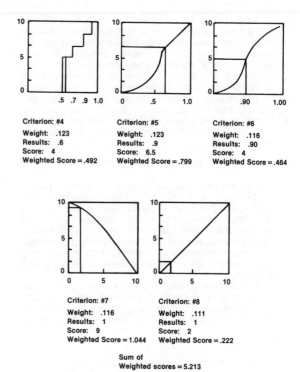

Criterion: #4
Weight: .123
Results: .6
Score: 4
Weighted Score = .492

Criterion: #5
Weight: .123
Results: .9
Score: 6.5
Weighted Score = .799

Criterion: #6
Weight: .116
Results: .90
Score: 4
Weighted Score = .464

Criterion: #7
Weight: .116
Results: 1
Score: 9
Weighted Score = 1.044

Criterion: #8
Weight: .111
Results: 1
Score: 2
Weighted Score = .222

Sum of
Weighted scores = 5.213

Figure 2.   Multi-Criteria Performance/Productivity Measurement Technique

The final computational step in this procedure is to add together all the weighted scores. The value for the computer example is 5.213 out of a maximum score of 10.0. The individual performance scores in addition to the total weighted score or overall performance indicator can be tracked over time and utilized to develop evaluation and control systems.

A general matrix format for this technique is presented in Figure 3. (Riggs and Felix, 1983) Note that column eight of the matrix represents the y-axis and columns 1-7; rows 3-13 represent the x-axis of the utility curves. Note also that it is possible to have sub-criteria for a given measure. In other words, there could be more than one way to operationalize a given performance/productivity measure. In that case, the weighting for the given measure or criteria is simply divided up among the sub-criteria. For instance, if customer satisfaction were operationalized with two independent measures rather than one, then the .137 weighting would need to be divided among the two sub-criteria. This is done by repeating the ranking, rating, and weighting procedure within the criteria itself.

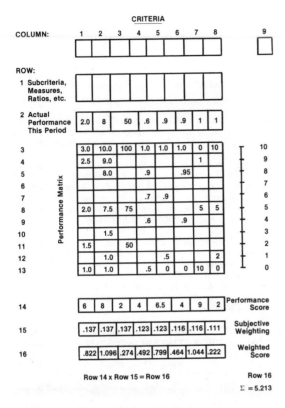

CRITERIA

COLUMN: 1 2 3 4 5 6 7 8 | 9

ROW:

1 Subcriteria, Measures, Ratios, etc.

2 Actual Performance This Period

	1	2	3	4	5	6	7	8		
2	2.0	8	50	.6	.9	.9	1	1		
3	3.0	10.0	100	1.0	1.0	1.0	0	10		10
4	2.5	9.0						1		9
5		8.0		.9		.95				8
6										7
7				.7	.9					6
8	2.0	7.5	75				5	5		5
9				.6		.9				4
10		1.5								3
11	1.5		50							2
12		1.0		.5				2		1
13	1.0	1.0		.5	0	0	10	0		0

Performance Matrix

14	6	8	2	4	6.5	4	9	2	Performance Score
15	.137	.137	.137	.123	.123	.116	.116	.111	Subjective Weighting
16	.822	1.096	.274	.492	.799	.464	1.044	.222	Weighted Score

Row 14 x Row 15 = Row 16

Row 16

Σ = 5.213

Figure 3. Productivity/Performance Measurement Matrix General Format

CONCLUSION

This approach and technique for performance and productivity measurement and evaluation for organizational systems has tremendous potential. The roots of this technique lie in multi-attribute decision analysis which is at least twenty-five years mature. The ideas have been there; American managers and researchers have simply failed to innovate with the basic ideas, theories, concepts and techniques. This is changing. The Japanese have taught us that things don't have to be complex to work.

Current developments with this technique are in the area of validation, methodology refinements and documentation, and creation of a computerized decision support package for the IBM-PC. The IBM-PC software for this technique will be available from our Center in approximately six months. Those interested in learning more about this technique are urged to call our Center and talk to either myself or one of our research associates.

OKLAHOMA PRODUCTIVITY CENTER
Serving the Oklahoma Region and the Nation to Strengthen America's Productivity

OKLAHOMA STATE UNIVERSITY
CENTENNIAL DECADE
1980 · 1990

Oklahoma Productivity Center

The Oklahoma Productivity Center is located in the School of Industrial Engineering & Management at Oklahoma State University. The Director is Dr. D. Scott Sink, associate professor in Industrial Engineering Management. Other faculty in the School engaged in productivity-related activities are: Dr. Wayne C. Turner, Director of the Oklahoma Industrial Energy Management Program; Dr. Kenneth E. Case, Quality Management; Dr. Philip M. Wolfe, Computer Systems; Dr. C. Patrick Koelling, Management Sciences; Dr. James E. Shamblin, Director of the Center for Local Government Technology; Dr. John W. Nazemetz, Manufacturing Systems; Dr. Palmer Terrell, Management Sciences; Dr. Earl J. Ferguson, Management Systems.

Much of the Oklahoma Productivity Center activities and services are coordinated by OSU Engineering Extension. Dr. William C. Cooper is the Director of CEAT extension.

Productivity Management, the Oklahoma Productivity Center's newsletter is published quarterly (Sp, Su, Fa, Wi). The newsletter is available by subscription. The rate is $25 per year (4 issues). A one-year complimentary subscription is provided with attendance at an Oklahoma Productivity Center short course.

For a current copy of OPC's five page listing of available programs, goods and services, please check the appropriate box on the subscription form.

DIRECTORY OF MAJOR PRODUCTIVITY AND QUALITY OF WORKING LIFE CENTERS

The National Productivity Network (NPN) is a consortium of nongovernmental, nonprofit organizations that devote their efforts to productivity and quality of work life research, development, and dissemination/extension. The NPN has been in existence since 1978, but there is a lot of active work going on now to publicize and promote the importance of the network. The organizations within the NPN work on a direct, one-to-one basis to help business and government clients identify and implement management and technical systems that can improve productivity and quality of work life. In this issue, we present a list of major U.S. centers and productivity related publications.

Major U.S. Productivity Centers

Dr. Keith McKee
Manufacturing Productivity Center
Illinois Institute of Technology
10 W. 35th Street
Chicago, IL 60616
(312) 567-4808

American Productivity Center
123 North Post Oak Lane
Houston, TX 77024
(713) 681-4020

Dr. Thomas Tuttle
Maryland Center for Productivity
and Quality of Working Life
University of Maryland
College Park, MD 20742
(301) 454-6688

Mr. LeRoy Marlow, Director
PENNTAP
Penn State University
501 J. Orris Keller Bldg.
University Park, PA 16802
(814) 865-0427

Dr. Rudy Yobs
Georgia Productivity Center
Georgia Tech. Eng. Exp. Station
Atlanta, GA 30332
(404) 894-3404

Dr. James Riggs
Oregon Productivity Center
100 Merryfield Hall
Oregon State University
Corvallis, OR 94331
(503) 754-3249

Dr. Barry Macy
Texas Center for Productivity
and Quality of Working Life
College of Business
Texas Tech University
Lubbock, TX 79409

Dr. William Smith
Productivity Research and
Extension Program
P.O. Box 5192
North Carolina State University
Raleigh, NC 27650

Mr. Joel Goldberg
National Center for Public Productivity
John Jay College
445 W. 59th Street
New York, NY 10019
(212) 489-5030

Dr. Gary Hansen
Utah State Center for Productivity
and Quality of Working Life
Utah State University
Logan, UT 84322

Dr. Scott Sink
Oklahoma Productivity Center
School of Industrial Engineering
and Management
322 Engineering North
Oklahoma State University
Stillwater, OK 74078

Productivity Related Publications

MANUFACTURING PRODUCTIVITY FRONTIERS: This monthly journal is published by the Manufacturing Productivity Center at the Illinois Institute of Technology at 10 West 35th St., Chicago, IL 60616. Annual subscriptions are $100. This is probably the best of the monthlies as it contains about 50 commercial free pages of up-to-date information on topics such as: construction, government, manufacturing, book reviews, seminar schedules, etc. (contact: Keith McKee, 312-567-4808).

THE MARYLAND WORKPLACE is the bi-monthly newsletter of the Maryland Center for Productivity and Quality of Working Life, University of Maryland, College Park, MD 20742. It is a short newsletter focusing on labor, government and management news on productivity issues. (contact: Tom Tuttle, 301-454-6688).

NATIONAL PRODUCTIVITY REPORT is published twice monthly and is four pages long. It is a provocative newsletter, typically covering one topic in some depth. $48 per year. Information and samples are available from NPR, 1110 Greenwood Rd., Wheaton, IL 60187. (contact: Bill Schleicher, 312-668-5146).

PRODUCTIVITY IN ACTION is a monthly newsletter published by the Georgia Productivity Center, Georgia Tech Eng. Experiment Station, Atlanta, GA 30332. A very informative, well done newsletter. (contact: Rudy Yobs, 404-894-3404).

PRODUCTIVITY is one of the newest monthlies and the most expensive at $126. It is 12 pages long and describes itself as "management's tool for improving productivity through worker satisfaction and innovation." Information and samples from: Productivity Inc., P.O. Box 3831, Stamford, CT 06905. (contact: Norman Bodek, 203-322-8388).

PRIMER is a four page newsletter of the Oregon Productivity Center, 100 Merryfield Hall, Oregon State University, Corvallis, OR 97331. It is a very well done and very informative newsletter. (contact: Jim Riggs, 503-754-3249).

THE PRODUCTIVITY LETTER is a bi-monthly publication of the American Productivity Center, 123 North Post Oak Lane, Houston, TX 77024. The objectives are to develop a useful and comprehensive newsletter and to expand national awareness of the need for productivity improvement. Available to Founders, Sponsors and Members as well as to other interested parties. (contact: Carole Roper, 713-681-4020).

PRODUCTIVITY MANAGEMENT is a quarterly newsletter publication of the Oklahoma Productivity Center, School of Industrial Engineering and Management, Oklahoma State University, 322 Engineering North, Stillwater, OK 74078. The newsletter reviews: techniques, methodologies, and approaches being developed and applied for improved management of productivity. Annual subscriptions are $25 U.S., Canada and Mexico; $35 all others.

INDUSTRIAL ENGINEERING is the monthly periodical of THE INSTITUTE OF INDUSTRIAL ENGINEERS. It is impossible to talk about productivity and not include *INDUSTRIAL ENGINEERING*. IIE has provided state of the art insights into productivity management for well over 32 years. *INDUSTRIAL ENGINEERING* is recognized internationally as being a premier source of information about productivity management.

NATIONAL PRODUCTIVITY REVIEW: THE JOURNAL OF PRODUCTIVITY MANAGEMENT is a journal in the style of *Harvard Business Review, California Management Review,* and *Sloan Management Review.* It is a quarterly journal focusing specifically on productivity-related issues such as: effective utilization of human resources, effective utilization of capital, effective uses of materials and energy resources, and organizational technical systems and processes that relate to productivity. The intended audience is executives and managers as well as consultants, faculty, and students. For subscription information, write to: 33 W. 60th St., N.Y., N.Y. 10023. Annual subscriptions are $96. This journal may well emerge as being a significant source of quality information on the topic of productivity.

PREP NEWS is the newsletter for the Productivity Research and Extension program at North Carolina State University. It is an informative newsletter focusing upon communicating activities at NC State and in North Carolina in general. The newsletter is very well done. (contact: Bill Smith, 919-733-2370).

References

Riggs, J.L. and Felix, G.H. Productivity by Objectives: Results-Oriented Solutions to the Productivity Puzzle, Prentice-Hall, Inc.: Englewood Cliffs, NJ, 1983.

Felix, G.H. Productivity Measurement with the Objectives Matrix, Oregon Productivity Center Press: Corvallis, OR, 1983.

Sink, D.S. Productivity Management: Planning, Measurement, Evaluation, Control, and Improvement, John Wiley & Sons, Inc.: New York, NY 1984.

Stewart, W.T. "A Yardstick for Measuring Productivity", Industrial Engineering, Feb. 1978.

Archer, B.L. Technological Innovation: A Methodology, Royal College of Art, London, 1970 (unpublished working paper/manuscript)

Sink, D.S. The Essentials of Productivity Management: Strategic Planning, Measurement and Evaluation, Control and Improvement. (Shortcourse Notebook) LINPRIM, Inc., Stillwater, OK, 1984.

Keeney, R.L. and Raiffa, H. Decision with Multiple Objectives: Preferences and Value Tradeoffs, John Wiley and Sons, NY, 1976.

Huber, G.P. "Multi-Attribute Utility Models. A Review of Field and Field-Like Studies" Management Science, 20., 1974.

Morris, W.T. Decision Analysis, Grid Publishing, Columbus, OH. 1977.

Morris, W.T. Implementation Strategies for Industrial Engineers, Grid Publishing, Columbus, OH. 1979.

MacCrimmon, K.R. and Toda, M. "The Experimental Determination of Indifference Curves", The Review of Economic Studies, 36, 1969.

Sink, D.S. "Initiating Productivity Measurement Systems is MIS" 1978 ASQC Technical Conference Transaction, Chicago, 1978.

Swalm, R.O. "Utility Theory--Insights into Risk Taking" Harvard Business Review, 44, 1966.

BIOGRAPHICAL SKETCH

D. SCOTT SINK, P.E., is an associate professor in the School of Industrial Engineering and Management at Oklahoma State University and director of the Oklahoma Productivity Center at OSU. He received his BSIE, MSIE and Ph.D. degrees from Ohio State University. His areas of interest include productivity management and measurement, work measurement and improvement and organizational behavior. Sink is a senior member of IIE and is currently director for the management division. Sink is a recipient of the Haliburton Award of Excellence, the Dow Outstanding Young Faculty Award and the Oklahoma Society of Professional Engineers Outstanding Young Engineer of the Year award. He is a registered professional engineer in the state of Oklahoma.

(This paper is excerpted from an upcoming book entitled Productivity Management: Planning, Measurement, Evaluation, Control and Improvement by Dr. Sink to be published by John Wiley in Fall, 1984.)

Reprinted from 1984 Annual International Industrial Engineering Conference Proceedings.

HOW TO MEASURE AND IMPROVE PRODUCTIVITY IN
PROFESSIONAL, ADMINISTRATIVE, AND SERVICE ORGANIZATIONS

William F. Christopher
Counsel for Management

Measurement makes productivity performance visible, and managable. This paper deals with productivity measurement techniques for professional, administrative, and service organizations. It describes, first, the need for productivity improvement, and then explains concepts from which effective measurement techniques have developed. These techniques are then explained, with case examples.

Recent competitive history, and the 1981-82 recession, have dramatized some important lessons:

(1) Productivity in plant operations must be improved.
(2) Productivity techniques can improve professional, administrative, and service functions as well as plant operations.
(3) In many companies, the people costs in functions other than direct plant production have ballooned, and productivity improvement is needed to control these costs.

No longer is our concern for productivity limited to factories, production, and direct labor. Today more than 50% of U.S. workers process information rather than material goods. In 1981, there were 51,848,000 white collar workers in the United States:

Professional & technical workers	16,055,000
Managers & administrators	11,315,000
Sales workers	6,291,000
Clerical workers	18,187,000
Total	51,848,000

Employment costs for these workers in 1981 totalled more than one trillion dollars. In the typical company workforce, 53% are white collar workers, with this white collar payroll accounting for 70% of the total company payroll.

No longer can we limit our measures of labor productivity performance to plant operations. Productivity performance for professional, administrative, and service functions can also be measured, monitored, and improved.

Concepts and techniques are now available and experience-proven for applying a disciplined input-output way of thinking about and structuring our professional, administrative, and services organizations. No longer can we grow these organizations according to the perceived amount of work to be done, so long as the cash flow can pay for it. This approach seemed to work satisfactorily...until the cash dried up. To prevent such expand/cutback, boom/bust management, it is now time to think about "output" rather than "function," not only in our plants, but also in our white collar organizations.

Successful productivity measures for white collar workers can be built on this very simple conceptual framework:

Inputs → Function(s) performed → Output(s) for use by "customer"

Inputs for the most part will be person/hours of work by the employees in the unit. There may also be a significant capital input employed by the unit. And there may be inputs received from other internal or external units. Inputs of this type are analogous to the "materials" inputs for manufacturing operations.

Inputs are used by the unit in the functions it performs to produce "outputs." Industrial engineering, management engineering, and work simplification methods can be used for studying and improving the functions performed. Best results will be obtained if members of the unit are trained in these methods, and participate in applying them to improve performance.

"Outputs" for white collar units are not easy to define. The most measurable or countable outputs often are not those that best measure successful achievement of the purpose of the unit. The best measures can only be developed by going outside the unit and standing in the shoes of the "customer," who, for the typical administrative unit, will be another unit or units of the company. This "customer" orientation is essential for the successful development of productivity measures.

The point of view of the unit, and the point of view of the "customer" unit, share a common perspective. Both aim, through the outputs they produce, to contribute to the success of the company as specified in company objectives.

It is from the perspective of unit purpose related to company objectives that "customer" needs and "customer" values determine definition of the output. When we take this approach for professional, administrative, and service productivity measurement, we will find often that we develop new measures of successful performance: →

Typical	A Better Way
Corporate R&D	
Function: Conducts research	Output: Developed or acquired technology needed for the success of the company.
Measure: Expenses vs. budget	Measure: Successful innovation
Corporate Marketing	
Function: Provides functional guidance to sales and marketing operating units	Output: The information system needed in the company to achieve its market objectives
Measure: Expenses vs. budget	Measure: The achievement of company marketing objectives
Accounting	
Function: Produce financial reports	Output: (1) Financial reports; (2) Economic information needed for decision-making
Measure: (1) Meeting report schedules; (2) Expenses vs. budget	Measure: Reports and information provided
Corporate Engineering	
Function: Specify and contract plants and equipment	Output: Plant and equipment performance to meet company objectives
Measure: Project costs vs. budget	Measure: Plant and equipment performance

The above are simplified highlights, but a difference in the two columns is apparent. In the traditional approach, emphasis is on function performed, and the most important measurement is cost in comparison with budget. This focus does not provide the conceptual framework needed for getting at the problem of productivity. We must take a very important further step and define function in terms of output to be produced. This output will be what most contributes to "customer" units and company success. Defining this output is the major task for productivity measurement. Measurement, then, will be a count or measure of this output, related to inputs. We will continue to watch and control expenses vs. budget, but the key measurement of performance now becomes the production of the desired outputs.

Outputs, when defined to satisfy "customer" needs, provide both the quality and quantity considerations needed for measurement. Adding the "customer" to our conceptual framework, we can now fill in the "feedback loops" provided by measurement:

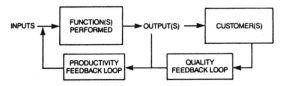

We can define and measure the inputs
We can define and measure the outputs
We can understand customer needs
Customer need will define the output quality required
Quality will be part of the output measure
Feedback loops will provide information needed for self control to achieve objectives

With measurements developed through dialog and participation, with prompt feedback on performance, and with the tracking of performance trends, the unit will have both the information and the motivation needed for self-control. Improvement in both quality and productivity can be substantial for performance measures on both quality and productivity are integrated in the feedback loops.

Some practitioners treat productivity and quality as different, separate objectives. I consider them both together as the key performance measure for work output. Quality is part of the measure of output. Quantity is part of the measure of output. To secure quality and productivity improvements, we must eliminate waste and minimize error. When we improve productivity, we will improve quality. When we improve quality, we will improve productivity. So...when we define outputs, we will include in our definition the appropriate quality attributes.

Was rework caused by faulty process? A management error.

Was scrap caused by operator error? Training and supervision a management responsibility.

Was the returned shipment a result of incorrect or unknown customer specification? A marketing error.

Was business lost due to a quality problem? Faulty SQC. ...or to unreliable delivery? Faulty scheduling and control.

How much time is spent on unneeded but internally required paperwork? A Quality Cost.

How many times was the strategic plan or profit plan done, redone, reviewed, revised, recalculated, restated, retyped, and resubmitted? A major Quality Cost in most large companies.

The list could go on and on. As in the plant, in white collar, professional, and management work we can aim to "do it right the first time." We just haven't thought about it enough. We haven't applied a disciplined input/output way of thinking about and structuring our white collar, professional, and management jobs. We think about functions and responsibilities instead of outputs. We leave the outputs undefined and unmeasured; the customers for our outputs unrecognized and unsatisfied.

Without measures, we continue error and waste, in alarming amounts. What an opportunity for operations improvement! What an opportunity for employee and company growth! What an opportunity for profit improvement! Quality and productivity methods learned and implemented pays off in the plant. It can pay off even more in management and in white collar and professional departments.

Plant downtime we analyze and fix. But white collar downtime we ignore. How much white collar and professional time is spent on activity other than producing or creating the desired output?

In the plant we target scrap and rework. What about white collar, professional, and management scrap and rework? How much time is spent correcting error? How much effort goes into maintaining and fixing processes that failed to produce quality output? How many times is something done more than once? How often is something produced that is not used or not needed?

Every meaningful job has outputs that can be defined and measured. Every output is produced for a customer who has needs to be satisfied by that output. Very often the customer will be internal, within the organization. My output is your input, in a process aimed toward achievement of company goals.

An MIS group outputs information (information, not data) needed by marketing, by accounting, by employee relations, by sales, and by other company groups to achieve their objectives. Defining customer needs–the needs of the receiving groups—made quality a part of the MIS output measure.

A corporate engineering department measured performance by comparing project actual cost with approved AFEs. After evaluating Quality Costs, the key performance measure was changed to facility performance in comparison with requirements.

The Management Task

Error and waste are found in two areas:

(1) The system, or process, may produce the error, may cause the waste. When this is the case, we must change the system to prevent error and waste.

(2) The employees operating the system or process may produce the error or cause the waste. When this is the case, some aspect of work structure, supervision, or training must change.

In both instances—fault of the system or fault in the operation of the system—the needed remedy is a task for management. Employees operate the system according to prevailing norms. Both fixing the system and fixing the operation are management responsibilities, but all employees can participate in the process. Employees know much about where changes are needed. Training will increase their contributions. Always, new learning will be an important part of the change process...learning at all levels. Participation and motivation are not enough, if needed skills are lacking.

I recently visited twelve companies in Japan to study their quality and productivity programs. In answer to my question at each company, "Where will future improvements come from?" the answer, almost always, was "training." And they described training for everyone, from CEO to production line workers.

Today, new knowledge and new skills for improving quality are needed at all levels—knowledge and skills in statistical quality control, work structure and participation, process technology, managerial engineering, microprocessor technology, software and systems, managerial economics. There is an information base for quality and productivity improvement that requires all of us to be keen students at the same time we are results-producing "doers."

To begin a program to improve administrative productivity, the following strategies can be helpful:

STRATEGIES FOR IMPROVING ADMINISTRATIVE PRODUCTIVITY

(1) Involve employees in the productivity effort.

(2) Provide job security, assuring that no employee will lose his or her job because of improvements in productivity.

(3) Define outputs to be produced, supportive of company goals.

(4) Identify client groups for the outputs, and user value measures of the outputs.

(5) Define inputs (typically labor is the only major input) and measures for the inputs (labor hours).

(6) Develop and monitor productivity measures (output/input), providing the appropriate feedback for each work group.

(7) Employ office technology in the appropriate ways.

(8) Provide leadership committed to achievement, and supportive of subordinates.

(9) Provide for ongoing education and learning.

For white collar workers, much of the initial work in productivity improvement was done in the areas of motivation, team building, employee participation, and such techniques as employee involvement teams and quality circles. This work has led to an increasing amount of quantification and measurement. Unless we have useful measurement we cannot set meaningful objectives. Unless we have measurement, we cannot monitor our performance toward those objectives.

The Query Form outline at the end of this Commentary can be used for assembling information needed to establish productivity measures for professional, administrative, and service units.

Companies undertaking to develop measures of productivity performance, and using the methods recommended in this Commentary, will find that the process of developing measures provides many benefits:

(1) Many people become involved, this participation improving motivation and identification with the company, and providing a better productivity measurement and productivity performance result.

(2) The purpose of the unit is clarified, and understood by members of the unit.

(3) The important outputs that add up to achievement of purpose are identified, and understood by those responsible for producing them.

(4) Relationships with other company units are clarified.

(5) The process is a training experience for those who participate, upgrading their skills.

(6) With common understanding of output measures, and monitoring of these measures, output will improve in relation to the inputs used to produce them.

(7) "Effectiveness" as well as "efficiency" will be improved.

(8) The measurement system will continue to be improved over time.

(9) Both productivity and quality will improve.

Two general types of productivity measures have been applied successfully in professional, administrative, and service organizations to monitor and improve productivity performance:

(1) A single, total unit "Administrative Productivity Indicator."

(2) Multiple Output measures.

The Administrative Productivity Indicator (API) method most closely approximates the techniques used in plant productivity measurement, and can be used in those units where a single output can be defined as the measure of the successful performance of the unit. Time, thought, dialog, and testing are needed to develop API measures. With the needed homework, APIs can often be developed...even for units where the "going in" attitude is that such measures are impossible.

An alternative approach is the Multiple Output method. This method requires also that outputs be defined, but allows: (1) the inclusion of several outputs in the measure; and (2) both qualitative and quantitative measures.

DEVELOPING AN "ADMINISTRATIVE PRODUCTIVITY INDICATOR" [1]

$$API = \frac{Work\ Output}{Labor\ Hours\ Input}$$

Work output: the physical, measurable, or countable thing which describes what the administrative unit does in achieving what it was organized to do.

How to find an API:

(1) State the key purpose for the unit. Some helpful questions are:

- What was the unit organized to do?
- What is the major transaction flow?
- What is the end product or service?
- Who are the "customers" for what is produced?
- What is value to the "customer"?

(2) Identify the physical output which determines how successfully the unit has achieved its purpose. An output can be defined as a physical product or service that has been created or processed. Examples:

Physical Thing	Action Performed
Employees	Hired
Invoices	Paid
Analyses	Documented
Work Order	Completed

(3) Test the selected work output. Does it show that work is done? Does completion of the work assure that purpose is achieved?

(4) Define the input measure or measures. For most administrative units the predominant input is person/hours of work.

Materials, energy, and capital inputs typically are minor. For this reason, productivity measures for administrative units can usually best be expressed in terms of work output in relation to person/hours of work.

Basic Concept — An Administrative Productivity Indicator is the single overall measure that quantifies how successfully the unit achieves its purpose.

By monitoring the API over time, you know:
...Where you were
...Where you are
And you can set meaningful objectives for:
...Where you intend to be in the future.

An API provides a measure of change in productivity performance over time. For administrative productivity measures, the traditional formula can be reversed, with labor in the numerator:

$$API = \frac{Labor\ Hours\ Input}{Units\ of\ Output}$$

This calculation provides an "Hours Per Unit" (HPU) measure that:

(a) Makes labor cost calculations easy
(b) Aids staffing and capacity planning
(c) Provides an HPU benchmark for monitoring productivity change

HPU is a number the administrative unit will aim to reduce, rather than to allow to increase...emphasizing that improving administrative productivity is principally a matter of managing the way hours (inputs) are combined and used. The base period or "going in" HPU can be used as a base for determining future productivity trends, with the API then being expressed as HPU. For its HPU measures, Intel uses the reporting format shown in Figure 1.

---

[1] This API method was developed by Keith A. Bolte, Corporate Director–Administrative Productivity, Intel Corporation.

Figure 1

	J	F	M	A	M	J	J	A	S	O	N	D
1. HEADCOUNT Cur/Goal	/	/	/	/	/	/	/	/	/	/	/	/
2. AVAILABLE HOURS												
3. UNITS OF WORK OUTPUT												
4. HOURS PER UNIT (HPU) (No. 2 ÷ No. 3)												

Explanation:
1. Headcount: Cur/ —Number of people on payroll at end of period
                /Goal —Number of people from staffing table
2. Available hours: Number of hours paid minus holidays and all personal absences
3. Units of work: Total volume of the primary output produced (the API measure)
4. Hours per unit: Available hours divided by units of work

The experience in using API measures at Intel shows that their use in an organization with open communication and participation can improve both productivity performance and attitudes.

Intel Corp. has made extensive use of the API method for measuring and improving administrative productivity. After APIs are developed, work simplification and workload management techniques are applied, and APIs are monitored. Employee training and participation are used throughout the process. Job security is assured — no employees lose their jobs as a result of improvements achieved. In the Intel experience, goals and measures of administrative productivity have improved both performance and employee attitudes. In general, Intel has found that a 30% improvement in unit productivity can be achieved in most administrative units.

Intel has successfully developed and applied the API method using person/hours of work as the input measure. In my experience, this also has been the most usual measure. However, there are some administrative and service units where the capital input is large, and for these I have found that a measure of output in relation to the capital input can help improve the efficiency with which capital is employed. Among administrative and service units that are capital intensive as well as labor intensive are:

Data processing
Analytical laboratory
Some kinds of service business (airline, bank)

Where capital input is significant, the Administrative Productivity Indicator can include a measure of this input in the denominator:

$$API = \frac{Work\ Output}{Labor\ Input + Capital\ Input}$$

For administrative or service units where more than one input is included in the denominator, partial productivity measures can be calculated and monitored for each of the inputs:

$$Labor\ API = \frac{Work\ Output}{Labor\ Input} \qquad Capital\ API = \frac{Work\ Output}{Capital\ Input}$$

DEVELOPING MULTIPLE OUTPUT PRODUCTIVITY MEASURES

In many professional, administrative, and service units a single output measure is not considered adequate. Instead, several outputs are defined as representing the successful achievement of the purpose of the unit. Some of these outputs may be quantifiable; others require a subjective appraisal. In such situations, multiple output measures can be used. A rating scale technique can then be used to combine these measures into a single, overall measure — a Multiple Output Productivity Indicator (MOPI).

The procedure is as follows:

(1) Define the purpose of the unit.

(2) Test the definition of purpose by relating it to the purpose and objectives of the company, or of the next higher level organization of which the unit is a part.

(3) Identify the outputs which represent the successful achievement of purpose.

(4) Determine how each of these outputs can be measured.

(5) Use the "going in" period for a base, and calculate the measures for each output for this base period.

(6) Establish a rating scale (1 to 10, or 1 to 20). Define performance levels for each output along the scale, with base period performance generally around the midpoint of the scale to permit performance trend measurement up or down from the base period levels.

(7) Prepare a rating form listing each output, and showing current period rating and position on the scale.

(8) Determine weights for each output, and combine the several outputs.

(9) Identify inputs and measures for them.

(10) Monitor performance trends for each output, the combined outputs, inputs, and for output related to input (MOPI).

Olin Corporation has made extensive use of the multiple output measurement method for measuring and improving administrative productivity. The Olin method is to establish Employee Involvement Teams for administrative functions in plant operations, and in headquarters staff departments. Each of these teams provides a forum for:

(a) Training and education
(b) Dialog
(c) Developing and monitoring productivity measures
(d) Problem solving

Olin began its work in measuring, monitoring, and improving administrative and service productivity with the administrative functions in manufacturing plants. Productivity measures had been developed for plant operations, and extending the productivity improvement efforts to plant administration was a logical next step. With successful experience in plant administration productivity improvement, the work was extended to divisional and headquarters professional, administrative, and service units.

The corporate director of productivity, in collaboration with the corporate training department, motivates and coordinates program efforts, with the actual work done through the Employee Involvement Teams established in each unit. Work began with a pilot group, and was then extended gradually to additional groups. Participating professional, administrative, and service groups were selected on the basis of:

    (a) Resources employed (people, capital)
    (b) Importance of outputs
    (c) Potential leverage from productivity improvement
    (d) Management and employee interest in the program

One of the first units to participate in the Olin program was the Information Services Division — computer operations and systems development. A project team organized the work, following the general approach of the Query Form outline. The project team, working under the direction of an overall ISD Steering Committee and in collaboration with the various organizational units of the Division:

(1) Developed a written statement of purpose for the Division.

(2) Defined four Division performance areas, and the purpose for each:

      Batch Processing Operations
      Time Share
      Transaction Processing
      Systems Development

(3) Defined purpose for subsets of the four Division performance areas.

(4) Defined outputs for each of the performance areas and subsets.

(5) Determined appropriate measures for the outputs.

(6) Worked out how data for measurement could be collected.

(7) Established productivity measures and a feedback reporting system.

(8) To the extent practical, developed a commonality in measurements used in the four performance areas.

A measure of "service units" was developed as a standard that could be used as an output measure for several performance units. Using, where possible, this standard unit of measure provides the capability for analyzing the changes over time of each unit, of one unit in relation to another, and the impact on each other's performance.

Outputs and productivity measures were defined for each of the four Division performance areas. The outputs for Time Share, for example, were defined as:

      System Availability
      Network Load
      Response Time
      Log-Ons
      Number of Service Units

Productivity measures were defined for each, and rating scale values established. Figure 2 illustrates the reporting format used.

Rating scale values were determined for each measure. For example, 98% was considered "par" for system availability, and the following rating scale values established:

		%			%
1	94	%	6	98.5	%
2	95.5		7	98.75	
3	97		8	99	
4	97.5		9	99.25	
5	98		10	99.5	

Figure 2

TIME SHARE	Productivity Measure	Rating 1 2 3 4 5 6 7 8 9 10	Rating	Weight	Points
System Availablility	___	___	___	___	___
Network Load	___	___	___	___	___
Response Time	___	___	___	___	___
Log-ons	___	___	___	___	___
Connect Hours	___	___	___	___	___
Service Units	___	___	___	___	___
		Total Earned Points			
		Maximum Points			1,000

Productivity Index: $\dfrac{\text{Total Earned Points}}{1000} \times 100 = PI$

Note: Weights total to 100

In establishing rating scale values, the midpoint of the range, 5, was selected for the current level of performance, or the standard for performance. Changes up or down would then represent relative change from "going in" performance levels, or performance up or down from an agreed-upon standard. The significant consideration in the measurement system is not the absolute rating, or number, but the performance change over time. Monitoring of performance trends, feedback of information, dialog among the Employee Involvement Teams, and problem solving are all important contributors to improving productivity.

Weights were determined for each output so that a total productivity performance measure can be determined for Time Share for each reporting period. A similar measurement system was developed for each of the other three Division performance areas.

To calculate a total Information Services Division productivity indicator, weights were assigned to each of the four performance areas — Batch Processing Operations, Time Share, Transaction Processing, and Systems Development — so that a combined productivity measure for total Division performance can be calculated and monitored.

In addition to the multiple output and rating scale measures, selected measures of output in relation to input of person/hours and capital are monitored in each of the four performance areas, and for the total Division. Person/hours of work is measured and monitored in relation to a major output of the unit. Capital input is calculated for the total Information Services Division and measured as an input in relation to service units. Capital input measures are made in the same way as for plant productivity measures. For the Information Services Division, person/hours of work is the major input, but capital input is substantial:

ISD Inputs

Person/Hours	61%
Capital	39
Total	100%

Several important conclusions developed from the Olin experience:

(1) Employee Involvement Teams can successfully improve professional and administrative productivity performance.

(2) The process is a practical, "hands on" kind of learning experience.

(3) "Effectiveness" as well as "efficiency" can be improved.

(4) Measures that are not strictly productivity measures will be included in the measurement and monitoring program.

(5) With experience from using the measures, the "going in" system of measures will be changed and improved over time.

(6) Measurement is an essential element for productivity improvement, for:

　(a) Setting goals and providing feedback on achievement

(b) Motivating performance improvement
(c) Sustaining the productivity effort over time

(7) An administrative productivity improvement program contributes to planning and budgeting procedures by clearly defining purpose, outputs and performance measures.

(8) In administrative and service organizations, and in organizations of individually contributing technical and professional employees, the process gives everyone a voice and a means to contribute to the success of the organization.

PRODUCTIVITY ACHIEVEMENT REPORTING

Once we have productivity measures, we can then establish a system of information flow that will help us achieve productivity improvement objectives. I call such an information system "Achievement Reporting." It's not a system to evaluate performance. It's a system to help people achieve productivity objectives. Productivity Achievement Reporting is a system of productivity measures and a structure of information flow that provides the feedback information at all decision points needed for the self-control that enables the enterprise and its operating, administrative, and service groups to achieve their productivity objectives.

The system is based on three disciplines:
(1) Managerial economics, which provides the needed dollar measures
(2) Behavioral science, which provides us an understanding of motivation
(3) Cybernetics, which provides us an understanding of how control enables a complex system (like a business enterprise) to achieve its purpose (objectives)

Managerial economics is the economics of the decision making process. A future issue of *Commentary* will describe basic concepts and applications.

Behavioral science has studied human motivation, and defined the characteristics of self-motivated achievers:
(1) They set their own goals
(2) Goals are meaningful and achievable
(3) They get prompt, reliable feedback on the results of their work showing progress toward goal achievement
(4) These characteristics apply in all activities, as well as in work
(5) They have experience over time in the satisfactions of achievement

These characteristics can be applied in the way we structure work, to increase on-the-job achievement motivation:
(1) Add acceptable risk to the job in the form of attainable, but meaningful objectives
(2) Establish objectives through participation and dialog
(3) Provide prompt and reliable feedback on progress toward the achievement of objectives
(4) Give recognition for achievement
(5) Make sure the achievement is significant to the individual
(6) Relate individual achievement to group objectives

Achievement motivation is learned, and it is learned from experience with setting and achieving objectives. So the achievement motivation of individuals, and of groups, can be increased in the right kind of environment, and with the right kind of experience. Achievement reporting provides this environment, and this experience.

The starting point is the setting of objectives and performance measures. Objectives may be individual objectives or group objectives. In my experience, productivity objectives are usually group objectives, with dialog among the group the method for developing objectives and measures, and effecting control through use of the feedback measures.

The prompt and reliable feedback important to self-motivated achievers is the basic principle of control as described for us by cybernetics. Control is not so much what is done to us from higher levels. It is more what we do for ourselves using feedback information we get from the results of our work. Higher level control is most effective in the dialog process by which objectives and measures are determined. Feedback from the performance provides the information we need to help us achieve the objectives. Self-control is what produces results. Feedback is the information fuel for self-control, and our measures provide this feedback.

With Productivity Achievement Reporting, feedback measures are more than specific measures for a reporting period. Each measure relates to past measures to show trends and changes in trends. It is direction — trends and changes in trends — that matters. Each measure is one more measure of a trend, a direction. The task then is to take actions that will influence the trend toward our objectives.

This "looking ahead" kind of management — which can be used at all levels — is "predictive control." We don't wait for a result and then try to correct it if it is bad. We look at where performance is taking us. We look ahead to where we are going. We then take actions to move the trend toward our objectives. We manage productivity improvement much like we drive a car — looking ahead to where we're going and controlling the car to get us there. Productivity Achievement Reporting helps us find our way to our productivity objectives by providing us the information we need for our next decisions and actions, as we are driving the car. And "us" is everyone, at all levels...not just top management.

Each company and each unit will have to develop the professional, administrative, and service unit productivity measures most suitable for them, using methods such as those described in this *Commentary*. Using the measures they can then established the productivity achievement reporting that will motivate the self-control to achieve the objectives. The results can be:

(1) Significantly improved productivity
(2) Improved quality
(3) Improved profitability
(4) High morale and commitment

QUERY FORM OUTLINE

The following outline can be used to assemble information needed to define outputs, to develop productivity measures, and to implement a productivity improvement program in professional, administrative, and service units. Use dialog. Include those doing the work in the process. Organize orientation and training sessions. The process is an important part of achieving performance improvement. Query form questions:

(1) Name of organizational unit, and structure:

(2) How will participation and dialog be used in developing answers to the questions posed by this Query Form?

(3) What is the purpose (mission) of this unit?

(4) How does this purpose relate to company purpose and objectives?

(5) When the purpose of this unit is successfully accomplished, what is the output(s)?

(6) Who are the customers (users) for the output(s)?

(7) What are the measures of value to the customer(s) (the quality requirements) for each of the outputs?

Output   User Values

(8) What are useful measures of the output(s)? Include the appropriate quality considerations in the output definition.

(9) How can quality of outputs be assured?

(10) What are the inputs required to produce the outputs?
Person/hours of work:
Materials:
Capital:
Other:

(11) What are useful measures of the inputs?

Inputs   Measure, or Measures

(12) What single "Administrative Productivity Indicator" can be defined that provides a useful measure of performance?

(13) What "Multiple Output Productivity Indicators" can be defined that provide useful measures of performance?

(14) What measure(s) best describe performance that successfully accomplishes the purpose of the unit?

(15) How will measures be reported, trends evaluated, and actions taken?

William F. Christopher

Mr. Christopher's executive career includes thirty years with General Electric Company, Hooker Chemical Corporation, and Occidental Petroleum Corporation in marketing, management, and internal consulting positions. In 1977 Mr. Christopher formed a consulting company, William F. Christopher Counsel for Management. Consulting areas include: strategic and operations planning, productivity, new products and new ventures, sales and marketing, profit improvement, organization and motivation, and performance measures. He has worked with more than seventy-five businesses, in seventeen countries.

Mr. Christopher is a frequent speaker and seminar leader before business audiences. During the Tokyo Round trade negotiations he was an advisor to government, serving as chairman of the Industry Advisory Committee for Rubber and Plastics. Mr. Christopher's publications include a technical book on plastics, Polycarbonates, Reinhold, 1962; two books on business management, The Achieving Enterprise, Amacom, 1974, (winner of the 1976 James A. Hamilton Award), and Management For The 1980's, published in 1980 by Amacom (hardcover) and Prentice-Hall (paperback); a "how to" manual on productivity measurement, Productivity Measurement Handbook, Productivity, Inc., and numerous articles on management, marketing, and international trade.

Mr. Christopher's address is P. O. Box 3085, Stamford, CT 06905.

PRODUCTIVITY MEASUREMENT IN HIGH TECHNOLOGY AREAS

Harvey H. Smerilson and Patricia D.S. Fields
Martin Marietta Orlando Aerospace
Orlando, Florida

The Industrial Engineer has always recognized the need to measure productivity in order to control cost. Setting standards by use of time studies and predetermined times has long been a method employed to measure performance. This technique supplies management with a tool to judge the effectivity of the total area of responsibility. With a changing trend toward high technology production controlled by computer and robots, man is playing a smaller role and machine a greater role in the actual cost of manufacturing. The authors will attempt to show how productivity should be estimated and then measured in today's manufacturing environment.

Productivity improvement is usually accomplished by means of method changes or work habit improvements. In most cases, this was done by increasing the efficiency of the operator, which in turn, increased output. In a high technology machine controlled operation, increasing the efficiency of the operator may not result in added productivity if the cycle time of the machine is the limiting factor. Increasing the efficiency of the machine may not be an alternative since the cost of new or improved tooling may be prohibitive as may be the case of new equipment. It now becomes necessary to explore all the factors that affect productivity and the possible alternatives that can be used to increase output.

In today's age of computerized equipment the human factor is actually becoming a smaller part of the run time. By studying the human operator, we may actually increase productivity by 20 to 30 percent. If the operator's time is limited by the machine run time, we have the problem of what to do with the operator during the time span when the machine is busy and the operator is idle. What actually may occur is an increase of idle time, as illustrated below:

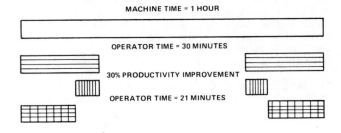

On longer runs, this newly found "operator free time" can be used to increase productivity if the actual operator attention time was significant to begin with. For example, a run time of one hour with an operator attention time of 30 minutes could be increased to free time of 39 minutes with a productivity increase of 30 percent.

MACHINE TIME = 1 HOUR

OPERATOR TIME = 30 MINUTES

30% PRODUCTIVITY IMPROVEMENT

OPERATOR TIME = 21 MINUTES

We now have 9 additional minutes of idle time at this point. What has been termed a productivity improvement is in fact nothing more than an increase of idle time for the operator, unless of course, a productive use can be found for the 39 minutes. For one additional example, consider the case of a 30 percent productivity improvement on a free time of 5 minutes. The increase would be to 6-1/2 minutes or an additional available operator time of 1-1/2 minutes. The second set of statistics identifies the cut off point when increasing productivity becomes feasible by increasing operator free time. This is an often overlooked detail to even the most experienced Industrial Engineer.

INSTITUTE OF INDUSTRIAL ENGINEERS
1984 Annual International Industrial Engineering Conference Proceedings

Now that the case has been built for analyzing the operator slack time in relation to machine time, a scheme to measure performance and identify problem areas is necessary. At this point it becomes necessary to enter another variable known as "yield". Yield can be defined as the number of good parts out compared to the number of good parts in. In the modern high technology fields of today there is planned yield and actual yield. Plan yield refers to the "expected loss" while actual yield refers to the experienced loss. The importance of yield plays a big role in determining the overall cycle time of the machine. A typical high technology operation flow is as follows:

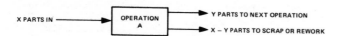

A combination of predetermined standard times and stop watch study can be used to determine the overall time for operation A. Often it is the stop watch study that shows the total machine cycle time is greater than the operator attention time, thus harder to control, measure, or improve.

In high technology areas with a low planned yield it is often the practice to build the yield into predetermined standard time. To evaluate the man versus machine time a raw standard for both, without a yield factor, must be used. The next approach is to determine what the true idle or free time really is, to decide how to best balance the operator's time.

At Martin Marietta Orlando Aerospace in the Optical Component Center, a high technology area, it was determined that different rates of learning were applicable for man controlled operations and machine controlled operations.

Looking first at a man controlled operation, it was determined that learning showed an initial bump at the beginning due to factors such as changes, experimentation, and cross-training, and then smoothed to follow an 85% experience curve. A plot of the data studied is shown in Figure 1.

A similar study was done on machine times with the results shown in Figure 2. Figure 2 is an operation which is entirely machine controlled and shows no improvement possible due to the controlling factor that nothing can be accomplished to reduce the machine time.

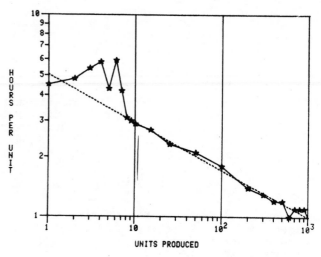

Figure 1. Man Controlled Operation

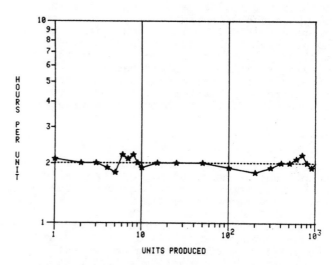

Figure 2. Machine Controlled Operation

Another case is shown in Figure 3. This is the typical case where the operation is still machine controlled. Method improvements, tooling, some operator habits and experience will all have a total effect on the outcome of productivity. A point to be stressed here is that learning occurs per cycle and not per part. This is an important factor when more than one part is processed during each machine cycle as is often the case in a high technology area (the title of the X axis of Figure 3 is really referring to each cycle as one unit produced).

After defining and choosing the proper learning curve, the rest of the problem becomes easier to solve mathematically.

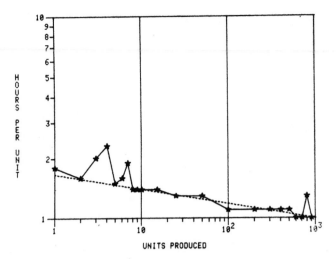

Figure 3. Machine Controlled
Operation/With Changes

The first step is to study the task and
set a predetermined time standard on the
operator dependent portion of the task.
A total machine dependent cycle time
should also be determined. If the task
is presently being performed the relationship between machine and operator time
will be observed during the study. If
the task is in a development stage a
caution flag must be flown until it can
be determined if man controls machine
or if machine controls man. The machine
time, as studied, can be the present time
and an improvement could possibly be expected. The operator time, could be a
standard time and would have to be calculated on an experience curve for starting performance and slope. We now have
the following situations:

    Machine time at standard - starting
    operator time = shortest idle time

    Starting machine time - operator time
    at standard = longest idle time

The time we are interested in is the time
between the above two times. Mathematically, where CAU is the cum average unit
value from the learning curves used, this
time can be considered as:

    Average CAU of machine time - average
    CAU of operator time = average idle
    time

In the case where average idle time is
calculated to be a negative number the
problem becomes simplified. The operator
is an operator controlled one rather than
a machine controlled one and can be treated
as such.

Now that the average idle time is calculated, there are several options open for
proceeding with measurement of productivity.

By using Industrial Engineering techniques
such as line and machine balancing, the
idle time can be minimized in many areas.
One measure of productivity is the assignment of man and machine which produces
the less possible idle time for both.

When actual idle time exceeds calculated
idle time the result is longer machine
time, therefore, lower productivity. As
Industrial Engineers are involved with
the measurement functions, we often go
to great extremes to measure the performance of an operator. A step by step
approach based on the ideas presented
will clearly show how easy productivity
measurement is. Remember, our first
step was to calculate what the average
idle time should be. Next, the reduction of this idle time to the minimum
possible was accomplished. If machine
time is still the controlling factor, it
can be correctly assumed that the following formula will produce a correct measure of performance:

$$\text{Performance} = \frac{\text{Number parts per run x expected time per run}}{\text{Expected parts per run x length of time per run}}$$

The above formula measures performance
in relationship to both idle time and
yield which are the two variables most
commonly associated with a high technology
area. It can be determined from our prior
mathematical expression that an increase
in the length of idle time will result in
an increase in the length of the run time.
It can also be readily determined that a
decrease in yield will result in a decrease in performance. Tracking performance based on the above formula, will not
tell without further investigation the
exact cause of a productivity decline.
It will signal the need to investigate
the contributing factors. Performance
can, at the desire of management, be
reported as yield performance and productivity performance. However, productivity performance is further dependent on both man and machine as is
yield. It is therefore, the recommendation of the authors, to report and
track performance as a whole to encourage
productivity improvement in all areas
without giving false emphasis to any one
factor. This also will prevent neglecting one or more important factors.

BIBLIOGRAPHIES

Harvey H. Smerilson is the Supervisor of
the Optical Component Center's Industrial
Engineering group at Martin Marietta
Aerospace, Orlando, Florida. He is also
an elected Commissioner of the City of
Longwood, Florida. Mr. Smerilson is a
past officer of Region IV and the Central
Florida Chapter of AIIE. He is a graduate

of Western New England College, BSIE with
graduate studies at the University of
Central Florida. Mr. Smerilson is an active
author and has made numerous presentations
on Industrial Engineering subjects.

Patricia D.S. Fields is an Associate
Industrial Engineer at Martin Marietta
Aerospace, Orlando, Florida. Presently
assigned to the Optical Component Center,
Patricia is responsible for the measurement and control of the factory touch
labor. Prior to her present assignment,
she worked in the Microelectronic Center
of Martin Marietta Aerospace, Orlando,
Florida. She received a BA in Business
Administration at Rollins College.

Reprinted from 1984 Annual International Industrial Engineering Conference Proceedings.

PRODUCTIVITY MEASUREMENT IN THE SERVICE SECTOR: A HOTEL/MOTEL APPLICATION OF THE MULTI-FACTOR PRODUCTIVITY MEASUREMENT MODEL

by

Mr. Jeff Swaim
Associate, **Oklahoma Productivity Center**

D. Scott Sink, Ph.D., P.E.
Director, **Oklahoma Productivity Center**
and, Associate Professor
School of Industrial Engineering and Management
Oklahoma State University

INTRODUCTION

The transition from a predominantly manufacturing oriented society to a greatly service oriented society is by now well discussed. (Toffler, 1980; Naisbitt, 1982) Computer assisted systems utilization increases this service, white-collar, indirect labor, professional employee oriented trend. That is, within every private sector, manufacturing firm there are service oriented functions, positions, etc. and this domain and the numbers of people in this domain are increasing substantially. Hence we are seeing decreasing numbers of direct labor and increasing numbers of indirect and service oriented labor both within our society as a whole as well as within traditionally direct-labor dominated industries and firms.

The indirect labor force, service related professions, and "professional" employees including management have traditionally been labelled as "hard to measure" and in many respects they do present added difficulties. Increasing numbers of these type of employees coupled with measurement and evaluation difficulties have created considerable attention and interest. "White Collar Productivity" books, articles, and seminars are selling very well. Unfortunately, we are convinced that much of this interest is misdirected.

This paper is an attempt to accomplish several objectives. **First,** we would like to examine performance related concepts relative to service oriented (white collar, indirect, professional, managerial, etc.) positions and functions. Building upon other recent papers (Sink, 1983a, 1983b), we would like to specifically examine the relationship between various performance criteria in service related organizations, function, and jobs.

Secondly, we would like to present a technique called Performance/Productivity Process Modelling or Input/Output Analysis. This technique is, in our opinion, a necessary first step to the development of effective measurement and evaluation systems in any type of organizational system. This approach is particularly beneficial in service oriented organizations, functions, and jobs because of the lack of crystalization that often exists as to what outputs are, what goals are, etc.

Thirdly, we will very briefly review a variety of strategies and approaches that can be taken to develop measurement and evaluation information for service related organizational systems. The Oklahoma Productivity Center has recently completed a DoD sponsored study that cataloged productivity measurement and evaluation theories and techniques that exist in the literature and in practice. Results from this study (Sink, Tuttle, et al, 1984) are presented in a paper by Sink and DeVries (1984) and are only highlighted here in this paper.

Finally, we will present a case study of an application of the Multi-Factor Productivity Measurement Model for a hotel/motel operation. The multi-factor productivity measurement model and similar approaches are described in some detail in other sources so we will focus primarily on the case study application. (Swaim and Sink, 1983; Sink, 1983, 1984; Sink and Keats, 1983; APC, 1978; Craig and Harris, 1973; Sumanth, 1984; van Loggerenberg and Cucchiaro, 1981-1982; Bain, 1982; Davis, 1955)

BASICS

The term organizational system is used to apply to a work group, a department, a function or division, a plant, a firm, or a corporation. Before performance can be defined and measured, the unit of analysis must be specified. The unit of analysis identifies the boundaries of the system being studied and therefore actually defines and delimits the specific organizational system being examined.

Perhaps the most critical test of management is to explicitly operationalize what performance means for the organizational system(s) they are responsible for and to communicate those impressions clearly to the people in the system(s). From a measurement, evaluation, and control standpoint we have likened this process to the development of an airplane control panel. (see Sink, 1983a) The measurement, evaluation, and control system must be an effective information system that is frequently a decision support system.

Organizational system performance is comprised of seven basic criteria. These criteria are: effectiveness, efficiency, quality, productivity, quality of work life, innovation, and profitability or budget accountability. These seven terms are defined only briefly in figure 1.

(1) Effectiveness = Accomplishment of Purpose (Barnard, 1938). Accomplishing the "right" things: (1) on time, (i.e., Timeliness); (2) right, (i.e., Quality); (3) all the "right" things, (i.e., Quantity)

Where: "things" = goals, objectives, activities, etc.

(2) Efficiency = (a) Satisfying individual motives, (Barnard, 1938), success at securing necessary personal contributions

(b) $\dfrac{\text{Resources Expected to be Consumed}}{\text{Resources Actually Consumed}}$

(3) Quality = Conformance to specifications, (Crosby, fitness for use.)

where "specifications" can be identified as: timeliness, various quality attributes, customer satisfaction, etc.

(4) Productivity =

$$\dfrac{\text{Quantities of Output from an Organizational System for some period of time}}{\text{Quantities of Input Resources Consumed by that Organizational System for that same period of time}}$$

or

$$\dfrac{\text{Quality Quantity}}{\text{Resources Actually Consumed}}$$

Hence, productivity is, by definition, a ratio and is a measure of effectiveness in the numerator divided by the denominator of the efficiency equation. Note that quite often productivity statistics for the nation are presented in terms of rates of change. These statistics are actually productivity indexes. A productivity index is a particular productivity ratio for one period in time divided by that same productivity ratio for an earlier period in time. For example:

Productivity Index =

$$\dfrac{\dfrac{\text{Quality Quantity 1982}}{\text{Resources Actually Consumed 1982}}}{\dfrac{\text{Quality Quantity 1981}}{\text{Resources Actually Consumed 1981}}}$$

or, rate of change of GNP to labor input, etc.

(5) Quality of Work Life =

Human beings' affective response to working in and living in organizational systems. Those attributes of organizational systems that "cause" positive affective responses. Often, the focus is on ensuring the employees are "satisfied", safe, secure, etc.

(6) Innovation =

The creative process of adaptation of product, service, process, structure, etc., in response to internal as well as external pressures, demands, changes, needs, etc. The process of maintaining fitness for use from the customer's eyes.

(7) Profitability =

A measure or set of measures of the relationship between financial resources and uses for those financial resources.

$\dfrac{\text{Revenues}}{\text{Costs,}}$

Return on Assets,
Return on Investments,
etc.

Figure 1. Organizational System Performance Criteria

These seven performance criteria will vary in importance from organizational system to organizational system. These seven criteria are causally related in roughly the following fashion.

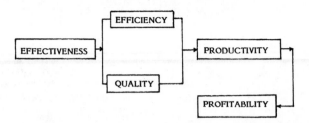

Figure 2. Relationship Between Effectiveness, Efficiency, Quality, Productivity, and Profitablility

Innovation and quality of work life are pervasive performance criteria influencing the other criteria directly as well as influencing the relationships between the criteria shown in figure 2.

In service related systems, in particular, management can achieve profitability partially and perhaps most importantly by being productive. Productivity will be acheived if the system is efficient and maintains high quality. If an organizational system, say a marketing department is efficient and achieving high quality but is ineffective (ie. doing the wrong things) it cannot be very productive. Therefore, in service oriented systems it is essential to first address effectiveness related questions and issues. Only once those issues have been resolved should the efficiency and quality criteria be addressed. In short, "White Collar Productivity" is a premature question. The term sells books and attracts large crowds to short courses but the real issues are effectiveness, efficiency, and quality.

PERFORMANCE/PRODUCTIVITY PROCESS MODELLING
(Input/Output Analysis)

Prior to the development of a measurement, evaluation, and control system, we recommend that appropriate persons/members from the focal organizational system work through the following exercise.

1. Define the focal organizational system.
   What are its boundaries?
   Who are the people inside?
   Which key organizational systems are in its task environment?
   What is its mission?
   What are its goals and objectives (2-5yrs)?
   What are the critical challenges facing it internally as well as externally during the next 2-5 years?

INPUTS
2. What resources (data, information, labor, energy, capital, materials) come into the organizational system?
   Who do they come from?
   What do they look like (quality wise)?
   What should they look like?

What do they cost?
How much comes in?

TRANSFOR- 3. What do the people in the organizational
MATION system do to these incoming
PROCESSES resources?

4. What comes out of the organizational
OUTPUTS system? (ie. information, people,
energy, goods, service, capital,
materials, etc.)
Who do they go to?
What do & should they look like?
What do they cost?
How much comes out?

OUTCOMES 5. What do we expect the distribution of
these to cause? (ie. desired outcomes
such as sales, profits, customer
satisfaction, etc.)

6. How can we measure, evaluate, and
control the performance of this
system?
What does performance mean?
What role can, should, or does
productivity play in this system?

7. How can we best control and improve
performance of the organizational
system?

8. Develop a 2-5 year, strategic plan for a
productivity management effort for
this organizational system.

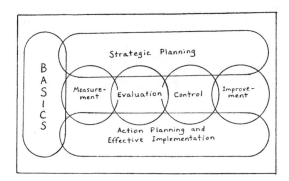

Figure 3. Productivity Management Process

Forms can be and have been developed and utilized to
facilitate this exercise.

The Productivity and Quality Center at Honeywell
Aerospace and Defense utilizes this process to assist
managers and supervisors with the performance
management process. They link this input-output analysis
closely to the measurement, evaluation, and control
process. They utilize this process to crystallize basic
systems understanding of goals, objectives, resource flows,
performance criteria, and the general management
process. They utilize the Nominal Group Technique to

identify priority and consensus ideas for performance
improvement within specific organizational systems.
Knowledge from the performance process modelling
exercise assists analysis of how to capitalize on the top
priority ideas. Proposals are developed and reviewed prior
to implementation. The Normative Productivity or
Performance Measurement Methodology (NP/PMM, see
Sink, 1984a, 1984b; and Sink and DeVries 1984), a
structured, participative approach to developing
productivity and performance measures follows these
improvement oriented steps. The end results for
Honeywell are decoupled (decentralized) productivity and
performance measurement, evaluation, control, and
improvement systems. The logic for these systems
resides within the organizational system itself (usually at
the department level). This approach has obvious benefits
and costs.

PRODUCTIVITY MEASUREMENT AND EVALUATION

If productivity is to become more than "Much Ado
About Nothing" (Sink 1983b) we will have to begin to
operationally differentiate between the seven performance
criteria. Moreover, we will have to be more disciplined
about measuring criteria as we have defined them.

"You cannot manage that which
you cannot measure.
You cannot measure that which
you cannot define."

Productivity ≠ Performance

Too often a manager will agree to define
productivity as output ÷ input and then operationalize it in
an entirely different fashion. If productivity is an
important performance criteria, then we will have to begin
to measure it correctly.

Productivity is quite simply a relationship between
what comes out of an organizational system during some
given period of time divided by what resources were
consumed during that same period of time to create those
outputs. As we saw in figure 1, productivity essentially is:

$$\frac{\text{quality quantity (effectiveness)}}{\text{resources actually expended}}.$$

Therefore, productivity nicely integrates the concepts
effectiveness, quality, and part of the efficiency
equation. This is the strength of productivity as a
performance criteria, it allows one to simultaneously view
the relationship between these three critical criteria.

Productivity measures include the following:

	STATIC	DYMANIC
TOTAL FACTOR	$\dfrac{\text{Total Outputs 1983}}{\text{Total Inputs 1983}}$	$\dfrac{\dfrac{\text{Total Outputs 1984}}{\text{Total Inputs 1984}}}{\dfrac{\text{Total Outputs 1983}}{\text{Total Inputs 1983}}}$
MULTI- FACTOR	$\dfrac{\text{Outputs 1983}}{\text{Labor \& Materials \& Energy 1983}}$	$\dfrac{\dfrac{\text{Outputs 1984}}{\text{Labor \& Materials \& Energy 1984}}}{\dfrac{\text{Outputs 1983}}{\text{Labor \& Materials \& Energy 1983}}}$
PARTIAL FACTOR	$\dfrac{\text{Outputs 1983}}{\text{Labor 1983}}$	$\dfrac{\dfrac{\text{Outputs 1984}}{\text{Labor 1984}}}{\dfrac{\text{Outputs 1983}}{\text{Labor 1983}}}$

The goal in productivity measurement is to develop one or more of these measures for the purpose of evaluation, control, and eventual improvement. The technique we will be presenting a case example for primarily develops multifactor, dynamic indexes for productivity.

Note, however, that one does not necessarily have to measure productivity to improve it. It is, for instance, possible to develop excellent measurement and evaluation, control, and improvement systems for effectiveness, efficiency, and quality and to manage productivity quite well. We suspect, in fact, that in service oriented applications, management would be better off focussing upon these three causal criteria directly and initially prior to paying direct attention to productivity measurement and evaluation.

ALTERNATIVE STRATEGIES AND APPROACHES TO MEASURING AND EVALUATING PRODUCTIVITY IN SERVICE ORIENTED ORGANIZATIONAL SYSTEMS

In our recent DoD sponsored study, we were able to locate only four generic approaches to measuring and evaluating productivity. Note that we differentiated between measuring productivity, as strictly defined and measuring, evaluating, and verifying productivity improvement has taken place. Productivity measurement techniques can measure, evaluate, and verify that improvement has taken place but so can cost accounting systems, engineering economic analysis and evaluation procedures, and cost benefit procedures. (see Schmidt, 1984)

The four generic productivity measurement techniques we identified are:

(1) The Nomative Productivity / Performance Measurement Methodology (NP/PMM): a structured participative approach utilizing the Nominal Group Technique (NGT) to develop decoupled or decentralized measurement and evaluation systems.

(2) The Multi-Criteria Performance / Productivity Measurement Technique (MCP/PMT): a multi-attribute, decision analysis/utility theory approach to measurement and evaluation. (also called the Objectives Matrix)

(3) The Multi-Factor Productivity Measurement Model (MFPMM): an accounting based, computerized approach that examines changes in price-recovery, productivity and profitability from period to period.

(4) Surrogate Performance / "Productivity" Measurement Approaches: Measurement and evaluation approaches that measure other elements of the other six criteria and which assume high correlation with productivity.

A review of the DoD study and these techniques is provided by Sink and DeVries. (1984) We will focus the remainder of this paper on a case application, in the service sector, of the MFPMM.

MULTI-FACTOR PRODUCTIVITY MEASUREMENT MODEL

The Multi-Factor Productivity Measurement Model (MFPMM) is a computerized approach to measuring productivity and was developed by the Oklahoma Productivity Center at Oklahoma State University. The MFPMM and other similar approaches essentially require two periods of data for quantities and prices of the outputs and inputs of the organizational system being analyzed. Using this data, the MFPMM can provide change ratios, cost/revenue ratios, productivity ratios, performance indexes, and most importantly, the dollar effects of these measures on profits. Figure 4 shows the basic structure of the output generated by the MFPMM.

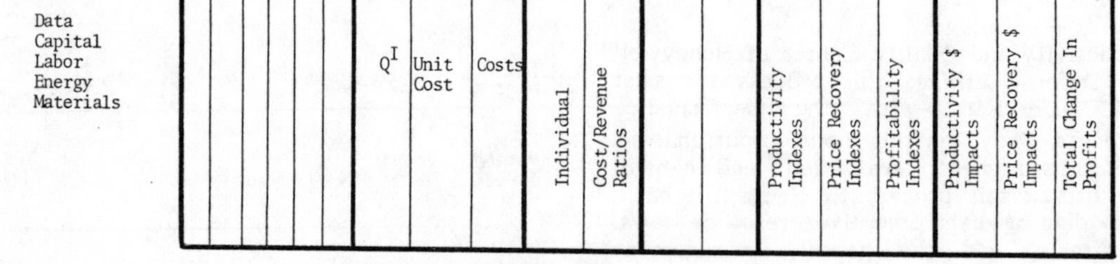

Figure 4. MFPMM Basic Structure

The MFPMM is based on the premise that profitability is a function of productivity and price recovery. That is, an organization can generate profit growth from productivity improvement and/or from price recovery. As mentioned earlier, productivity is a relationship between quantities of outputs and quantities of inputs. Price recovery, on the other hand, is the degree to which input prices (costs) are recovered by adjusting output prices. The relationship between productivity, profitability, and price recovery is depicted in Figure 5 (adapted from van Loggerenberg and Cucchiaro, 1981-1982).

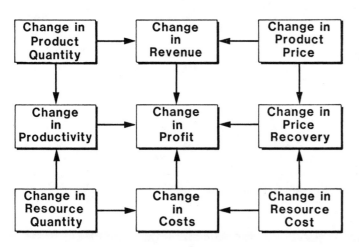

Figure 5. Relationship Between Productivity, Profitability and Price Recovery
(adapted from van Loggerenberg and Cucchiaro, 1981-1982)

MOTEL CASE EXAMPLE

The OPC has recently completed a one-year MFPMM pilot study with a local motel. The motel manager has been supplying the OPC with monthly data and in turn, the OPC has been running the model and providing the manager with the results. Typically, the manager first supplies the OPC with base period (period 1) and current period (period 2) data and the MFPMM is run comparing period 2 to period 1. Then, the manager reviews the MFPMM results along with other performance measurement data to which he has access and projects a future scenario (period 3) of quantities and prices (or costs). Using this projected data, the MFPMM is run again comparing period 3 to either period 1 or period 2. In order to assist the manager in his efforts, the OPC supplied him with data collection forms, data projection forms, and instructions. Base and current period data and projections for all twelve months were collected and run, but only one month will be presented here.

The remainder of this paper will utilize actual motel data from May 1980, May 1983, and projections for May 1984 to exemplify how the model was used by the manager to quantitatively view: 1) the motel's performance in comparison to historical performance, 2) the potential effects of planned management interventions and forecasted quantities and prices (costs) on future performance, and 3) the leverage afforded by specific input resources.

Comparison of Current Performance to Past Performance

The pilot study occurred during 1983 so the May 1983 data represents current period data (period 2). The year 1980 was selected as a base year against which the current and projected periods would be compared; therefore, the May 1980 data represents base period data (period 1). Due to the seasonality of the motel industry, the current month (period 2) was compared to the same month in the base year (period 1) and projections were made for the same month of the next year (period 3).

The first six columns of the MFPMM are data input (Table 1). Column 1 represents the quantities of outputs generating revenue and the quantities of input resources used to provide those outputs in period 1, the base period (May 1980). The outputs of the motel were defined as occupied rooms. The inputs were broken down into the following classes: Labor, Supplies, Utilities, and Capital. From Table 1, it can be seen that in May 1980, 714 Singles, 181 Doubles, 27 Triples, and 6 Meeting Rooms were occupied. (In other words, for example, since May consists of 31 days, an average of 714/31=23 Singles were occupied each day during the month.) The Restaurant and the Carpet Store are each leased out by the year so they constitute occupied rooms which generate monthly revenue. In the form of labor inputs, 3 employees labeled Personnel were utilized, and employees labeled Desk Clerks, Housekeepers, Maintenance, and Laundrymen worked 400 hours, 330 hours, 158 hours, and 118 hours, respectively. (It should be pointed out that the units of measure for outputs and inputs do not have to be consistent among categories.) Since each type of supplies (cleaning, office, and maintenance) included so many different items, they were simply lumped together and considered one unit of Cleaning Supplies, one unit of Office Supplies, and one unit of Maintenance Supplies. Regarding Utility inputs, 185 BTU's of Natural Gas, 17,725 KWH's of Electricity, and 372 thousand gallons of Water & Sewage were used during the month. The Capital inputs, like the Supply inputs, were each considered one unit because that is how they were treated by the motel's accounting system.

Column 2 represents the unit price for outputs and the unit cost for inputs for period 1, the base period (May 1980). From Table 1, it can be seen that Singles were $18, Doubles were $25, Triples were $27, Meeting Rooms were $40, and the Restaurant and Carpet Store were leased out for $1500 and $300, respectively. Personnel labor was paid $1000/person, Desk Clerks were paid $3.50/hour, Housekeepers were paid $3.00/hour, Maintenance labor was paid $4.00/hour, and Laundrymen were paid $3.00/hour. (All of these salaries are average salaries for May 1980.) Cleaning Supplies cost $489, Office Supplies cost $120, and $457 were spent on Maintenance Supplies. Natural Gas cost $2.55/BTU, Electricity cost $0.05/KWH, and Water & Sewage cost $0.70/thousand gallons. The amount paid on the Note Premium was $6500, Insurance cost $475, $200 was spent on Advertising, Phone Service cost $500, and $250 was spent on TV, Air Conditioning, and other various equipment.

Column 3 represents the period 1 (May 1980) value for each output and input. Since value equals quantity times price (V = Q x P), column 3 represents revenues for outputs and costs for inputs. For example, it can be seen from Table 1 that Total Outputs generated revenues of $20,146 and Total Inputs incurred costs of $16,985.

Table 1. Base Period and Current Period Data

	PERIOD 1			PERIOD 2		
	(1) QUANTITY	(2) PRICE	(3) VALUE	(4) QUANTITY	(5) - PRICE	(6) VALUE
SINGLE	714.0	18.00	12852.00	803.0	18.00	14454.00
DOUBLE	181.0	25.00	4525.00	198.0	25.00	4950.00
TRIPLE	27.0	27.00	729.00	22.0	27.00	594.00
MEETING ROOM	6.0	40.00	240.00	5.0	50.00	250.00
RESTAURANT	1.0	1500.00	1500.00	1.0	2000.00	2000.00
CARPET STORE	1.0	300.00	300.00	1.0	400.00	400.00
TOTAL OUTPUTS			20146.00			22648.00
PERSONNEL	3.0	1000.00	3000.00	2.0	1200.00	2400.00
DESK CLERKS	400.0	3.50	1400.00	360.0	4.00	1440.00
HOUSEKEEPERS	330.0	3.00	990.00	346.0	3.60	1245.60
MAINTENANCE	158.0	4.00	632.00	169.0	4.50	760.50
LAUNDRYMEN	118.0	3.00	354.00	123.0	3.50	430.50
TOTAL LABOR			6376.00			6276.60
CLEANING SUPP.	1.0	489.00	489.00	1.0	520.00	520.00
OFFICE SUPP.	1.0	120.00	120.00	1.0	136.00	136.00
MAINTENANCE SUPP	1.0	457.00	457.00	1.0	512.00	512.00
TOTAL SUPPLIES			1066.00			1168.00
NATURAL GAS	185.0	2.55	471.75	162.0	3.70	599.40
ELECTRICITY	17725.0	0.05	886.25	18423.0	0.08	1400.15
WATER & SEWAGE	372.0	0.70	260.40	402.0	0.90	361.80
TOTAL UTILITIES			1618.40			2361.35
NOTE PREMIUM	1.0	6500.00	6500.00	1.0	6500.00	6500.00
INSURANCE	1.0	475.00	475.00	1.0	532.00	532.00
ADVERTISING	1.0	200.00	200.00	1.0	358.00	358.00
PHONE SERVICE	1.0	500.00	500.00	1.0	775.00	775.00
TV AC ETC. EQMT.	1.0	250.00	250.00	1.0	400.00	400.00
TOTAL CAPITAL			7925.00			8565.00
TOTAL INPUTS			16985.40			18370.95

Breaking down the Total Inputs by the four major classes, it can be seen that Labor cost $6376, Supplies cost $1076, Utilities cost $1618.40, and Capital cost $7925. Column 3 is automatically calculated in the MFPMM by multiplying column 1 times column 2.

Columns 4 through 6 are the same as columns 1 through 3 except that they represent data for period 2, the current period (May 1983). Again, columns 4 and 5 represent quantity and price (or cost), respectively, and column 6, value, is obtained by multiplying column 4 times column 5. Column 4 of Table 1 shows, for example, that more Singles were occupied in May 1983, but less Personnel and less Desk Clerk hours were required. This already gives the manager some indication that labor productivity is improving. On the other hand, his unit prices for Singles, Doubles, and Triples stayed the same while his unit costs for all inputs went up.

Columns 7 through 9 provide weighted change ratios and are shown in Table 2. These change ratios represent the percentage increase (or decrease) of an item from the base period (May 1980) to the current period (May 1983). Column 7 reflects the price-weighted and base period price-indexed changes in quantities; column 8 reflects the quantity-weighted and current period quantity-indexed changes in unit prices and unit costs; and column 9 reflects the simultaneous impact of changes in quantities and prices (or costs) from period 1 (May 1980) to period 2 (May 1983) for each output and input. Essentially, column 9 indicates the rate of change of revenues and costs in period 2 relative to period 1.

Table 2. Weighted Change Ratios (Base Period to the Current Period)

	(7) QUANTITY	(8) PRICE	(9) VALUE
SINGLE	1.1246	1.0000	1.125
DOUBLE	1.0939	1.0000	1.094
TRIPLE	0.8148	1.0000	0.815
MEETING ROOM	0.8333	1.2500	1.042
RESTAURANT	1.0000	1.3333	1.333
CARPET STORE	1.0000	1.3333	1.333
TOTAL OUTPUTS	1.0919	1.0295	1.124
PERSONNEL	0.6667	1.2000	0.800
DESK CLERKS	0.9000	1.1429	1.029
HOUSEKEEPERS	1.0485	1.2000	1.258
MAINTENANCE	1.0696	1.1250	1.203
LAUNDRYMEN	1.0424	1.1667	1.216
TOTAL LABOR	0.8380	1.1747	0.984
CLEANING SUPP.	1.0000	1.0634	1.063
OFFICE SUPP.	1.0000	1.1333	1.133
MAINTENANCE SUPP	1.0000	1.1204	1.120
TOTAL SUPPLIES	1.0000	1.0957	1.096
NATURAL GAS	0.8757	1.4510	1.271
ELECTRICITY	1.0394	1.5200	1.580
WATER & SEWAGE	1.0806	1.2857	1.389
TOTAL UTILITIES	0.9983	1.4615	1.459
NOTE PREMIUM	1.0000	1.0000	1.000
INSURANCE	1.0000	1.1200	1.120
ADVERTISING	1.0000	1.7900	1.790
PHONE SERVICE	1.0000	1.5500	1.550
TV AC ETC. EQMT.	1.0000	1.6000	1.600
TOTAL CAPITAL	1.0000	1.0808	1.081
TOTAL INPUTS	0.9390	1.1518	1.082

Columns 10 and 11 represent cost to revenue ratios for period 1 (May 1980) and period 2 (May 1983), respectively. These ratios reflect the percentages of reported revenue consumed by a particular input in a given period. From Table 3, it can be seen that in May 1980, Total Labor costs represented 31.65% of total revenues, Total Supplies costs represented 5.29% of total revenues, Total Utilities represented 8.03% of total revenues, and Total Capital represented 39.34% of total revenues. Of all the inputs included in this model, Total Input costs consumed 84.31% of total revenues in May 1980. Note that since this is a "multi-factor" model rather than a "total factor" model, only the motel manager knows whether the 15.69% of the remaining revenues is all profits or attributable to other input costs not included in the model. Similarly, in column 11 of Table 3, 81.12% of reported revenues is due to Total Input costs, leaving 18.88% unaccounted for. These two columns provide the manager with insights as to the leverage afforded by specific resources. In other words, columns 10 and 11 show the manager where he will get the "biggest bang for his bucks."

Table 3. Base Period and Current Period Cost/Revenue Ratios

| | COST/REVENUE RATIOS | |
	(10) PERIOD 1	(11) PERIOD 2
TOTAL LABOR	0.3165	0.2771
TOTAL SUPPLIES	0.0529	0.0516
TOTAL UTILITIES	0.0803	0.1043
TOTAL CAPITAL	0.3934	0.3782
TOTAL INPUTS	0.8431	0.8112

Columns 12 and 13 represent productivity ratios for period 1 (May 1980) and period 2 (May 1983), respectively (Table 4). These are absolute price-weighted partial factor productivity ratios for each input item in each particular period. As you can see in Table 4, the productivity ratios of all of the input factors were higher in period 2 than they were in period 1. The next three columns (columns 14, 15, 16) of the MFPMM show just how much higher (in percentages) these ratios were.

Table 4. Base Period and Current Period Productivity Ratios

| | PRODUCTIVITY RATIOS | |
	(12) PERIOD 1	(13) PERIOD 2
TOTAL LABOR	3.16	4.12
TOTAL SUPPLIES	18.90	20.64
TOTAL UTILITIES	12.45	13.62
TOTAL CAPITAL	2.54	2.78
TOTAL INPUTS	1.19	1.38

Columns 14 through 16 of the MFPMM are called "Weighted Performance Indexes." Column 14 reflects price-weighted productivity indexes; column 15 reflects quantity-weighted price recovery indexes; and column 16 reflects profitability indexes. These columns are actually output over input change ratios from period 1 (May 1980) to period 2 (May 1983). Table 5 shows, for example, that Labor productivity increased by 30.3%, the unit cost of Supplies rose 6% faster than output prices, and the simultaneous changes in Total Inputs' quantities and costs contributed to a 3.9% increase in profitability.

Table 5. Weighted Performance Indexes (Base Period to Current Period)

| | WEIGHTED PERFORMANCE INDEXES | | |
	(14) CHANGE IN PRODUC-TIVITY	(15) CHANGE IN PRICE RECVRY	(16) CHANGE IN PROFIT-ABILITY
TOTAL LABOR	1.303	0.876	1.142
TOTAL SUPPLIES	1.092	0.940	1.026
TOTAL UTILITIES	1.094	0.704	0.770
TOTAL CAPITAL	1.092	0.953	1.040
TOTAL INPUTS	1.163	0.894	1.039

Columns 17 through 19 represent the dollar value effects on profits of columns 14 through 16. That is, columns 17 through 19 indicate what impacts (in dollars) are caused by changes in productivity, price recovery, and profitability, respectively. Table 6 presents the actual values for columns 17 through 19. For example, from May 1980 to May 1983, the increase in Total Inputs' productivity contributed $2597.20 to profits but the decrease in Total Inputs' price recovery affected profits adversely by $1873.27 resulting in a change in profitability (from May 1980 to May 1983) due to total reported inputs of $723.93. This does not mean that the motel made $723.93 in May 1983; it simply means that the motel was $723.93 more profitable in May 1983 than it was in May 1980.

Table 6. Dollar Effects on Profits (Base Period to Current Period)

| | DOLLAR EFFECTS ON PROFITS | | |
	(17) CHANGE IN PRODUC-TIVITY	(18) CHANGE IN PRICE RECOVERY	(19) CHANGE IN PROFIT-ABILITY
TOTAL LABOR	1619.14	-727.88	891.26
TOTAL SUPPLIES	98.00	-67.61	30.39
TOTAL UTILITIES	151.53	-693.48	-541.95
TOTAL CAPITAL	728.54	-384.30	344.23
TOTAL INPUTS	2597.20	-1873.27	723.93

Projections

Unique to the MFPMM is a sensitivity analysis routine which allows the user to project the impact of planned interventions and forecasted quantity and price (or cost) changes on profits. The planned interventions will obviously be aimed at controllable factors, but the user must seriously consider uncontrollable factors and attempt to offset any negative impacts they might have. This feature essentially gives the user an opportunity to play "what if" with the model by comparing a "projected" period (period 3) to either the current period (period 2) or the base period (period 1). But first the user must enter a projected "desired value" for Total Inputs' effect on change in profitability (last row of column 17). In this pilot study, the motel manager decided to compare the projected period to the base period and he wanted to be $1000 more profitable than he was in the base period.

In order to create projected period scenarios, the manager had to supply the OPC with three point estimates (pessimistic, most likely, optimistic) for each category he chose to vary in the projected period. These estimates were based upon his reaction to the results of the current period to base period comparison and his perceptions concerning forecasted quantities and prices (or costs). Since the manager chose to compare a projected period to the base period and the MFPMM uses a base period indexing mechanism, the effect of inflation was removed. The motel manager's quantity estimates for the projected period are shown in Table 7.

Table 7. Projected Quantity Estimates

NO.	CATEGORY NAME	QUANTITY		
		PESS.	M.LIKELY	OPTM.
1	SINGLE	650.	800.	875.
2	DOUBLE	100.	150.	170.
3	TRIPLE	20.	25.	30.
4	MEETING ROOM	5.	7.	9.
5	RESTAURANT	1.	1.	1.
6	CARPET STORE	1.	1.	1.
7	TOTAL OUTPUTS	0.	0.	0.
8	PERSONNEL	3.	2.	1.
9	DESK CLERKS	470.	400.	350.
10	HOUSEKEEPERS	360.	325.	295.
11	MAINTENANCE	300.	205.	160.
12	LAUNDRYMEN	123.	123.	123.
13	TOTAL LABOR	0.	0.	0.
14	CLEANING SUPP.	1.	1.	1.
15	OFFICE SUPP.	1.	1.	1.
16	MAINTENANCE SUPP	1.	1.	1.
17	TOTAL SUPPLIES	0.	0.	0.
18	NATURAL GAS	280.	250.	240.
19	ELECTRICITY	18000.	17000.	16500.
20	WATER & SEWAGE	385.	350.	330.
21	TOTAL UTILITIES	0.	0.	0.
22	NOTE PREMIUM	1.	1.	1.
23	INSURANCE	1.	1.	1.
24	ADVERTISING	1.	1.	1.
25	PHONE SERVICE	1.	1.	1.
26	TV AC ETC. EQMT.	1.	1.	1.
27	TOTAL CAPITAL	0.	0.	0.
28	TOTAL INPUTS	0.	0.	0.

In contrast to productivity performance, the motel's price recovery performance in May 1983 was poor (less than 1.0). Poor price recovery does not, however, imply trouble. It simply means that the motel manager has a competitive edge. He was able to increase profitability without increasing prices and probably gained a larger share of the market; therefore, he is now in a position to react to higher input costs by raising prices. His price estimates for the projected period are shown in Table 8.

Table 8. Projected Price Estimates

NO.	CATEGORY NAME	PRICE		
		PESS.	M.LIKELY	OPTM.
1	SINGLE	18.00	20.00	22.00
2	DOUBLE	24.00	26.00	28.00
3	TRIPLE	26.00	28.00	30.00
4	MEETING ROOM	40.00	55.00	60.00
5	RESTAURANT	2000.00	2000.00	2000.00
6	CARPET STORE	450.00	450.00	450.00
7	TOTAL OUTPUTS	.00	.00	.00
8	PERSONNEL	1500.00	1400.00	1000.00
9	DESK CLERKS	4.15	4.15	4.15
10	HOUSEKEEPERS	3.75	3.75	3.75
11	MAINTENANCE	4.75	4.75	4.75
12	LAUNDRYMEN	3.75	3.75	3.75
13	TOTAL LABOR	.00	.00	.00
14	CLEANING SUPP.	500.00	450.00	400.00
15	OFFICE SUPP.	120.00	100.00	80.00
16	MAINTENANCE SUPP	450.00	400.00	360.00
17	TOTAL SUPPLIES	.00	.00	.00
18	NATURAL GAS	4.20	4.20	4.20
19	ELECTRICITY	.08	.08	.08
20	WATER & SEWAGE	1.05	1.05	1.05
21	TOTAL UTILITIES	.00	.00	.00
22	NOTE PREMIUM	5000.00	5000.00	5000.00
23	INSURANCE	540.00	540.00	540.00
24	ADVERTISING	420.00	360.00	300.00

Simulation

After all quantity and price (or cost) projections have been made, a histogram plot and a probability statement such as those shown in Figure 6 are displayed. The histogram depicts the results of a Monte Carlo simulation which has generated 100 random outcomes. The 100 data points on the histogram represent the 100 simulated values for Total Inputs' effect on change in profitability based upon the projections made by the user. The probability statement tells the user how many of the 100 simulated values were greater than or equal to the desired value entered earlier. The statement at the bottom of Figure 6 tells the motel manager that, based upon his projections, he has a 75 % chance of equalling or exceeding his desired value of 1000. The histogram and probability statement give the manager a very good indication as to whether or not the projected scenario he created will result in his desired change in profitability before he ever looks at the new tableaus.

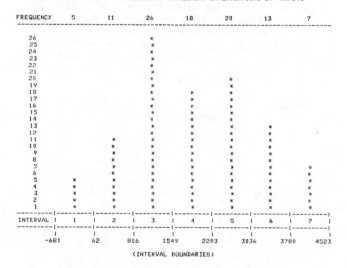

Figure 6. Histogram and Probability Statement

Table 9. Base Period and Projected Period Data

| | PERIOD 1 | | | PERIOD 3 | | |
	(1) QUANTITY	(2) PRICE	(3) VALUE	(4) QUANTITY	(5) PRICE	(6) VALUE
SINGLE	714.0	18.00	12852.00	775.0	20.00	15500.00
DOUBLE	181.0	25.00	4525.00	140.0	26.00	3640.00
TRIPLE	27.0	27.00	729.00	25.0	28.00	700.00
MEETING ROOM	6.0	40.00	240.00	7.0	51.67	361.67
RESTAURANT	1.0	1500.00	1500.00	1.0	2000.00	2000.00
CARPET STORE	1.0	300.00	300.00	1.0	450.00	450.00
TOTAL OUTPUTS			20146.00			22651.67
PERSONNEL	3.0	1000.00	3000.00	2.0	1300.00	2600.00
DESK CLERKS	400.0	3.50	1400.00	406.7	4.15	1687.67
HOUSEKEEPERS	330.0	3.00	990.00	326.7	3.75	1225.00
MAINTENANCE	158.0	4.00	632.00	221.7	4.75	1052.92
LAUNDRYMEN	118.0	3.00	354.00	123.0	3.75	461.25
TOTAL LABOR			6376.00			7026.83
CLEANING SUPP.	1.0	489.00	489.00	1.0	450.00	450.00
OFFICE SUPP.	1.0	120.00	120.00	1.0	100.00	100.00
MAINTENANCE SUPP	1.0	457.00	457.00	1.0	403.33	403.33
TOTAL SUPPLIES			1066.00			953.33
NATURAL GAS	185.0	2.55	471.75	256.7	4.20	1078.00
ELECTRICITY	17725.0	0.05	886.25	17166.7	0.08	1373.33
WATER & SEWAGE	372.0	0.70	260.40	355.0	1.05	372.75
TOTAL UTILITIES			1618.40			2824.08
NOTE PREMIUM	1.0	6500.00	6500.00	1.0	5000.00	5000.00
INSURANCE	1.0	475.00	475.00	1.0	540.00	540.00
ADVERTISING	1.0	200.00	200.00	1.0	360.00	360.00
PHONE SERVICE	1.0	500.00	500.00	1.0	775.00	775.00
TV AC ETC. EQMT.	1.0	250.00	250.00	1.0	390.00	390.00
TOTAL CAPITAL			7925.00			7065.00
TOTAL INPUTS			16985.40			17869.25

Comparison of Projected Performance to Past Performance

The three new tableaus are presented in the same format as the previous tableaus but this time they reflect period 3 (projected) and period 1 (base) data rather than period 2 (current) and period 1 data. The period 3 data is the result of expected value calculations on the estimates entered by the manager. If he did not enter new estimates for a particular item, then that item retained its current period (period 2) value.

Table 9 shows the first six columns of the new tableaus. Columns 1 through 3 provide base period (May 1980) data again and columns 4 through 6 provide the new projected period (May 1984) data.

The remaining columns of the new tableaus provide the motel manager with valuable information concerning the aggregate effects of his planned interventions and projected changes in quantities and prices. Columns 7 through 9 provide weighted change ratios (changes from the base period to the projected period) and are shown in Table 10. From this table, the manager can see the projected percentage increases (or decreases) in quantities and prices (or costs) in the future period. For example, column 9 shows that, based upon his projected estimates, revenues from Total Outputs will be 12.44% higher and costs from Total Inputs will be 5.20% higher than they were in the base period (May 1980).

Columns 10 and 11 do not show any significant variation (from May 1980 to May 1984) in the percentage of revenues consumed by various resource costs (Table 10). The motel operation is still heavily labor and capital intensive with the Note Premium representing the largest single percentage of revenues. However, the cost/revenue ratio of the Note Premium is expected to decrease from .3226 (32.26% of revenues) in May 1980 to .2207 (22.07% of revenues) in May 1984. This will help reduce the percentage of total revenues consumed by Total Inputs' costs from 84.31% to 78.84%.

Table 10. Base Period and Projected Period Cost/Revenue Ratios

| | COST/REVENUE RATIOS | |
	(10) PERIOD 1	(11) PERIOD 3
TOTAL LABOR	0.3165	0.3102
TOTAL SUPPLIES	0.0529	0.0421
TOTAL UTILITIES	0.0803	0.1247
TOTAL CAPITAL	0.3934	0.3119
TOTAL INPUTS	0.8431	0.7889

Columns 12 and 13 reflect the productivity ratios of the various inputs in period 1 (May 1980) and period 3 (May 1984) and are shown in Table 11. Overall, the multi-factor productivity (Total Inputs) is expected to increase slightly from 1.19 to 1.23.

Table 11. Base Period and Projected Period Productivity Ratios

	PRODUCTIVITY RATIOS	
	(12)	(13)
	PERIOD 1	PERIOD 3
TOTAL LABOR	3.16	3.57
TOTAL SUPPLIES	18.90	18.95
TOTAL UTILITIES	12.45	11.47
TOTAL CAPITAL	2.54	2.55
TOTAL INPUTS	1.19	1.23

Columns 14 through 16 are shown in Table 12 and they reflect the changes in productivity, price recovery, and profitability expected as a result of the projected May 1984 estimates. If you recall from the comparison of current performance to past performance, you will remember that the productivity of all the categories increased in May 1983 relative to May 1980. As a result, the motel manager paid little attention to improving productivity in the projected period.

Most of the motel manager's efforts were focused on improving price recovery and from the data shown in column 15 of Table 12 , it appears that he will be successful.

Table 12. Weighted Performance Indexes (Base Period to Projected Period

	WEIGHTED PERFORMANCE INDEXES		
	(14)	(15)	(16)
	CHANGE IN		
	PRODUC-TIVITY	PRICE RECVRY	PROFIT-ABILITY
TOTAL LABOR	1.130	0.903	1.020
TOTAL SUPPLIES	1.003	1.254	1.257
TOTAL UTILITIES	0.922	0.699	0.644
TOTAL CAPITAL	1.003	1.258	1.261
TOTAL INPUTS	1.038	1.030	1.069

Column 16 of Table 12 shows the percentage change in profitability of various inputs from the base period (May 1980) to the projected period (May 1984). For example, based upon projected estimates, the following changes will occur in profitability: Total Labor will increase by 2%, Total Supplies will increase by 25.7%, Total Utilities will decrease by 35.6%, Total Capital will increase by 26.1%, and Total Inputs (Multi-factor Profitability) will increase by 6.9%. In other words, the motel is projected to be 6.9% more profitable in May 1984 that it was in May 1980.

Columns 17 through 19 reflect the dollar effects on profits of projected changes in productivity and price recovery (Table 13). The aggregate effect of the projected change in productivity (relative to May 1980) is an increase in profits of $623.81. Column 18 of Table 13 shows that projected decreases in the price recovery of Total Labor and Total Utilities will affect profits negatively by $593.49 and $866.20, respectively. The total dollar impact on profits from projected changes in productivity and price recovery is indicated in column 19 of Table 13. Overall, the motel is expected to be $1228.71 more profitable in May 1984 (period 3) than in May 1980 (period 1). $623.81 of this increase in profitability is attributable to improved productivity and $604.90 is attributable to increased price recovery. The projected increases in productivity and price recovery indicate a definite increase in short term profits. However, if this projected performance actually transpires, the motel manager must then proceed cautiously. Consistently increasing price recovery can result in consumer resistance and increased competition. It is crucial to the survival of the motel that the manager have increasing productivity to fall back on if strong competition occurs.

Table 13. Dollar Effects on Profits (Base Period to Projected Period)

	DOLLAR EFFECTS ON PROFITS		
	(17)	(18)	(19)
	CHANGE IN PRODUC-TIVITY	CHANGE IN PRICE RECOVERY	CHANGE IN PROFIT-ABILITY
TOTAL LABOR	735.67	-593.49	142.18
TOTAL SUPPLIES	3.12	242.13	245.25
NATURAL GAS	-181.37	-366.21	-547.58
ELECTRICITY	30.51	-407.37	-376.86
WATER & SEWAGE	12.66	-92.63	-79.96
TOTAL UTILITIES	-138.19	-866.20	-1004.39
NOTE PREMIUM	19.04	2289.40	2308.44
INSURANCE	1.39	-7.31	-5.92
ADVERTISING	0.59	-135.71	-135.12
PHONE SERVICE	1.46	-214.28	-212.81
TV AC ETC. EQMT.	0.73	-109.64	-108.91
TOTAL CAPITAL	23.21	1822.47	1845.67
TOTAL INPUTS	623.81	604.90	1228.71

After seeing the histogram plot of simulated column 17 values, the motel manager could see that his projected scenario had a 75% chance of achieving his target of $1000 change in profits in May 1984 (period 3) relative to May 1980 (period 1), but he did not yet know precisely which factors would affect the motel's performance positively and which would affect it negatively. Now, he not only knows which factors were positive and which were negative but also the magnitude of each factor's dollar impact on profits in terms of productivity and price recovery. The trend charts in Figure 7 graphically depict the base, current, and projected performance of each of the motel's major input classes as well as the combined effect of all of the inputs included in the analysis. It should be pointed out that the projected period will eventually become the current period so the MFPMM can also be used to track the accuracy of projections and the effectiveness of planned action.

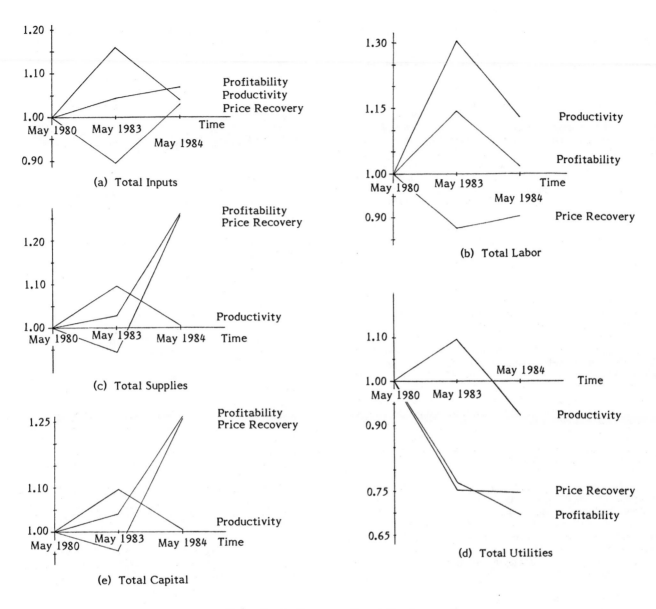

Figure 7. Performance Trend Charts

The performance trend chart for Total Inputs (Figure 7a) shows the motel manager that even though his projected productivity (1.038) will be higher than base period productivity (1.0), it will be a significant decline from current period productivity (1.163). The MFPMM could be run comparing the projected period (May 1984) to the current period (May 1983), but then the model's built-in base period indexing process would be negated. Furthermore, the results of a projected period to current period comparison would still reflect the declining productivity exhibited by the trend chart. Although productivity is projected to decline, profitability is expected to steadily increase because of rising price recovery. This would put the motel in a potentially dangerous situation. New competitors would be attracted (because of higher prices), market share would be reduced, and productivity would become an even more serious concern because of underutilized capacity.

If the motel manager had unlimited access to the MFPMM (e.g., a microcomputer in his office), he could try other scenarios until he was satisfied with the projected results. Then, he could proceed by developing specific action plans which would hopefully lead to the desired projected scenario. Although the manager's previous scenario was capable of achieving his desired change in profits, it would have initiated a downward trend in productivity covered up by the increase in price recovery. In this pilot study, the motel manager only submitted one projected scenario to be run by the OPC. The information provided by the MFPMM results and the performance trend charts should not be taken out of context; they simply monitor one particular month (May) of the motel's performance from year to year. Nevertheless, it is never advisable to overemphasize price recovery at the expense of productivity. That strategy is indeed a simple approach to short-term profitability but it will inevitably erode the long-term success of any organization.

Naturally, the motel manager will have more than just one month of data from each year but since his business is seasonal, it is advantageous to compare similar months of different years. However, since data from other months does exist, another capability being added to the MFPMM is a year-to-date analysis. The development of this feature is a direct consequence of the motel pilot study. During the study, the motel manager pointed out to the OPC that it would be beneficial to him to compare his year-to-date performance in the current year to the year-to-date performance in a base year and to project year-to-date performance in a future year. For example, with this feature he could not only compare May 1983 to May 1980, but he could also compare January-May 1983 to January-May 1980. A comparison of May 1983 to May 1980 might show acceptable performance when in fact the year-to-date performance of 1983 (relative to 1980) might be extremely poor.

CONCLUSION

The motel case example has shown how the MFPMM can be used to measure and monitor performance. The only requirements are that quantity and price (or cost) data be obtained for each output and input of the entity being analyzed. As evidenced by this example, output does not have to be something manufactured. (The motel does not manufacture rooms each month.) Output can be services provided as long as those services can be quantified and input data can be matched to output data. Depending upon the needs of the user, period length can be a week, a month, a quarter, a year, or any other length of time as long as the collection of data is feasible. Also, if the appropriate data exists, the base period could be used for standards or even a budget.

It is very important that the MFPMM be closely interrelated with an organization's existing control systems. Managers should be able to use the MFPMM to complement other sources of financial data when making business decisions. Also critical to the successful application of the MFPMM is the degree to which management feels comfortable using it and accepts the information it provides.

The likelihood that managers view the MFPMM as a valuable decision support system (DSS) is an ongoing concern of the OPC. Even though the MFPMM is presently a valuable management tool, it still can be improved. Enhanced interactive capabilities, added graphics, and improved flexibility are among the planned developments. These developments should significantly increase the degree to which the MFPMM can help organizations strategically plan for, measure, evaluate, and control sustained profit growth and strong competitive positioning.

REFERENCES

American Productivity Center (1978). How to Measure Productivity at the Firm Level. Short Course Notebook and Reference Manual. Houston, Texas.

Bain, D.F. (1982). The Productivity Prescription. New York: McGraw-Hill.

Craig, C.E. and Harris, R.C. (1973, Spring). Total Productivity Measurement at the Firm Level. Sloan Management Review.

Davis, Hiram (1955). Productivity Accounting. Philadelphia: The Wharton School, Industrial Research Unit, University of Pennsylvania (reprint 1978).

Naisbitt, John (1982). Megatrends. New York: Warner Books.

Schmidt, Robert (1984). Cost Justification for Advanced Manufacturing Technology. Institute of Industrial Engineers 1984 Spring Conference Proceedings. Chicago, Illinois.

Sink, D.S. and Keats, J.B. (1983). Quality Control Applications in Multi-Factor Productivity Measurement Models. Institute of Industrial Engineers 1983 Spring Conferenece Proceedings. Louisville, Kentucky.

Sink, D.S. (1983a). Organizational System Performance: Is Productivity a Critical Component? Institute of Industrial Engineers 1983 Spring Conference Proceedings. Louisville, Kentucky.

Sink, D.S. (1983b, October). Much Ado About Productivity: Where Do We Go From Here? Industrial Engineering.

Sink, D.S. (1984a). The Essentials of Productivity Management: Strategic Planning, Measurement, Evaluation, Control and Improvement. Short Course Notebook. LINPRIM, Inc. Stillwater, Oklahoma.

Sink, D.S. (1984b). Productivity Management: Planning, Measurement, Evaluation, Control and Improvement. To be published by John Wiley & Sons, New York, New York.

Sink, D.S. and DeVries, S.J. (1984). An In-Depth Study and Review of "State-of-the-Art and Practice" Productivity Measurement Techniques. Institute of Industrial Engineers 1984 Spring Conference Proceedings. Chicago, Illinois.

Sink, D.S., Tuttle, T.C., DeVries, S.J. and Swaim, J.C. (1984). Development of a Taxonomy of Productivity Measurement Theories and Techniques. AFBRMC Contract Number F33615-83-C-5071. Wright-Patterson Air Force Base, Ohio.

Sumanth, David J. (1984). Productivity Engineering and Management. New York: McGraw-Hill.

Swaim, Jeff and Sink, D.S. (1983). Current Developments in Firm or Corporate Level Productivity Measurement and Evaluation. Institute of Industrial Engineers 1983 Fall Conference Proceedings. Toronto, Ontario, Canada.

Toffler, Alvin (1980). The Third Wave. New York: Bantam Books.

van Loggerenberg, B.J. and Cucchiaro, S.J. (1981-1982, Winter). Productivity Measurement and the Bottom Line. National Productivity Review, 1 (1).

BIOGRAPHICAL SKETCHES

JEFF SWAIM is a graduate associate for the Oklahoma Productivity Center at Oklahoma State University. He is currently editor of "Productivity Management", an OPC Publication, and the IIE Management Division newsletter. Jeff received his B.S. and M.S. degrees in Industrial Engineering and Management from Oklahoma State University and is a member of IIE, OSPE, and NSPE.

D. SCOTT SINK, P.E., is an associate professor in the School of Industrial Engineering and Management at Oklahoma State University. His areas of interest include productivity management and measurement, work measurement and improvement and organizational behavior. Sink is a senior member of IIE and is currently director for the management division. Sink is a recipient of the Haliburton Award of Excellence, the Dow Outstanding Young Faculty Award and the Oklahoma Society of Professional Engineers Outstanding Young Engineer of the Year award. He is a registered professional engineer in the state of Oklahoma.

Reprinted from 1984 Annual International Industrial Engineering Conference Proceedings.

AN IN-DEPTH STUDY AND REVIEW OF STATE-OF-THE-ART AND PRACTICE PRODUCTIVITY MEASUREMENT TECHNIQUES

by

D. Scott Sink, Ph.D., P.E.
Associate Professor, and Director
Oklahoma Productivity Center
and
Sandra J. DeVries
Associate, Oklahoma Productivity Center

School of Industrial Engineering & Management
Oklahoma State University

ABSTRACT

The Department of Defense has initiated numerous productivity improvement programs relating to contractor productivity. One such program is the Industrial Modernization Incentives Program (IMIP). A component of IMIP is contractor productivity measurement. The Army Procurement Research Office was charged to sponsor and coordinate a study to investigate and develop a taxonomy of productivity measurement theories and techniques. The Oklahoma Productivity Center was awarded a contract through the Air Force Business Research Management Center to execute this research. The Maryland Center for Productivity and Quality of Working Life was a subcontractor to the Oklahoma Productivity Center on this project.

The project was completed in late Fall of 1983. A final report and briefing was made to the DoD in December, 1983. The result of the project was a comprehensive classification scheme of productivity measurement techniques. A very thorough literature search was made (both automated and manual) and site visits to organizations leading the field of productivity measurement were carried out to ensure that the taxonomy was valid and complete.

This paper presents the highlights of that study. The taxonomy or classification system is presented and described. Specific productivity measurement techniques that are either in use or should be in use are also outlined. Focus is placed on attempting to prescribe which techniques are applicable to specific situations. The paper, therefore, represents a very current look at what is being and can be accomplished through productivity measurement in organizations today.

INTRODUCTION

In a cost plus contract business such as exists between the DoD and their major contractors and subcontractors the issue to performance is critical. Performance means many things to many people, however, as discussed in previous papers (Sink; 1982, 1983) it is primarily comprised of at least seven criteria. Those criteria are: effectiveness, efficiency, quality, productivity, quality of work life, innovation, and of course, profitability/budgetability. The Federal Government and the DoD in particular have been concerned about the "value" of goods and services they receive from contractors and subcontractors in the private business sector. The industrial base that the DoD relies upon to deliver necessary goods and services is extremely complex, at times impossibly cumbersome, and often ineffective, inefficient, and nonproductive. As a result, the costs of goods, systems, and services being delivered to DoD contractors from subcontractors and ultimately to the DoD's various branches have grown greatly. There have been various attempts to create incentives for contractors and subcontractors to improve productivity (actually overall performance) and, therefore, ultimately reduce costs.

A simple example will serve to clarify the basic goals of these productivity improvement programs developed and being developed by the DoD. Suppose that a contractor has received a contract to provide component A for some new defense related system. The contractor would have gone through some estimating procedure to come up with expected total costs based upon existing technologies, anticipated volumes, service life of component market, etc. In cost plus systems, the plus is a percentage figure that is also negotiated and tacked on top of the contract's total costs. Let's assume that the estimate is for $10.00 per component plus margin. The relatively few number of multi-year contracts, which occurs for a number of reasons, adds to the complexity and costs inherent in this system.

Next year the contract is renegotiated and awarded again to the same contractor. Let's assume that some new technology (hardware, software, methods, etc.) is available which would allow that contractor to reduce the cost of the component to $8.00 per unit. Will the contractor be motivated to implement that technology? Is there currently an incentive for him to do that? Figure 1 presents the scenario such that you may be able to answer this question.

Note that since the plus in cost plus contracts is relatively fixed from year to year, total margins (total revenue -

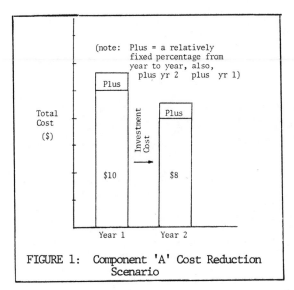

FIGURE 1: Component 'A' Cost Reduction Scenario

total costs) will actually shrink in year 2 as a result of this cost reduction. Additionally, there will inevitably be an investment cost associated with the improvement which reduces the yield even further.

Now, granted this is a micro analysis of a macro problem. And, for most contractors there are a great number of products and services being produced and each one is at various stages in its life cycle. Margins will obviously vary as a product or service moves through the life cycle. However, in the quasi-free market environment such as exists in DoD contracting the pressures to remain effective, efficient, to produce quality, and to maintain acceptable levels of productivity are not as dominant or clear cut as they are in a truly competitive market. The incentives for reducing a component's costs from $10.00 to $8.00 through some application of new technology are simply not very strong in a cost plus contracting environment.

The Industrial Modernizations Incentives Program (IMIP) is an attempt to create incentives for contractors and subcontractors to proactively search for ways to lower costs. The primary incentive mechanism is in the form of cost and benefit sharing. For instance, in our component example the investment required to reduce unit costs from $10.00 to $8.00 would be shared. The negotiation process is detailed and somewhat complex but still very realistic. The second component of the incentive mechanism is the cost reduction or benefit sharing. In other words, the $2.00 unit cost savings would be shared with the contractor. For example a 50% sharing might take place whereby a new contract would be given in year 2 at $8.00 per unit but the contractor would receive $1.00 per unit benefit sharing. Both sides win, a win-win strategy has been executed.

We have much simplified a fairly complex program and set of issues, however, the reader now has an understanding of the basic concepts of the program. A critical element and concern surrounding this DoD effort is that of measurement and evaluation. This concern is what prompted the recent DoD/APRO/WPAFB study that we are reporting in this paper.

DoD STUDY

The actual evolution of the DoD/APRO study has its roots in a document produced in 1981 entitled the "Carlucci Initiatives". Deputy Secretary of Defense Frank C. Carlucci, directed a 30-day study assessment of the Defense acquisition system on 2 March 1981. The report delivered to him on 31 March 1981 provided 24 specific recommendations and 8 issues for decisions. Of these initiatives, several were directed towards the need for increased contractor productivity (i.e. encourage capital investment to enhance productivity, contractor incentives to improve reliability and support, management principles, budget most likely costs, etc.).

As a result of the Carlucci Initiative and other related impetuses, a number and variety of Defense Industry, productivity related programs and efforts have been initiated (see the February 1983 issue of Industrial Engineering for a review and update on these activities). In particular, the DoD initiated a contractor productivity study under the direction of the Army Procurement Research office in Fort Lee, Virginia. Mr. Monte Norton is directing that project. The goals, objectives, and activities for that project are presented in Figure 2 in the form of a project timetable. Our project and the focus of this paper was on measurement of contractor productivity (Task #3 from Figure 2). This component of the overall APRO project was deemed as being the linchpin since any incentive plan for investment and cost savings would necessarily require a strong measurement, evaluation, and verification component. The contractor has to be able to verify that he can and does provide the component at $8.00. And, the DoD would obviously like to ensure that their investment reaped expected benefits.

PROPOSED TASK ASSIGNMENTS

0. DEVELOP PROJECT DEFINITION - APRO WITH INPUT FROM OTHER PARTICIPANTS.

1. DESIGN AND DISTRIBUTE CONTRACTOR SURVEY - APRO

2. DEVELOP INCENTIVES AND REWARD MECHANISMS - IMIP*

3. INVESTIGATE PRODUCTIVITY MEASUREMENT THEORY - AFBRMC

4. ANALYZE RESULTS AND SURVEY RESPONSES - APRO/AFBRMC

5. SYNTHESIZE PROPOSED SYSTEM - APRO

6. DRAFT REPORT - APRO WITH INPUT FROM OTHER PARTICIPANTS

7. DESIGN SYSTEM TEST - NCAR/DSMC

8. CONDUCT TEST AND ANALYZE RESULTS - NCAR WITH INPUT FROM OTHER PARTICIPANTS

9. DRAFT GUIDE - NCAR/DSMC

10. MANAGE & COORDINATE EFFORT/IMPLEMENTATION - ADUSD-IP

*Separate ongoing activity supported by this effort.

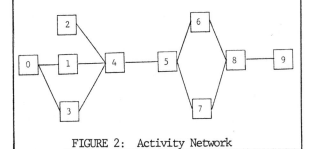

FIGURE 2: Activity Network

The Oklahoma Productivity Center received the contract to develop a taxonomy of productivity measurement theories and techniques in August 1983. We subcontracted with Dr. Tom Tuttle of the Maryland Center for Productivity and Quality of Work life to ensure breadth of knowledge. Our goal was to uncover and present a review of current theories and techniques for measuring productivity from the work group level up to the firm level. We were also expected to develop a taxonomy or classification system with which to conceptually categorize each theory and technique.

FINDINGS: PRODUCTIVITY MEASUREMENT TECHNIQUES

We made an extensive study of the literature via automated data searches and utilizing conventional search techniques. We accessed government, public, and private data bases. The Oklahoma Productivity Center resource center containing over 1300 productivity related books, research documents, articles, etc., was also tapped. In addition to this comprehensive literature review, we made a number of site visits to organizations having exemplary productivity measurement efforts. We identified these organizations by obtaining a top ten list of organizations from productivity center directors in the National Productivity Network. Carl Thor, of the American Productivity Center, was particularly helpful. Site visits were made to; the Honeywell Productivity and Quality Center, Hershey Foods, Lockheed Georgia, Westinghouse, and IBM. Previous knowledge and experience of the researchers bolstered these site visits in terms of breadth of coverage.

As a result of our search, we concluded that there are four generic and primary techniques available and to some extent in use for measuring productivity. They are:

1. **Multi-Factor Productivity Measurement Model** (MFPMM) (also called the Total Productivity Measurement Model, Total Factor Productivity Measurement by Product in other related forms and variations)

2. **Normative Productivity Measurement Methodology** (NPMM)

3. **Multi-Criteria Performance/Productivity Measurement Technique** (MCP/PMT) (also called the Objectives Matrix)

4. **Surrogate Approaches**
   a. Common Staffing Study
   b. Benefit/Cost Tracking
   c. Checklists, Audits, etc.
   d. Managing Productivity by Objectives

We will report on the three major approaches and the surrogate approaches briefly in this section of the paper.

MULTI-FACTOR PRODUCTIVITY MEASUREMENT MODEL

The Multi-Factor Productivity Measurement Model (MFPMM) is a dynamic, aggregated, indexed and computerized approach to measuring productivity. It is capable of blending the major inputs of a particular organizational system together and relate the resulting aggregate input to the total output of that same system. MFPMM can be utilized to measure productivity change in labor, materials, energy and capital. It also measures the corresponding effect each one has on profitability. With essentially the same accounting data used to track revenues and costs, the MFPMM can provide additional insight into the individual factors that are most significantly affecting profits.

The MFPMM is based on the premise that profitability is a function of productivity and price recovery; that is, an organizational system can generate profit growth from productivity improvement and/or from price recovery. Productivity relates to quantities of output and quantities of inputs, while price recovery relates to prices of output and costs of inputs. Price recovery can be thought of as the degree to which input cost increases are passed on to the customers in the form of higher output prices. The relationships between productivity, profitability and price recovery is depicted in Figure 3 (adapted from van Loggerenberg and Cucchiaro, 1981-1982).

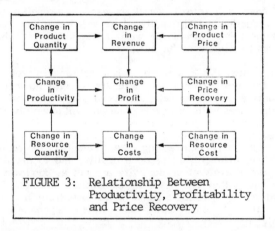

FIGURE 3: Relationship Between Productivity, Profitability and Price Recovery

The data required for the MFPMM are periodic data for quantity, price and value of each output and input of the organizational system being analyzed. Since value equals quantity times price, having two of the quantity, price and value variables obviously yields the third algebraically. Quantity, price and/or value of the various outputs produced and most of the inputs consumed are straight forward and should be provided by most basic accounting systems.

The MFPMM compares data from one period (base period) with data from a second period (current period). This comparison forms the basis of the productivity/price recovery/profitability analysis. The choice of a base period is a critical decision since it establishes the period against which the current period will be compared. Therefore, the base period should be as representative of normal business conditions as possible. If the data exists the budget or "standards" could be used as the base period data. Depending on the needs of the user, the availability of data, product cycle time, etc., period length could be a week, a quarter, a year or any other period for which input data can be matched to output data.

From the base and current period data, the MFPMM generates a series of ratios and indexes, each communicating different information about the system under study. The ratios and indexes, as depicted in Figure 4, include weighted change ratios, cost/revenue ratios,

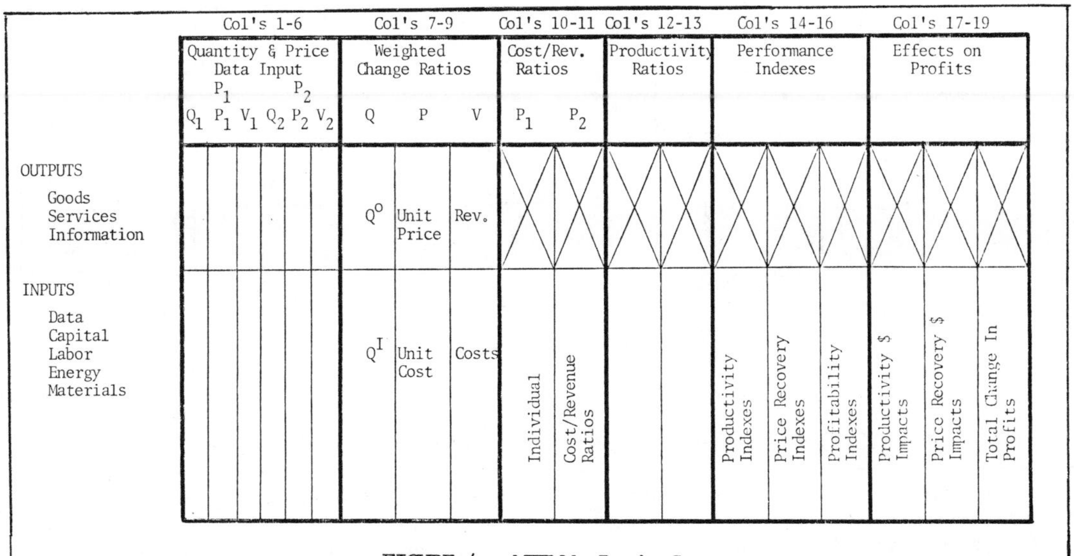

	Col's 1-6						Col's 7-9			Col's 10-11		Col's 12-13	Col's 14-16			Col's 17-19		
	Quantity & Price Data Input						Weighted Change Ratios			Cost/Rev. Ratios		Productivity Ratios	Performance Indexes			Effects on Profits		
	P_1			P_2														
	Q_1	P_1	V_1	Q_2	P_2	V_2	Q	P	V	P_1	P_2							
OUTPUTS Goods Services Information							Q^O	Unit Price	Rev.									
INPUTS Data Capital Labor Energy Materials							Q^I	Unit Cost	Costs	Individual Cost/Revenue Ratios			Productivity Indexes	Price Recovery Indexes	Profitability Indexes	Productivity $ Impacts	Price Recovery $ Impacts	Total Change In Profits

FIGURE 4: MFPMM Basic Structure

productivity ratios, weighted performance indexes and total dollar effects on profits. Weighted change ratios depict the percentage increase (or decrease) of an item from the base to current period. Both price and quantity weighted change ratios are generated by the model to show the percentage changes from period to period. Cost/Revenue ratios reflect the percentages of reported revenue consumed by a particular input in a given period. This information provides the user with insights as to where leverage exists. The most common method of productivity improvement is cost reduction and these ratios show exactly where cost reductions will pay the biggest dividends. Productivity ratios depict absolute productivity values in the base and current period. These absolute values are used in calculating the price weighted productivity indexes which show increases or decreases in productivity for the overall system and each component. Weighted performances indexes are actually output over input change ratios from period 1 to period 2. The final set of indexes are the dollar effects on profits. In other words, these indexes indicate what impact (in dollars) are caused by changes in productivity, price recovery and profitability. The ratios and indexes, alone or together, provides management with information on their systems. The ratios and indexes identify areas that need improvement and they also identify areas that are operating at an acceptable level. If the information is used correctly, productivity can improve which in turn should increase profits.

The MFPMM is most appropriate at the firm and plant level and would be most useful to senior management. It could be used at the cost center level as a separate accounting system for an assembly line, individual product line, etc; however, at lower levels of organizations, managers do not normally need the kind of detail offered by the model. The MFPMM has been most often applied in manufacturing settings but it can be used anywhere the necessary data exists.

It is estimated that somewhere between 50 and 100 organizations in the United States are utilizing the multi-factor productivity measurement approach. Among these are: Phillips Petroleum Company, Anderson Clayton, General Foods, Hershey Foods, Sentry Insurance, John Deere and Federal Express.

NORMATIVE PRODUCTIVITY MEASUREMENT METHODOLOGY

The Normative Productivity Measurement Methodology (NPMM) was designed, developed and tested at The Ohio State University by the Productivity Research Group of the Industrial and Systems Engineering Department during the period 1975-1978 (Sink, 1981). The basic and early methodology, as tested at Ohio State, incorporated structured group processes to identify appropriate productivity measures for such work groups as engineering, marketing and personnel. The structured group processes are used as mechanisms for shaping consensus and for developing a commitment for further follow-through. Once the productivity measures are identified, it becomes the task of the group to operationalize and implement the productivity measurement system. The final and perhaps most important task of the NPMM is to provide feedback to the workers in hopes of identifying productivity improvement opportunities.

The Normative Productivity Measurement Methodology, as a component of a productivity measurement system, is shown in Figure 5. As illustrated, NPMM is not implemented until several necessary preconditions are fulfilled. Without support from all levels of management and labor, the NPMM cannot be successful. Once it is determined that NPMM is to be used in the productivity measurement system, there are five distinct stages to be followed.

Stage 1 of the NPMM involves execution of a structured group process such as the Nominal Group Technique (NGT) or the Delphi Technique (DT). This process is used as a mechanism for generating a prioritized list of productivity measures, ratios, and/or indices for each specified unit of analysis. The basic reason for using a "normative" or participative approach is to ensure adequate motivation commitment, and accountability on the part of key participants for implementation and acceptance of the resultant productivity measurement system (Sink, 1984).

Stage 2, the design of the productivity measurement system, requires intervention from one or more productivity analysts, who work at operationalizing measures obtained from the structured group process. The

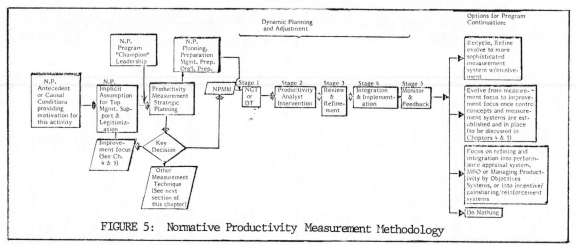

FIGURE 5: Normative Productivity Measurement Methodology

analysts must also determine how and where to collect the data for the productivity measures previously suggested by the structured group processes. Eventually, it becomes the responsibility of the analysts and the group members to shape a workable productivity measurement system (Sink, 1984).

Stage 3 of the NPMM process requires a briefing, review, discussion, potential revision and eventual approval of the draft operating system of the productivity measurement program (Sink, 1984). This is also the stage that provides feedback to the participants in the organizational system for which the productivity measurement system is being developed. This feedback is important in keeping commitment from all participants involved in the process.

Stage 4 is the integration and implementation of the productivity measurement system with existing performance measurement systems within the organization. This integration and implementation requires consultative or managerial backing in order for the resultant productivity measurement system to be fully implemented.

The final stage, Stage 5, occurs once the measurement program has been implemented. Continuous monitoring of the productivity measurement system after implementation is recommended. This monitoring will provide feedback and validity for the program. The monitoring also provides flexibility to the method. If at any time the program developed by the NPMM does not accomplish its objectives, corrections can be made by repeating the NPMM process. Like any worker participation program the measurement program will eventually run its course. This is a natural occurrence and should not be viewed as a flaw in the methodology. Figure 5 lists several options for continuing a participative productivity measurement system.

The NPMM can be implemented in any organization that is interested in pursuing a participative method of developing productivity measures. A substantial body of literature and experience relative to participation and group processes suggests that there exists a great amount of untapped resources within organizations in the form of its employees (Sink, 1984). By tapping the resources of its employees, an organization can capture a group wisdom that is most likely to identify the real problems and opportunities involved in productivity improvement.

It has been determined that the NPMM process functions best when the measurements focus is not on the corporation as a whole, or the individual employee, but upon the individual department. The productivity emphasis; therefore, would be placed upon the "producing" unit.

The NPMM is most often implemented in organizations where white collar employees prevail. For example, Westinghouse uses a form of the NPMM to measure the productivity of their Research and Development group. Other companies such as the Shell Oil Company, Control Data Corporation, Babcock and Wilcox, Tektronix Inc., and Champion International also use a form of the NPMM to form a participative productivity measurement program.

MULTI-CRITERIA PERFORMANCE PRODUCTIVITY MEASUREMENT TECHNIQUE

An innovative, widely-applicable, and reasonably simple approach to measuring performance or productivity is being developed at the Oregon Productivity Center and at the Oklahoma Productivity Center. The technique can be nicely integrated with the Normative Productivity Measurement Methodology (Participative approach utilizing the Nominal Group Technique) in order to facilitate more effective use of the tremendous number of measures and ratios of productivity or performance that can be attained from groups in organizations. The technique will be briefly presented in this paper.

Studies beginning at The Ohio State University in 1975 developed a participative, yet highly structured methodology for identifying consensus productivity/performance measures for a given organizational system. The value of a participative approach lies in the creation of an "ownership" for the resulting measures. Successful implementation of ensuring measurement, evaluation, and control systems is more assured with effective participative approaches.

However, difficulties in operationalizing measurement systems that have origins in a participative process hindered early efforts. The question of how to evaluate performance against a list of measures that is often highly heterogeneous became critical to continued development. William Stewart addressed the issue of how to aggregate and hence evaluate performance against many measures or criteria in his dissertation effort at Ohio State. His approach was to develop a prioritized set of productivity/performance measures utilizing the NGT. (This approach evolved from the NSF sponsored Ohio State Studies of 1975-1977). He then developed a "utility" curve for each of the priority (top eight to ten) measures. A ranking and rating process was executed so as to weight the relative importance of each productivity/performance measure. The utility curve was utilized to transform actual performance against each specific measure or criteria to a common 0 to 1.0, was then multiplied by the relative value. The various performance values for each of the top priority measures are then added together to obtain a productivity/performance index. Stewart based the procedure upon the works of Morris (1977, 1979), The Ohio State Productivity Research Group (1977), and Keeney and Raiffa (1976).

Since those early developments in 1976-1978 at Ohio State, several other efforts have been made in this general area. In 1980, Stewart applied this approach to the common carrier industry. In 1981, William Viana, while a graduate associate in The Oklahoma Productivity Center at Oklahoma State University applied a hybrid design of this procedure in a fairly large, diversified manufacturing firm (gate valves, ball valves, etc.) in Brazil. More recently, Riggs and Felix have developed and published an analogous approach called the "objectives matrix", (1983).

MCP/PMT Procedure

Assume you have just generated a consensus and prioritized list of productivity/performance measures for a given organizational system utilizing the NGT (Sink, 1982). You have a list of heterogeneous measures (i.e., apples, oranges, peaches, etc.). You are interested in aggregating or evaluating performance against these criteria in an integrated fashion.

A common performance scale or utility scale needs to be developed that converts all the uncommon measures into some common denominator. The performance scale commonly is rather arbitrarily allowed to range from either 0 to 1.0, 0 to 10.0, or 0 to 100.

```
10.0 ┬ "Excellence"
     │
 5.0 ┼ "Acceptable"
     │
   0 ┴ "Lowest Possible"
```

Common Performance Scale for each Productivity/ Performance Measure

Level 0 represents the lowest level of performance possible for a given measure. Level 5 represents a minimally acceptable performance level (MAPL). And, level 10 represents the perception of best performance or excellence. Levels 0, 5, and 10 should be clearly defined and accepted benchmarks.

Each productivity/performance measure or criterion has at least one "natural" scale that it can be or is measured with. Often this "natural" scale is simply an industry consensus or norm. For example, measuring liquids in gallons or liters, measuring coal in tons, measuring profitability performance in terms of ROI, etc.

The objective in the MCP/PMT is to develop a valid set of natural scales used to measure performance against a given criteria and to match levels of performance on that scale to levels of performance on the common utility scale.

Customer Satisfaction
"Natural" Scale(s) (x-axis)

A utility curve, (such as curves a, b, or c) which can, and often will be, subjective, is developed and used to transform performance on one scale to a common scale. There exist techniques for developing these curves in a valid manner (Kenney & Raiffa, 1976).

The M in the MCP/PMT stands for multi, which signifies that there are many criteria or measures of productivity/ performance against which we are attempting to perform. Therefore, there will often be many utility function graphs as depicted above. The question becomes one of how to aggregate performance scores. Since these scores are all from a common scale we might be tempted to just add them up. However, the question of relative impact of performance against various criteria on overall performance becomes critical.

The NGT will have given you a ranked list of productivity/performance measures. The next step is to rate this list. Figure 6 depicts this process. The first step, once the criteria have been ranked, is to arbitratily assign 100 points to the top priority measures or criteria. Next, the relative importance of the second most important measure is assessed. In the example depicted in Figure 6, which is for a computer center, customer satisfaction is seen as being equally important to projects completed/constant value budget dollars. So, customer satisfaction is also assigned 100 points. This paired comparison relative assignment of points is done for each successive criterion. (i.e., the most important (1st) relative to the second most important (2nd); the 2nd to the 3rd; 6th to 7th, etc.) The total points allocated for all measures or criteria is summed. Relative weights are then determined by dividing the individual points assigned by the total points (i.e., 100/730,...80/730). One then has a sense for the relative importance or contribution of each measure or criterion to overall performance/productivity. There are some critical nuances to this procedure that are not described herein, however the approach is basically as straightforward as it seems. These weights can be determined unilaterally by the manager of the group or by an analyst or participatively by the same persons who identified the criteria and their rankings.

#	Criterion	Rank/Priority	Rating	Weight
1	Reports/Projects completed and accepted / Constant value Budget $	1	100	$\frac{100}{730} = .137$
2	Customer Satisfaction	2	100	$\frac{100}{730} = .137$
3	Quality of Decision Support From Systems Developed	3	100	$\frac{100}{730} = .137$
4	Meeting User Flexibility Requirements	4	90	$\frac{90}{730} = .123$
5	Existence of and use of work scheduling/project management	5	90	$\frac{90}{730} = .123$
6	Projects completed on time / Total projects completed	6	85	$\frac{85}{730} = .116$
7	# of requests for rework/ redoing a project	7	85	$\frac{85}{730} = .116$
8	Existence of and Quality of strategic planning for facilities, equipment, staffing, management processes, and operational systems	8	80	$\frac{80}{730} = .111$
			730	1.000

FIGURE 6: Ranking and Rating Procedure

The next step in the MCP/PMT is to integrate the performance (utility) graphs (scales and curves) with the criteria weightings. This will allow the development of one performance/productivity indicator which will indicate the overall performance of the organizational system. Figure 7 conceptually depicts what is happening in this step. Actual performance as measured against the scales represented on the x-axis is transformed into a performance score (0 to 10) on the y-axis. Those performance scores are then multiplied by the criteria weighting factors to obtain weighted scores. Note that these weighted scores all have common units while the x-axis reflects a variety of units. Note also from the computer center example that only one of the eight measures is a pure productivity measure (ratio) and that is criterion #1.

The final computational step in this procedure is to add together all the weighted scores. The value for the computer example is 5.213 out of a maximum score of 10.0. The individual performance scores in addition to the total weighted score or overall performance indicator can be tracked over time and utilized to develop evaluation and control systems.

A general matrix format for this technique is presented in Figure 8 (Riggs and Felix, 1983). Note that column eight of the matrix represents the y-axis and columns 1-7; rows 3-13 represent the x-axis of the utility curves. Note also that it is possible to have sub-criteria for a given measure. In other words, there could be more than one way to operationalize a given performance/productivity measure. In that case, the weighting for the given measure or criteria is simply divided up among the sub-criteria. For instance, if customer satisfaction were operationalized with two independent measures rather than one, then the .137 weighting would need to be divided among the two sub-criteria. This is done by repeating the ranking, rating, and weighting procedure within the criteria itself.

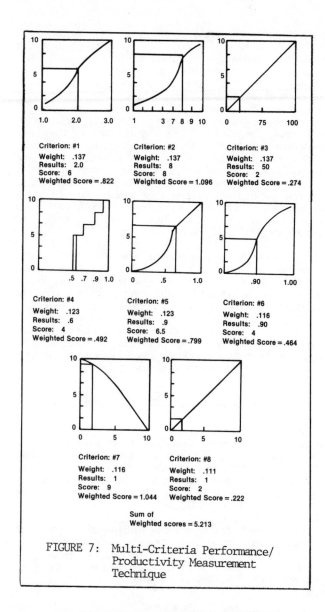

FIGURE 7: Multi-Criteria Performance/ Productivity Measurement Technique

This approach and technique for performance and productivity measurement and evaluation for organizational systems has tremendous potential. The roots of this technique lie in multi-attribute decision analysis which is at least twenty-five years mature. The ideas have been there; American managers and researchers have simply failed to innovate with the basic ideas, theories, concepts and techniques. This is changing. The Japanese have taught us that things don't have to be complex to work.

Current developments with this technique are in the area of validation, methodology refinements and documentation, and creation of a computerized decision support package for the IBM-PC. The IBM-PC software for this technique will be available from our Center in approximately six months. Those interested in learning more about this technique are urged to call our Center and talk to one of the OPC staff.

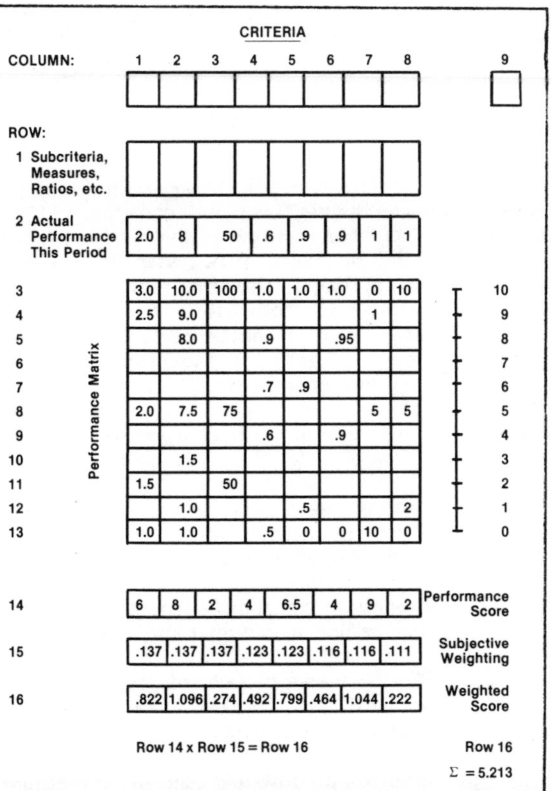

FIGURE 8: Productivity/Performance Measurement Matrix General Format

CRITERIA

COLUMN:	1	2	3	4	5	6	7	8	9	
ROW:										
1 Subcriteria, Measures, Ratios, etc.										
2 Actual Performance This Period	2.0	8	50	.6	.9	.9	1	1		
3	3.0	10.0	100	1.0	1.0	1.0	0	10		10
4	2.5	9.0					1			9
5		8.0		.9		.95				8
6										7
7				.7	.9					6
8	2.0	7.5	75				5	5		5
9				.6		.9				4
10		1.5								3
11	1.5		50							2
12		1.0			.5			2		1
13	1.0	1.0		.5	0	0	10	0		0
14 Performance Score	6	8	2	4	6.5	4	9	2		
15 Subjective Weighting	.137	.137	.137	.123	.123	.116	.116	.111		
16 Weighted Score	.822	1.096	.274	.492	.799	.464	1.044	.222		

(Performance Matrix spans rows 3–13, Column 1)

Row 14 x Row 15 = Row 16

Row 16 $\Sigma = 5.213$

SURROGATE PRODUCTIVITY MEASUREMENT TECHNIQUES

The authors include a few distinct techniques that are not specifically productivity measurement oriented but are directly relevant to this paper. A surrogate productivity measurement approach is one that does not measure productivity directly but measures something that is highly correlated with productivity. The implicit assumption behind these approaches is that if we measure, evaluate, control and improve upon these factors then productivity will also be managed.

Any measurement, evaluation, control and improvement technique that focuses on the broader issue of performance could; therefore, be labeled a surrogate productivity measurement approach since efficiency, effectiveness, quality, quality of working life, and innovation are highly interdependent with productivity. In this respect, the MFPMM, NPMM, and the MCP/PMM are partially surrogate approaches. More specifically, however, the following techniques are presented which typically do not measure productivity directly but which have productivity management implications.

The Common Staffing Study (CSS) was established on an experimental basis in a manufacturing division of IBM eleven years ago. The technique was designed as an attempt to measure/plan/improve productivity in the indirect labor areas. These areas tend to encompass work that is complex, non-repetitive, irregular in character, and often unpredictable. The CSS approach is based on the assumption that it is not feasible nor economical to measure most "indirect" manufacturing jobs in the sense of determining the minimum requirements, to accomplish a task with "100%" performance. Instead, the objective is to describe the level of productivity, whatever it is, at one point in time, and then to strive to continuously improve that productivity in all areas through future points in time. For further information on CSS see Olson, 1983 and Sink, et. al., 1984.

In addition, Price-Waterhouse has their own system of measurement. They measure and verify productivity improvement through a process called Benefit/Cost Tracking. This tracking is project based and continues through the life of the project. For further reference see Schmidt, 1984.

Productivity Audits and checklists (Mali, 1978; Hughes, 1978; Productivity, Sept. 1983; and Bain, 1982) can serve as starting points toward the design of productivity measurement systems or be used to evaluate the effectiveness of existing productivity efforts. They are listed here because they fit the definition of surrogate measurement techniques and assist in the productivity management process.

One final surrogate measure that is finding success in American Organizations is Managing Productivity by Objectives (MPBO). MPBO was developed by Paul Mali and is based upon the management strategy, Management By Objectives, (MBO). MPBO provides a format for creating a management system for improving productivity. The whole point of the management system, such as MPBO, is to coordinate all of the available resources and efforts of people toward agreed-upon goals and objectives. MPBO does not measure productivity directly, but like the other surrogate techniques listed, it provides measures which are highly correlated with productivity.

PRODUCTIVITY MEASUREMENT TAXONOMY

Taxonomy Development

Two assumptions guided the taxonomy development process. One assumption was to adopt a narrow definition of productivity. Adopting this narrow definition of productivity ruled out consideration of other performance dimensions such as efficiency, effectiveness, or quality from the txonomy. As a result, the taxonomy development centered on productivity measurement, not performance measurement. A second assumption which guided the taxonomy development process was that the taxonomy should have functional value as well as descriptive value. The taxonomy should have functional utility for a manager who wants to select a productivity measurement method or technique. If the manager can define certain parameters, the taxonomy should point to one or more techniques which correspond to the selected parameters (e.g. organizational level, etc.). A descriptive taxonomy on the other hand would not necessarily be as concerned with utility. It would give more concern to accuracy and precision of classification. In the domain of productivity measurement, which is poorly defined, such a descriptive taxonomy would most likely be more complex, and have more dimensions than a functional taxonomy which would lean toward greater simplicity and ease of use.

Given these two assumptions, the researchers considered a wide range of possible dimensions which could be used to classify productivity measurement techniques. The most seriously considered dimensions are the following:

1. **Unit of analysis** - refers to the level of the target system being measure.
2. **Scope of Measurement** - consists of discrete "snapshots" of productivity at certain time intervals. This dimension simply refers to the length of time between these intervals.
3. **Functional Discipline** - this is a nominal variable and might have values such as economist, manager, cost accountant, industrial psychologist or industrial engineer.
4. **Type of Technology** - this dimension ranges from a manufacturing technology where inputs and outputs have low variability to a service technology where inputs and outputs may vary considerably over time.
5. **Degree of Measurability** - Measurability refers to the extent to which inputs and outputs in the target organizational system lend themselves to quantification.

Following an analysis of these dimensions and in light of the assumptions stated earlier, two dimensions were selected to form the taxonomy. The dimensions selected were "unit of analysis" and "scope of measurement". The dimensions selected lead to a two dimensional taxonomy as depicted in Figure 9. Shaded in cells indicate combinations of the two dimensions which are theoretically possible but operationally make little sense. As a functional taxonomy, it is possible to indicate the appropriate cell or cells in which particular measurement techniques fall. It will then be possible for a manager to select a technique or techniques which meet his/her specifications regarding unit of analysis and scope of measurement. For example, if a manager desires a measurement technique which is appropriate for the plant level and which covers a monthly time period, then the cell indicated in Figure 9 would contain the appropriate techniques.

This is quite straight forward. However, the selection of a measurement technique is more complex. Using the taxonomy will put the manager is the "ballpark", however, some additional "fine tuning" is necessary in order to settle on a particular approach. The "fine tuning" takes place by considering a range of moderator variables. Ech of the moderator variables represent considerations which will alter the content or form of the productivity measure depending on the circumstances.

A wide range of moderator variables were considered and those listed below were felt to be the most significant.

1. **Output variability** - extent to which the physcial characteristics of the system outputs changes over time.

2. **Process cycle time** - length of time it takes for one unit of output to be produced.

3. **Resource as a percent of costs** - in selecting a productivity measurement methodology, feasibility and costs are major concerns.

4. **Intended purpose and user of the measure** - the selection of a measurement method is in a large part a function of what the measure is supposed to do and who will use it.

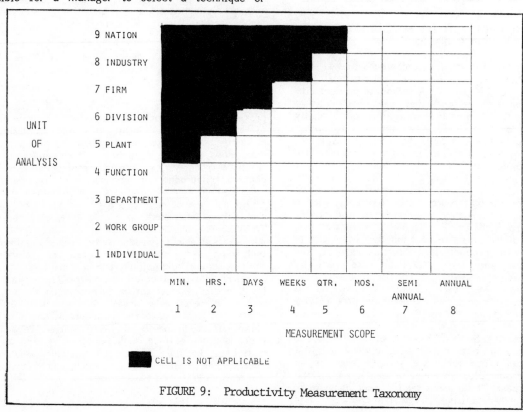

FIGURE 9: Productivity Measurement Taxonomy

5. **Controllability of inputs** - extent to which management can "manage" or control input levels affects what is measured.

6. **Control system maturity** - extent to which measurement and control systems are part of the organizational culture.

7. **Management style** - measurement techniques are most effective when they augment and compliment the existing management style.

8. **Commitment to measurement** - extent to which an organization sees productivity measurement as a critical part of its effort to remain competitive.

9. **Decentralization/centralization issue (control)** - extent to which measurement is a decentralized/centralized function.

10. **Management understanding/awareness** - extent of management understanding/awareness of productivity measurement systems.

Through the use of these moderator variables and the original taxonomy dimensions it is possible to identify more precisely the type of measurement technique most appropriate to a particular organizational system.

Taxonomy Validation

As a means of validating the taxonomy, the researchers conducted field investigations of measurement systems in use by a wide range of organizations. These actual measurement techniques in current use by industry provide a means of assessing the utility of the taxonomy. Taxonomy validation refers to the ability of the taxonomy dimensions and moderator variables to adequately explain major characteristics of the measurement systems.

The field investigations were used as a vehicle to define the unit of analysis, scope of measurement and moderator variables inherent in each organization. Once these variables have been classified the taxonomy can be used to identify the measurement system most appropriate for them. As an example, one of the companies visited exhibited the following characteristics:

1. **Unit of Analysis:** work group, department, function, plant and division level.

2. **Scope of measurement:** weeks, quarters and annual periods of measurement.

3. **Moderator Variables:**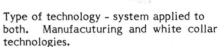

a. Type of technology - system applied to both. Manufacuturing and white collar technologies.

b. Output variability - low.

c. Process cycle time - short.

d. Resources as percent of costs - all resources are significant and measured.

e. Purpose and audience - multifactor model used for overall tracking, planning and evaluating trade-offs in operational

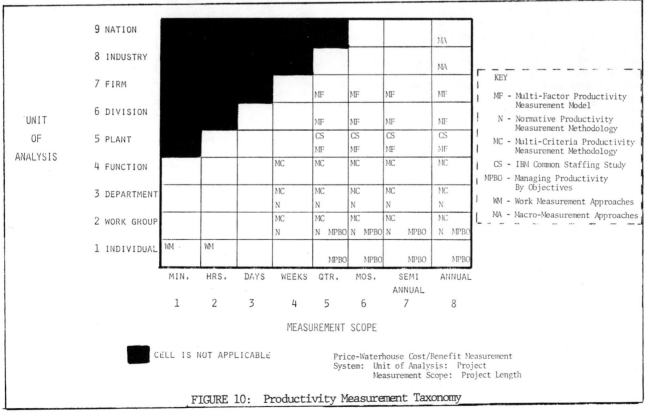

FIGURE 10: Productivity Measurement Taxonomy

decisions audience is top management. Surrogate and normative models used for diagnostic and improvement purposes by lower and middle management.

f. Controllability of inputs - all measured inputs are controllable.

g. Control system maturity - moderate to high.

h. Management style - not assessed.

i. Commitment to measurement - high commitment by top management.

j. Decentralization/centralization - muultifactor measurement model is centralized, not used for controls. Normative/surrogate models are decentralized and used for feedback and control at plant level.

k. Management understanding/awareness - not assessed but is a continuing stated priority of top management. An ongoing process.

The taxonomy depicted in Figure 10 illustrates the location of the measurement techniques described earlier in this paper. Using Figure 10 and the data from the example company identifies the measurement systems that are appropriate. Based on the information presented, this company would benefit most from using the multi-factor model when measuring productivity at the plant or division level on an annual basis. The Normative model and selected surrogate models are more suited to the smaller unit of analysis (work group, department or function) and the shorter measurement period (weeks and quarters).

Based on the limited trials conducted by the researchers, the taxonomy appears promising as a model for use in identifying suitable productivity measurement techniques for a given situation. It provides a reasonably objective basis to review various measurement schemes in light of a range of organizational characteristics (moderator variables). The question of the validity of the model for use by managers in selecting measurement techniques must await further testing and experience.

CONCLUSION

The purpose of this paper has been to present a succinct review of the findings of our recent study on productivity measurement techniques existant in theory and in practice. We began the study with a predisposition that one cannot manage what one cannot measure; and, that one cannot measure what one cannot define. We devoted substantial space in our final report to WPAFB/APRO/DoD on productivity basics (Sink; 1983, 1984). We systematically differentiated between effectiveness, efficiency, quality, productivity, quality of work life, innovation, and profitability. We presented and discussed a causal model linking these performance criteria together.

American managers seem to have lost sight of what drives what. Our sales, marketing, accounting, finance, MBA dominated upper management seems to have forgotten what we taught the Japanese. "We were so impressed with

the factories and organization of the 40's we decided to preserve them" (Richard Stimson, 1984).

We review, in this paper, the four generic productivity measurement techniques we were able to locate in the literature and in our site visits. We became poignantly aware of the operational difference between the phrase "measurement of productivity or productivity measurement" and "Measurement, evaluation, and verification of productivity improvement." There are a limited number of ways to measure productivity (output/input). While there are a great number of ways to measure whether productivity has improved and by how much. Productivity measurement can tell you this but so can a number of other approaches. Any measurement and evaluation approach that will indicate that input resource consumption changed favorably relative to outputs from the system can be utilized to verify productivity improvement. Cost/Benefit analysis, engineering economic analysis evaluation, etc., are all mechanisms which could be utilized to verify productivity improvement. A critical factor, however, is the ability of these techniques to measure in constant value terms. An advantage to utilizing the Multi-Factor Productivity Mesurement Model is that it automatically indexes prices and costs to some established base period value. Our conclusion is that a great deal of relative to developing productivity measurement systems needs to be accomplished. Most organizations we visited are just now in the early stages of development of productivity measurement approaches. We suspect there are very few organizations in this country with more than 3-5 years of experience with actual productivity measurement techniques. With respect to the needs of the DoD and in particular the IMIP program, it appears that the MFPMM holds the most promise. The next stage of the APRO directed study will be to field test the MFPMM along with several other approaches.

The final major comment we would like to make as a result of this study has to do with the distinction between a measurement system designed to control and verify versus a measurement system designed to improve. We found productivity measurement approaches that are effective at control, diagnosis, and evaluation (MFPMM). And, we found productivity measurement approaches that are effective at evaluation and improvement (NPMM, and MCP/PMT). We found no techniques that were effective at doing both. We also found techniques that are not accurately labelled as productivity measurement approaches but that can measure, evaluate, and verify the existence of and magnitude of productivity improvement (IMIP engineering economic alaysis and evaluation approach, cost benefit analysis mechanism--Price Waterhouse). These approaches can be effective at control and verification but are often less effective at linking to measurement successfully to improvement. We conclude from our investigation that organizations may need to develop a variety of productivity measurement approaches and develop mechanisms for integrating these systems together. Our investigations did not diminish our perception of the need to develop a strategic (2-5 year) plan for a productivity management, measurement, evaluation, control and improvement effort. This will be the focus of future efforts and papers from the Oklahoma Productivity Center.

REFERENCES

1. Bain, D. F. Productivity Measurement. Westmead, England: Gower Publication Company, 1982.

2. "The Carlucci Initiatives (For Improving the Defense Acquisition Process)". Briefing given by Sterling Institute (ROC), Washington, D. C. Fall, 1981.

3. Hughes Aircraft Company. R. & D. Productivity Study Report Culver City, CA, 1978.

4. Keeney R. L. and Raiffa, H. Decisions with Multiple Objectives: Preferences and Value Tradeoffs. New York: John Wiley and Sons, 1976.

5. Mali, Paul, Improving Total Productivity. New York: John Wiley and Sons, 1978.

6. Morris, W. T. Decision Analysis. Columbus, OH: Grid Publishing Company, 1977.

7. Morris, W. T., Implementation Strategies for Industrial Engineers. Columbus, OH: Grid Publishing Company, 1979.

8. Olson, Val., White Collar Waste, Gain the Productivity Edge. Englewood Cliffs, NJ: Prentice-Hall, Inc., 1983.

9. Productivity Measurement Systems for Administrative Computing and Information Services: An Executive Summary, (1977). The Productivity Research Group, The Ohio State University, NSF Grant No. APR 75-20561, Columbus, OH.

10. Productivity Measurement Systems for Administrative Computing and Information Services, (1977). The Productivity Research Group, The Ohio State University, NSF Grant No. APR 75-20561, Columbus, OH.

11. Riggs, J. L., and Felix, G. H., Productivity by Objectives. Englewood Cliffs, NJ: Prentice-Hall, Inc., 1983.

12. Robbins-Myers. "Productivity Checklist", Productivity, September, 1983.

13. Schmidt, Robert, Sr., "Cost Justification for Advanced Manufacturing Technology". Institute of Industrial Engineers Annual Conference, May, 1984.

14. Sink, D. S., 1982. "Building A Program For Productivity Management: A Strategy for IE's". Industrial Engineering, October, 1982, pp. 43-50.

15. Sink, D. S., "Organizational System Performance: Is Productivity a Critical Component?" Institute of Industrial Engineers 1983 Annual Conference Proceedings. Norcross, GA.

16. Sink, D. S., Productivity Management: Measurement, Evaluation and Improvement. New York: John Wiley & Sons, 1984.

17. Sink, D. S., Productivity Measurement and Improvement Strategies and Techniques. Seminar Notebook. Stillwater, OK: LINPRIM, Inc., 1981.

18. Sink, D. S., Tuttle, T. C., DeVries, S. J., and Swaim, J., Development of a Taxonomy of Productivity Measurement Theories and Techniques AFBRMC Contract No. F33615-83-C-5071. Wright-Patterson Air Force Base, Research Management Center, 1983.

19. Stewart, W. T., "A 'Yardstick' For Measuring Productivity". Industrial Engineering, February, 1978, pp. 34-37.

20. Stimson, Richard L., Keynote Address, Institute of Industrial Engineers Aerospace Division Conference, Clearwater, Florida, February, 1984.

21. Van Loggerenberg, B. J. and Cucchiaro, S. J., "Productivity Measurement and the Bottom Line." National Productivity Review 1(1), 87-99, Winter 1981-82.

BIOGRAPHICAL SKETCHES

D. SCOTT SINK, P.E., is an associate professor in the School of Industrial Engineering and Management at Oklahoma State University and director of the Oklahoma Productivity Center at OSU. He received his BSIE, MSIE and PhD degrees from Ohio State University. His areas of interest include productivity management and measurement, work measurement and improvement and organizational behavior. Sink is a senior member of IIE and is currently director for the management division. Sink is a recipient of the Haliburton Award of Excellence, the Dow Outstanding Young Faculty Award and the Oklahoma Society of Professional Engineers Outstanding Young Engineer of the Year award. He is a registered professional engineer in the state of Oklahoma. His book titled, "Productivity Management: Planning, Measurement, Evaluation, Control, and Improvement", will be released by John Wiley and Sons in late 1984.

Sandra J. DeVries is a graduate associate for the Oklahoma Productivity Center at Oklahoma State University. She has received a Bachelor's Degree in Industrial Engineering and Management and is currently pursuing a Master's Degree in the same field. As an undergraduate, Sandy worked as a summer engineer for Procter and Gamble in Cape Girardeau, Missouri and Texas Instruments in Sherman, Texas. Sandy is a member of IIE, OSPE and NSPE.

*Reprinted from **Industrial Management**,
May-June 1984.*

The Productivity Index:
A Total Factor Measurement
System that Works

William M. Brady

In 1976, General Foods Corporation adapted and installed a total factor productivity system in response to a management question about the status of productivity in the plants. Today, this system — known as the Productivity Index — is in use in the company's manufacturing plants in the U.S. and Canada. It is also being expanded to the next higher organization level — the Division.

The success of the Productivity Index does not rest solely in the fact that it gives management a measure and a trend of productivity decrease or increase. Rather, it lies in the many other advantages inherent in a total factor measurement. These benefits are more important than the measure iteself. While some of the benefits were envisioned when the system was first installed, most of them have been learned through experience.

Advantages of the Productivity Index

Improved productivity is an objective of most corporations today. Productivity improvement programs are widespread. For many of them, improved labor productivity is the main purpose. Most are aimed at the manufacturing plants only and not at the total organization. This is frequently the result of a lack of definition and understanding of the term productivity. But what is productivity and how should it be defined?

The Productivity Index defines productivity in a way which is easily understood by everyone in the organization.

"Productivity is a measure of how effectively an organization — a plant or a corporation — uses its total resources — management skills, materials, equipment and capital — in producing its products."

WILLIAM M. BRADY is currently a partner in the consulting firm of McNesby, Brady & Co., White Plains, New York. He is retired from General Foods Corporation where he served in Engineering and Operations positions for 35 years. He was Corporate Manager of Industrial Engineering for nine years before retirement. He has been active in the Institute of Industrial Engineers, past President of the Tappen Zee Chapter and for three years a member of I.I.E.'s Educational Policy Board.

The Productivity Index — a total factor measurement system -- results in a productivity trend over time which helps to measure the effectiveness of improvement programs. This, of course, is the fundamental objective of any measurement program.

But a total factor productivity measurement system — The Productivity Index — offers many other advantages, more important than the basic objective.

The Productivity Index provides:

- A basis for analysis by identifying important leverage points impacting productivity.

- A basis for developing action programs — such as Quality Circles — for productivity improvement.

- A basis for planning cost reduction programs and goals to meet productivity objectives.

- An outlook or philosophy that relates management decisions to the impact on productivity.

- A productivity consciousness that helps in motivating people at all levels of the organization.

- A basis, when applied at a division or corporate level, for determining what portion of increased (decreased) profitability is due to real productivity versus pricing action.

From another perspective, the Productivity Index provides a way for evaluating an organization's performance:

- In improving its profitability

- In demonstrating its management skill in adjusting to different factors impacting productivity

- In improving its quality performance

- In improving its production efficiency

- In its ability to innovate, as for example in product/process design to reduce costs

- In developing the Quality of Work Life and employee motivation to improve productivity.

Installing the Productivity Index

Our experience at General Foods and that of several other firms indicates that it is important to first consider at what organizational level the Productivity Index should be introduced.

The need to improve productivity applies to all levels of the Company — the manufacturing plants, operating divisions, the Corporation. However, the Productivity Index should first be installed at the manufacturing plants. Generally, the manufacturing cost is the largest single component of the total product cost. Starting at the plants also insures that the necessary input data is available in the proper form. After installation at the plant level, it is appropriate to move up the line to the division and then the corporate level. It is also essential that the basic principles and system concepts be understood.

Basic System Concepts

There are several basic concepts behind the Productivity Index —

- In its simplest form, Productivity equals Output ÷ Input.

- It includes all the cost inputs required to produce the products or services which are the organizational system outputs.

- The basic unit of measure for Output and Input is $.

- All costs are expressed in "base year" costs or constant dollars.

- The Productivity Index is designed to measure the effectiveness of the management of the organizational system or unit under consideration.

For a manufacturing plant, therefore, the Index is designed to measure the performance of the plant manager and his staff in producing quality products at a higher productivity level.

It is essential then that the organizational unit should have control over all the costs included. Therefore, allocated costs from the corporation should not be included. Direct Costs from a Corporate Staff function should be included particularly if the staff function is on a charge out basis. Where raw materials/packing materials are purchased at a division or corporate level and the plant has no control over price, the Productivity Index is calculated on a "Value Added" basis rather than on a total cost basis.

When higher management directs a plant to take an action not originally planned, the plant is able to calculate the effect of that action on productivity and so inform higher management.

How the Productivity Index Works

First it is necessary to define Output and Input and to describe the procedure for calculating each. The values used in Figures 3 and 6 are for demonstration only and are not related to the two actual examples which follow.

The first example is based on calculating the Productivity Index for an actual manufacturing plant. Since this plant was not responsible for purchasing its major raw and packaging materials, it is calculated on a Value Added basis.

The second example is based on the total division of which the first plant example was a part. Here the Productivity Index is calculated on a total cost basis.

The basic formula for calculating the Productivity Index is as shown in Figure 1.

HOW IT WORKS

BASIC FORMULA

OUTPUT ÷ INPUT = PRODUCTIVITY INDEX

Figure 1. Basic Formula.

Output (Figure 2) is the "value added" cost of converting purchased raw and packaging materials into finished goods at base year cost rates reflecting the base year operations efficiency.

HOW IT WORKS

OUTPUT

COST VALUE THAT SHOULD HAVE BEEN ADDED BY CONVERTING RAW AND PACKAGING MATERIALS INTO FINISHED GOODS IF PLANT OPERATIONS WERE CARRIED OUT AT BASE YEAR EFFICIENCY LEVELS. CONSISTS OF:

- RAW AND PACKAGING MATERIALS LOST IN PRODUCTION.

- ALL MANUFACTURING LABOR AND OVERHEAD.

Figure 2. Output Definition.

"Value added" cost includes:

- Raw and packaging material lost in production.

- All manufacturing labor and overhead.

The procedure for calculating output is shown in Figure 3. First, the base year is selected and for that year the value added cost per unit is established for each product. This is multiplied by the actual number of units for each product produced in any year by the base year cost. The output is always the current year volume at the base year cost and experience. In the example shown, the plant is producing two products at different base year cost at different volumes equaling output for each product for a total as shown.

HOW IT WORKS

OUTPUT PROCEDURE

- SELECT BASE YEAR.

- IDENTIFY BASE YEAR VALUE ADDED COST PER UNIT FOR EACH PRODUCT.

- ACTUAL NUMBER OF UNITS OF EACH PRODUCT PRODUCED X BASE YEAR COST RATE = OUTPUT (VALUE ADDED).

PRODUCT	BASE YEAR ADDED COST PER UNIT	CURRENT YEAR UNITS PRODUCED	OUTPUT
A	$5.00	10,000	50,000
B	2.00	50,000	100,000
TOTAL OUTPUT			$150,000

Figure 3. Output Procedure.

Input is defined (Figure 4) as the effort and resources expressed in constant base year dollars used to achieve output.

HOW IT WORKS

INPUT

EFFORT AND RESOURCES, EXPRESSED IN CONSTANT BASE YEAR DOLLARS, USED TO ACHIEVE OUTPUT. CONSISTS OF:

- RAW AND PACKAGING MATERIALS LOST IN PRODUCTION.

- ALL MANUFACTURING LABOR AND OVERHEAD (DIRECT LABOR, SUPPORT STAFF, REPAIR AND MAINTENANCE, PURCHASED SUPPLIES AND SERVICES, ETC.).

- INCREASE OR DECREASE IN COST OF CAPITAL COMPARED TO BASE YEAR. (CAPITAL IS RAW MATERIAL, PACKAGING MATERIAL AND FINISHED GOODS INVENTORIES, AND PLANT INVESTMENT IN LAND, BUILDINGS AND EQUIPMENT.)

Figure 4. Input Definition

The cost of capital used in our example is 15%. The hurdle rate within your company may differ.

To develop the input (Figure 5), the first step is to define quantitative resources expended, such as the number of man-hours worked, and multiply this by the base year cost. This eliminates the need for using government indices for most of the calculations.

By using the base year cost, you have essentially created your own deflation factor for the various elements. For minor costs such as miscellaneous operating supplies which will have relatively little impact on the final index, it is easier to use government-published indexes such as the GNP or the Wholesale Price Index.

HOW IT WORKS

INPUT PROCEDURE

- IDENTIFY QUANTITATIVE RESOURCES EX-PENDED (I.E., LBS. OF RAW MATERIALS LOST IN PRODUCTION, NUMBER OF HOURS WORKED, NUMBER OF SALARIED PEOPLE ON PAYROLL, KWH OF ELECTRICITY USED, ETC.) AND MULTIPLY BY BASE YEAR COST RATES.

- USE GOVERNMENT-PUBLISHED INDICES TO DEFLATE MINOR COSTS (MISCELLANEOUS SUPPLIES, MINOR INGREDIENTS, ETC.) TO BASE YEAR CONSTANT DOLLARS.

Figure 5. Input Procedure.

Input is always constant dollar cost for the current year to produce current year volume. It reflects, therefore, the current year experience such as hours worked at the base year cost per hour. This is shown in Figure 6.

In this example, we identify the amount of pounds lost in the subject year as 55,000 pounds at the base year cost per pound of $2.00 for an input value of $110,000. The example also shows how a deflator index is used for minor ingredients and supplies. In this example, it shows a net increase of $7,500 in the input value for capital; a mix between an increase in inventories and a decrease in the value of fixed capital. Incidentally, the total input dollars shown here are not related to the previous output example (Figure 3).

An Actual Plant Example

When we first started in Fiscal 1976, we were able to retrieve cost data at the initial plants going back to Fiscal 1973. This enabled us to establish a trend line using 1973 as the base year, followed by the data for the years 1974 and 1975.

If you are unable to go back three years to establish the base year Productivity Index, you should select the first previous year for which good data is available.

INPUT (CONT'D)

PROCEDURE EXAMPLE:

INPUT ITEM	(UNIT OF MEASURE)	BASE YEAR COST PER U/M	ACTUAL U/M'S	INPUT
MATERIAL X	LBS. LOST	$ 2.00	55,000	$110,000
HOURLY LABOR	HRS. WORKED	7.00	30,000	210,000
SALARIED LABOR	NO. OF PEOPLE	15,000	6	90,000
ELECTRICITY	KWH	.40	750,000	30,000
SUBTOTAL				$440,000

	CURRENT YEAR COST	DEFLATOR INDEX	
MINOR INGREDIENTS/SUPPLIES	$ 22,000	110%	$ 20,000

	CONSTANT $ INVESTMENT INCREASE (DECREASE) VS. BASE YEAR	% COST OF CAPITAL	
INVENTORIES	$100,000	15%	15,000
LAND, BUILDINGS, EQUIPMENT	(50,000)	15%	(7,500)
SUBTOTAL			$ 7,500
TOTAL INPUT			$467,500

Figure 6. Input Procedure Example.

	PLANT "A" $ MILLIONS BASE YEAR 1973	1974	1975	1976	1977	1978 PLAN
OUTPUT						
RAW MATERIALS LOST IN PRODUCTION	$42.6	$41.9	$33.2	$39.8	$35.5	$35.3
PACKAGING MATERIALS LOST IN PRODUCTION	.3	.3	.3	.3	.3	.3
LABOR AND OVERHEAD	25.4	25.5	19.2	22.7	19.9	19.4
TOTAL OUTPUT	$68.3	$67.7	$52.7	$62.8	$55.7	$55.0
INPUT						
RAW MATERIALS LOST IN PRODUCTION	$42.6	$41.0	$32.4	$37.9	$34.2	$32.7
PACKAGING MATERIALS LOST IN PRODUCTION	.3	.3	.3	.4	.3	.3
LABOR AND OVERHEAD	25.4	25.3	26.2	28.1	25.3	25.9
INCREASE IN COST OF CAPITAL	–	–	.4	(.9)	(1.0)	(1.4)
TOTAL INPUT	$68.3	$66.6	$59.3	$65.5	$58.8	$57.5
PRODUCTIVITY INDEX (OUTPUT ÷ INPUT)	1.000	1.017	.888	.958	.947	.956
VOLUME (MILLIONS OF UNITS)	18.1	17.6	14.5	16.2	13.5	12.5

Figure 7. Productivity Index – Plant Example.

As more and more plants came on line over the first three years, the more recent additions were forced to use Fiscal 1976 as the base year. At that point in time, we converted all plants to a Fiscal 1976 base year. In the future, changes or updating of the base year are expected to be made every 5 to 10 years. The example shown in Figure 7 is for one plant out of a four-plant division.

The data shown is real from the base year 1973 through the plan for Fiscal '78. The output data shows that raw materials lost in production is a major output factor reflecting the high cost of the materials used in this product. The basic variable in the total output dollars as shown is the variable in volume from each year versus the base. Therefore, this number shows the erratic volume picture at this plant over that period of time. Part of that volume variation is due to the cost problems in the plant, but volume fluctuation induced by the high cost of the raw materials is also a factor.

The input costs for the base year are identical to the output. Each year after that, however, they essentially reflect the difference in experience between any one year and the base year. The raw material lost in production line on the chart, when compared with the output number for that same line item, dramatically shows the effect of management's attention to improving raw material yields.

While the Productivity Index performance as shown is not good, the 1978 plan also does not reflect the planned cost reduction program at that plant. If that program met its objectives, the Productivity Index for 1978 would have been 1.01.

Since the time period shown on this chart, Plant A has increased its Productivity Index to 1.08 in Fiscal '81.

In the Productivity Index report for the division involved, the financial area summarized (Figure 8) their reasons for the changes in the Productivity Index that you have just seen.

Input/$ of Output

There are several ways of analyzing the data to pinpoint the trends for individual components or the line items used in the preceding example. For instance, under Labor and Overhead it would be desirable to know the productivity trend for direct labor or for maintenance labor and expense. Showing a more detailed breakdown of the basic line items in terms of Input/$ of Output makes this analysis possible.

In the example which follows (Figure 9) this approach is taken for a total division.

The line items, of course, are different; the costs are total costs, not value added. The total manufacturing costs line shows the total results for the division plants.

Here you can see that the major area of improvement over the base year is in Total Manufacturing Cost —the result of planned cost reduction activities based on the Productivity Index at this division's plants since 1976.

Overall, the division's productivity performance is good considering the volume problems it experienced due to high material costs during this period.

How the Productivity Index Is Used

More important than the mechanics of the system is how the Productivity Index is used within the organization. There are many factors which can affect how it is used. The specifics of the organization and the culture of the organization are two significant factors.

Consider, however, a fairly typical organization. A Vice President of Operations reports to the President of the company. Reporting to the Vice President of Operations are three plant managers. The initiative for installing the Productivity Index may come from the President but more likely from the Vice President of Operations. This initiative will be based on a desire or need:

- to know what the real productivity of the plants is and has been

- to be able to plan with his plant managers for higher productivity

- to evaluate the performance of each of his plant managers, and their staffs, on a more objective basis than may have previously been possible.

With these needs in mind, it is essential that the Vice President of Operations work with his plant managers, both individually and as a group, to develop a productivity improvement objective:

- for the operations area in total

- for each plant

PLANT "A" PRODUCTIVITY INDEX

THE PRIMARY REASONS FOR THE CHANGE IN THE PRODUCTIVITY INDEX ARE:

- VOLUME REDUCTIONS HAVE HAD A SIGNIFICANT IMPACT ON FIXED COSTS UNDERABSORPTION. IT IS ALSO FAIR TO CONCLUDE THAT THE UNFAVORABLE TREND IN VARIABLE COSTS PRODUCTIVITY IS PARTIALLY A RESULT OF LOW VOLUMES NECESSITATING INEFFICIENT LEVELS OF OPERATION.

- IT APPEARS THAT THE PLANT CAN BE MORE RESPONSIVE TO VOLUME DECLINES IN INCURRING FIXED COSTS.

- REDUCED RAW MATERIAL SHRINKAGE AND REDUCED COST OF CAPITAL HAVE BEEN THE BRIGHTEST SPOTS OVER THE LAST SIX YEARS.

Figure 8. Reasons for Change in Productivity Index.

DIVISION "X"
INPUT PER DOLLAR OF OUTPUT

	BASE YEAR F 1976	F 1977	F 1978	F 1979	F 1980	F 1981 AFP
VARIABLE MANUFACTURING COST	S .560	S .542	S .543	S .523	S .518	S .521
FIXED MANUFACTURING COST	.059	.054	.066	.052	.049	.052
TOTAL MANUFACTURING COST	.619	.596	.609	.575	.567	.573
TRANSPORTATION AND WAREHOUSING	.022	.021	.025	.023	.025	.023
SELLING EXPENSE	.018	.019	.025	.022	.022	.021
TOTAL DELIVERY COST	.040	.040	.050	.045	.047	.044
MEDIA ADVERTISING	.048	.046	.052	.078	.088	.084
CONSUMER PROMOTIONS	.076	.049	.060	.031	.041	.050
TRADE PROMOTIONS	.107	.077	.079	.087	.073	.065
OTHER ADVERTISING AND PROMOTIONS	.005	.003	.005	.003	.004	.004
TOTAL ADVERTISING AND PROMOTIONS	.236	.175	.196	.199	.206	.203
ADMINISTRATION AND PRODUCT MANAGEMENT	.002	.002	.003	.003	.003	.003
TECHNICAL RESEARCH	.007	.010	.013	.012	.013	.012
MARKET RESEARCH	.005	.005	.007	.005	.005	.005
TOTAL OTHER MARKETING	.014	.017	.023	.020	.021	.020
TOTAL MARKETING	.250	.192	.219	.219	.227	.223
GENERAL AND ADMINISTRATIVE EXPENSE	.003	.003	.004	.004	.004	.005
TOTAL OPERATING COSTS	.912	.831	.882	.843	.845	.845
COST OF CAPITAL	.098	.074	.098	.068	.065	.075
TOTAL INPUT	1.000	.905	.980	.911	.910	.920
PRODUCTIVITY INDEX	100.0	110.6	102.0	109.7	109.9	108.9

Figure 9. Input Per $ of Output – Division Example.

The productivity objective for each plant may not be the same. Individual situations can vary substantially and this must be recognized. The objective for the Productivity Index for each plant, and plant manager, becomes a major factor in the performance evaluation of each plant manager.

This process continues down the line. The plant manager with his staff develops the Cost Reduction objectives required to meet the Productivity Index objectives for their plant. Together and individually they put together the ideas and projects needed to affect the known price increases for goods and services for the next year. The total process provides a basis for motivating the entire plant organization, for suggesting specific improvement programs suited to the particular culture of each plant.

The line of communication from the Vice President of Operations to the plant manager and his staff is not one way. The Productivity Index process also allows for communication up-the-line when a decision is made affecting a plant's productivity, particularly short term. For example, a reduction of previously planned production values at one plant will have a negative affect.

The plant can assess the impact on the Productivity Index in dollars and inform higher management. Where such changes may have been accepted in the past and shrugged off, the effect of such reductions will be made known in a forceful manner.

Conclusion

The objective of this article has been to point out the value of a Total Factor Productivity Measurement system as an essential and fundamental part of any productivity improvement program. It also describes the basics of how it works and how it should be used.

The value of a total factor measure of productivity goes beyond showing whether productivity is improving or not.

Its main value to an organization is its use as an analytical and motivational tool — and its ability to measure the overall performance of the organization.

In eleven General Foods plants, the latest estimate for Fiscal '84 versus a Fiscal '81 base year shows an improvement in productivity of from 3.5% to 36%. When properly applied Total Factor Productivity Measurement works.

Reprinted from Industrial Engineering, July 1983.

Work Sampling And Measurement

'Unmeasurable' Output Of Knowledge/Office Workers Can And Must Be Measured

By Chester L. Brisley, P.E.
University of Wisconsin-Extension
and William F. Fielder, Jr., P.E.
Hughes Aircraft Co.

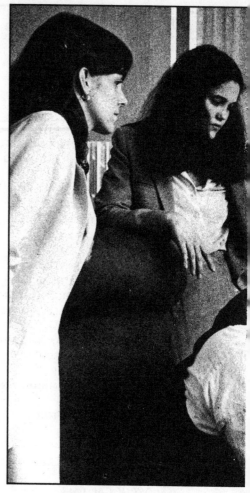

For decades, experienced industrial engineers have insisted that every job, without exception, can be measured. Today's proliferation of knowledge work provides a crucial challenge to industrial engineers who would make good on that conviction.

Knowledge work is an ever-increasing category of employment that includes the work of executives, managers and supervisors as well as a constantly growing number of professionals such as design engineers, accountants, market managers and industrial engineers themselves.

Peter Drucker coined the term "knowledge worker" in his book, *The Effective Executive.* Drucker claims that every knowledge worker is in fact an "executive" if "by virtue of his position or knowledge, he is responsible for a contribution that materially affects the capacity of the organization to perform and to obtain results."

But how do we measure these results, as well as the work that produces them? How much will it cost? And how will we benefit?

A prime motive for measuring knowledge work is the rapid increase in office costs. John J. Connell, executive director of the Office Technology Group, indicates that special studies are required to aggregate all of the costs associated with office operations and determine the direction in which these costs are moving (see "For further reading").

Companies that have undertaken such studies have turned up some startling results. Consistently the studies show that office costs are rising at a rate of 12 to 15% a year—and sometimes faster than that. With this rate of increase, the very existence of many enterprises is threatened.

We observe that a very rapid transition is taking place. As automation increases in our factories, the proportion of knowledge workers continues to increase. These people constitute the human-capital resource that makes an information-intensive economy viable.

As Vincent Giulano points out, "The benefits of an increase in the productivity of office workers are not always within the view of managers;" nevertheless, "the office is the primary focus of information work, which is coming to dominate the U.S. economy." Perhaps the benefits of increased productivity are not always recognized, but the rising costs of knowledge workers should soon stimulate that recognition.

What are the bottom-line results of increased effectiveness and reduced costs in the office? Measurement is absolutely essential to evaluate the true significance of the shift to electronics.

Improving productivity

Industry is becoming increasingly self-conscious. Analysis, interview, study, survey and measurement are adding an entirely new dimension to sales, finance, production and engineering. Everyone from sweeper to statistician feels the prod of the scientist's instrument.

Yet the effectiveness of the knowledge worker has remained little explored. No job analysis is done. No instructions are issued when a new name is painted on the door of a private office. The working day on the production line is prescribed down to the finest detail; how knowledge workers spend their time is left to their own judgment.

Carl D. Liggio of Arthur Young & Co., shown reviewing testimony with two associates, believes the office Factmaster computer system's $250,000 cost is insignificant compared to the increased productivity it has made possible.

Because white collar workers have no rules to guide their activities, personal factors may have a strong influence. Instead of putting in deliberate, preplanned days, such workers are subject to constant interruption and governed by chance occurrences; they fail to assign and carry out priorities; and too often they do not do first things first. They may favor that branch of the business from which they were promoted, or pay more attention to that aspect of the company with the most aggressive and vociferous spokesperson.

Because of the impact of compounding office costs, there is growing recognition of the need for knowledge workers to allow and even to insist on the application of job measurement and job improvement to themselves.

However, Randy J. Goldfield, president of the Gibbs Consulting Group, sees knowledge workers as having a tremendous resistance to such measurement. She asserts that trying to equate professional activities with the bottom line is extremely difficult. First the concept must be sold to management. This process involves complicated, yet very informal organizational procedures.

Goldfield emphasizes that there are many "products" (outputs) that are extremely difficult to measure in terms of lines of data, numbers or units, etc., in the professional sector. "You have to think not only about output-input, but also about the time that the knowledge worker expends thinking about his activity off the job as well as on the job. . . . Therefore, we are talking about tremendously difficult matters in justifying the value of the 'product.' How can you judge the value of your own time? How can you, as an outsider, judge the value of someone else's time? We are looking at state-of-the-art concepts—concepts of thinking—concepts of self-purpose."

Refining methodology

Marvin Mundel strongly asserts that the methodology for establishing and measuring knowledge work is available. It is a matter of applying our current knowledge and making refinements with those companies that are willing to participate.

Some of the principles Mundel advocates are answers to the following questions:

☐ What is the objective?
☐ What are the outputs to be counted?
☐ How can the outputs be counted?
☐ How much of what kind of resources is needed to produce the outputs?
☐ What is a feasible and desirable operating plan for the next time period?

Mundel emphasizes that the answers to these five questions provide quantitative controls for both the knowledge worker and the knowledge worker's product. Among his techniques for measurement are work sampling, multiple regression and linear programming.

W. B. Bumbarger offers another approach to the measurement and improvement of the knowledge worker's productivity. He presents four rules for successful productivity improvement in the professional/ knowledge/office sector. The four key elements of what he calls operation function analysis (OFA) are:

1.) *Demand orientation:* Demands drive the business and determine the productivity of work flow. An OFA ratio of man hours to customer orders (demand) is one such useful performance measure.

2.) *Interorganizational focus:* In many businesses, 80% of the potential improvement involves interorganizational work flow.

3.) *Creativity:* When old ideas are not working, new ideas are needed. This requires creativity. A creative attitude can be fostered through special training in creative problem solving as applied to productivity improvement work.

4.) *Do it yourself orientation:* Your own people can do it best. Real improvements come from within and cannot be imposed from the outside.

Bumbarger's method for employing OFA consists of three phases:

☐ Phase 1: Senior management planning and commitment.

☐ Phase 2: Middle management analysis and recommendation.

☐ Phase 3: Improvement implementation.

Robert E. Hoisington, director of Arthur Young & Co., Detroit, meets the challenge of measuring technical workers such as engineers, draftsmen and technicians using engineering operations analysis (EOA). He defines these jobs as involving long cycle work, considerable mental activity and much creativity. The long cycle work component of these jobs requires costly data collection; the knowledge work requires careful job content analysis; and the high skill work requires an analysis of the skill mix. Therefore, productivity measurement in this area must be multi-dimensional to determine:

☐ An optimal skill mix to segregate intellectual and routine tasks.

☐ Normal distribution of effort.

☐ Time standards for work "units."

The application requirements are that the activity be performed in a manner acceptable to management; that the work produce some tangible, countable output; and that work cycles/output be repetitive.

Work sampling and multiple regression data are Hoisington's usual measurement tools.

At Hughes Aircraft Co., measuring the "unmeasurable" knowledge work in the organization is seen as an opportunity for organizational productivity improvement. Over the past two decades the knowledge worker segment of that firm's professional work force has burgeoned, as is the common situation. A primary objective of measurement is to unleash the huge potential of the firm's greatest resource—its population of knowledge workers. Hughes Aircraft is actively experimenting with knowledge work measurement.

Measuring project coordination

To provide a maximum challenge to their techniques, Hughes management personnel chose a very difficult knowledge function for one of their initial studies. They felt that if they could measure their project coordination function, they could measure anything.

Project coordination is a functional group within the production control organization. Responsible for getting the right parts to the right place at the right time, this group had served as a "fire-fighting" operation, constantly resolving critical parts shortages to keep the production line producing efficiently.

The initial step in measuring the project coordination function was defining function tasks. The group's supervisor accomplished this with the active and influential involvement of the remaining eight employees in the group. After some additions, modifications and deletions, the group defined 32 different tasks which they performed.

Next, these employees logged their time each day according to the 32 task categories. Active participation in the definition of tasks facilitated a common understanding of them and realistic time recording.

The workers recorded only their productive time. If they individually felt that specific amounts of their time were given to personal activities, these were not recorded. The net difference between total reported time and gross payroll hours provided an overall measure of nonproductive time. This was significant and contributed to the analysis of the data.

The study continued for ten representative weeks. During this period a quality circle addressed needs and opportunities for improving the productivity of the function.

A number of facts jumped out from the study. For example, administrative activities, including work on routine and special reports of several kinds, comprised 38% of all activities. Employees and management felt that major reductions could be accomplished here.

The value added concept was an important consideration. "Value added" is generally applied to productive work that adds value to the final product or service being rendered. Certain administrative activities were recognized as adding little value to the end result.

Only 7% of the group's time involved auditing inventory and production control records. The early detection of such errors is extremely important to the prevention of critical shortages.

In addition, no time was reported for formal training. Given the great complexity of the total production and material control systems, many of the critical shortages had been the result of poorly or erroneously coordinated activities by many people.

To address the latter finding, the group formed a "SMART" (systematic, modular approach to relevant training) committee to concentrate on training with other interrelated functional groups, within and outside the production control department.

Excessive slack time within the eight-person group was another key finding.

Management recognized an urgent need for a substantial 38% reduction in the staff of the project coordination group in order to meet operating budgets. Despite this major staff reduction, morale actually improved within the group as a result of the cooperative participation of all involved. The two persons who left the group were promoted to other open positions within the division.

The lowest performer was laid off. Actually, all the members had been frustrated by the slack time and nonperformance within the group, so correction of this problem was a positive action. Additionally, the supervisor, Geneva Potepan, received a promotion to the next higher level of management.

As a direct result of measurement, the project coordination function now is seen differently, and is operating much more effectively, efficiently and productively. The project coordination function no longer serves as a fire-fighting outfit, but is now mainly a fire-prevention organization. Critical shortages have dropped approximately 50% over the last year. This has improved efficiency on the production line and also reduced the work load of the project coordination group itself. The employment of the simplest measurement tool, a time log, contributed significantly to a substantial improvement in office productivity.

Shaping future work

Mechanization and the employment of electronics in office work are bringing about an amazing transformation. Productivity improvements cited include a decrease in delays relating to information, the elimination of retyping, a decrease in the time needed for filing and information retrieval, and rapid communication to facilitate decision making. The timeliness of decisions or of a response to an inquiry can have immense consequences.

Since knowledge workers are concerned with information, their productivity should increase as the means of handling information are mechanized and computerized. Therefore, in this changing world, the need to measure knowledge workers to determine their accountability becomes apparent.

The most important element in office automation is people, because people alone will make or break any innovative system. Most offices are well managed and have established procedures. However, change intimidates people. Technological advancements come quickly and often involve computer products, which are difficult to comprehend.

In dealing with knowledge workers we must recognize that we are not analyzing a production line. Automation's single biggest obstacle has been the fear among employees that it would mean lost jobs. But as knowledge workers become involved, they not only accept computer technology, but are welcoming it and even insisting upon it.

Carl D. Liggio, general counsel, Arthur Young & Co., believes the $250,000 cost of the firm's computer system is insignificant compared to the time savings and the increased efficiency and productivity that have been achieved with it. "Not only has computer technology materially increased our productivity; it's cut down on the chance for a substantial error as well," Liggio notes.

From an industrial engineering viewpoint, obtaining the hardware and software is the easiest part, according to Liggio. "The key factor is developing the systems and methodologies for the efficient use of these systems," he says.

Industrial engineers have little choice: if business and industry are to survive and prosper in today's difficult economic climate, the performance of knowledge workers *must* be measured. As we have shown, successful measurement hinges on refining our existing knowledge and on creatively adapting our knowledge of measurement to the specific job function.

In application, the real issue goes beyond measurement to management. If we haven't measured our knowledge work, how can we possibly manage it?

The challenge that faces management is to introduce new technology to improve the productivity of the total enterprise. A very large part of the cost of the enterprise is the salaries of knowledge workers.

Measurement must be employed to properly staff the company with producers who contribute to the bottom line of the financial statement. That bottom line must consistently show a profit, and productive knowledge workers are essential to realizing this result.

For further reading:

Connell, John J., "The Office of the Future in Perspective," *The Word,* Word Processing Society Inc., April, 1982, page 7.

Giuliano, Vincent E., "The Mechanization of Office Work," *Scientific American,* August 1982, page 110.

Liggio, Carl D., "Upping Law Office Output," *Management Information Systems Week,* April 21, 1982, page 28.

Mundel, Marvin E., *Measuring and Enhancing the Productivity of Service and Government Organizations,* Asian Productivity Organization, Tokyo, Japan, 1975. **IE**

Chester L. Brisley, P.E., is professor and associate chairman of the department of engineering and applied science of the University of Wisconsin-Milwaukee. He is executive vice president, chapter operations, of the Institute and a fellow of IIE. He has served as vice president and as chairman of Region XI and as coordinator of divisions and has been president of the Milwaukee and Detroit chapters of IIE. He has also been chairman of Institute conference programming. Currently, he is the Institute director of membership and chairman of the Rating Film Feasibility Committee. He is a fellow of SAM and MTM and a member of NSPE/WSPE, ASEE and ASME. In 1977 Region XI named him Engineer-of-the-Year. Brisley holds a diploma from General Motors Institute, a BSIE from Youngstown University and MSIE and PhD degrees from Wayne State University. He is a well-known lecturer and author on work measurement.

William F. Fielder, Jr., P.E., holds BA and MBA degrees in business administration from the University of California at Los Angeles. He is an internal work measurement consultant with Hughes Aircraft Co. and is chairman of the Hughes corporate work measurement technology committee. Fielder has had experience with work measurement at U.S. Steel Corp. and McCulloch Corp. as well as Hughes. Prior to assuming his present position at Hughes, he served as head of production control and head of personnel development. He is a senior member of IIE and a fellow and secretary of the MTM Association for Standards and Research.

Reprinted from Industrial Engineering, December 1983.

Short Method Makes Frequent Evaluation Of Office Productivity A Realistic Objective

By John Kristakis
Kristakis Associates

An organization-wide productivity improvement program based on company objectives can be highly beneficial. But all too often, such an approach takes too long to implement and either loses its effectiveness or never really gets started.

A straightforward short method is proposed as an alternative. It employs a basic evaluation process that can be carried out without special training or coaching from a practicing management scientist. The method is applicable to both small and large operations.

Each individual office can independently perform its own productivity improvement activity, although cooperation between offices having interrelated operations is of course possible and encouraged. The scope and depth of involvement are at the discretion of the manager.

In an organization with a functioning overall productivity improvement program in place, an individual office manager can easily employ the method while participating in the organization's overall program.

Often, managers who fully intend to conduct productivity analyses find that the evaluation is always being postponed until another time. The simplicity of the short method described here should encourage the manager to actually sit down and perform a productivity improvement evaluation more frequently.

The method

The process involves basically the simplified instructional steps shown in Figure 1. Begin the process with the often bypassed step of writing down the major work processes from start to finish. Use a standard chart like the one shown in Figure 2 if the process involves paperwork movement between office work stations.

Writing the process down leads the manager logically through the flow pattern. It also establishes a tangible device for thinking through changes or improvement alternatives.

The office manager can continue the process with these inquiries: Who does what? When? How long does it take? What is the cost? It is not essential to have precise, accurate answers in order to use the method. Naturally, the more accurate the data, the more quantitative the analysis can be. Should accurate data or standards be available for alternative comparison purposes, by all means they should be used.

The manager should not hesitate to use the method for lack of accurate data or concise calculations, however. When capital expenditures and justification rate-of-return analysis are not required, the manager's intuitive "that's a better idea" feel is sufficient justification for a change that might improve productivity.

Step one:

The types of work processes performed in the office are listed. The types are classified according to general major categories of work.

The manager identifies and writes down what the office's major work categories produce as output.

The input and resources required to produce the outputs within the department are identified and written down. This should include both human and non-human resources: labor, materials, equipment and information. For offices that perform more than one major category of work, a selection routine is required (step two); otherwise, skip to step three.

Step two:

When more than one major category of work is involved, the work should be ranked in order of greatest productivity-improvement potential. A sophisticated analysis is not required. The manager's intuitive judgment can be used in conjunction with the following general questions to determine the ranking:
☐ What office work process is causing the costliest problems?
☐ What office process requires the greatest improvement?
☐ What office process requires preparation for anticipated increased volume, etc.?

The office work process which has the greatest potential for improving office productivity is selected, and the other work processes are evaluated in the order of their ranking.

Step three:

Using the standard chart shown in Figure 2, the manager breaks down the selected work process into detailed operations. The depth of detail is at the manager's discretion. The operation's sequence is maintained as much as possible.

The operations are identified according to who performs the work and how long it requires to complete. Precise figures for how much time the operation requires are helpful, but not necessary. Assigning time duration to an operation on a best-estimate basis is sufficient in most cases. Use of established time standards is the most desirable procedure if they are readily available.

In step three the supervisor identifies the following:

☐ Paperwork documents, information, etc., that flow through the work process internally from operation to operation and externally to and from other departments.

☐ Materials, supplies, etc., that are involved with the work process.

☐ Equipment or work station that is required to perform each operation in the work process.

☐ Operations which generate the most errors and cause the greatest frequency of rework.

The manager adds or notes the previously mentioned inputs or identifications adjacent to the applicable operation on the standard chart. Any other specifics on the operation which would aid in the evaluation are added.

Step four:

The work process operations are evaluated from the standpoint of how productivity can be improved. The following list is a general guide:

☐ Is the work process necessary to the office or organization's welfare?

☐ Is the work process end product

Productivity analysis asks if the materials used in a work process are necessary. (Photo courtesy of Steelcase Inc.)

satisfying the purpose?

☐ Are any operations candidates for elimination?

☐ Can any operations be combined to eliminate duplication?

☐ Can the frequency of performing the operation be reduced?

☐ Does an operation need to be strengthened or improved to handle an expected increased volume?

☐ Would the redistribution of labor assignments achieve performance or cost improvements?

a) Can lower classification (pay grade) personnel perform all or a portion of an operation?

b) Can an assistant or secretary be delegated a portion of the manager's responsibility for the operation?

Unloading the manager's operational responsibility is a desired objective. The manager can use the freed-up time to do better planning or perform other essential duties or to prepare the department for increased volume.

☐ Are the materials, supplies and information used in the work process necessary? Can less expensive items be substituted without a quality loss, such as generic items? If these items represent only a small percentage of the operation's cost, they would not be evaluated.

☐ Are there operations which could be handled better with added equipment, electronic or otherwise? For example, dictating equipment is a relatively inexpensive item that many offices are not fully utilizing.

Step five:

After the selected work process operations have been evaluated, the operations whose improvement offers the largest return can be isolated. Cost calculations involving time saved, equipment justification, etc., are beneficial if available. Otherwise, judgment is used.

The following steps complete the evaluation:

☐ Rank the operations in the order

of the largest return when two or more improvement possibilities are uncovered.

□ Select the operation with the largest potential return and the highest chance for successful implementation.

□ Determine the method, or alternative methods, for improving the operation; cost considerations are taken into account when comparing the alternatives.

□ Select a productivity improvement method that is within the office or organization's capability to implement.

□ Plan the implementation of the improvement method into the office work process.

□ Execute the implementation plan.

A follow-up evaluation is used to determine the productivity improvement method's effectiveness.

Summary

The short method provides an office manager with a checklist for performing a productivity-improvement evaluation within his or her sphere of influence. The depth of the evaluation is proportional to the amount of time the manager allocates to the effort. An important advantage of the short method is that it can be carried out frequently with a modest expenditure of time.

An office manager need not dismay if a productivity improvement has not resulted from the evaluation of work processes. The benefits of reviewing office work process operations may show up at some future time. Knowledge of what and how office operations are performed can prepare an office for future growth, changes in the business environment, internal organizational changes, etc., even if immediate productivity gains do not result from every evaluation. Frequent analyses enable a manager to spot productivity improvement opportunities as they come up. **IE**

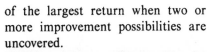

Figure 1: Productivity Analysis Process

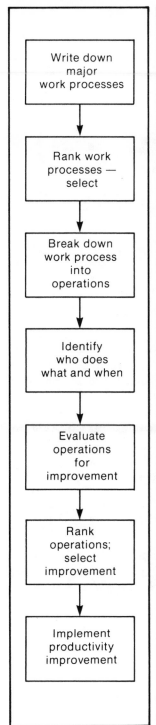

Figure 2: Paperwork Movement Chart Between Work Stations

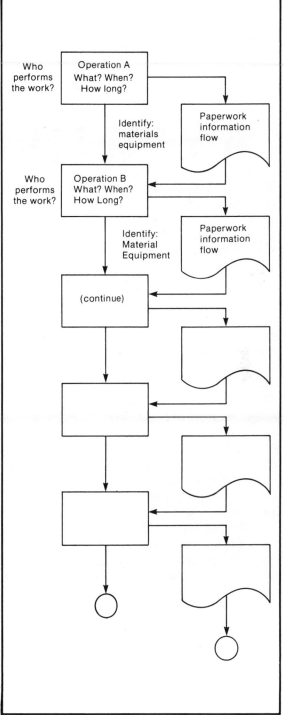

John Kristakis is president of Kristakis Associates and has performed consulting assignments in plant and office sites around the United States and Europe. He received BSIE and MSIE degrees from the University of Pittsburgh. A senior member of the Institute, Kristakis has served as president and vice president of Pittsburgh chapter No. 3 of IIE and as Region V chairman for the production planning and inventory control division.

Reprinted from Industrial Engineering, February 1983.

Simplified Method Of Measuring Productivity Identifies Opportunities For Increasing It

By Gary N. Brayton, P.E.
Touche Ross & Co.

Although American business managers have become increasingly aware of the necessity of improving productivity, they are seriously hampered in their efforts to do so by a lack of effective methods for determining how efficiently they use their productive resources.

A recent survey (see "Productivity Indicators Used by Major Manufacturing Companies: The Results of a Survey," by D.J. Sumanth, *Industrial Engineering,* May 1981, pp. 70-73) showed that less than 3% of U.S. businesses have systems for measuring total productivity. Moreover, many business-management professionals do not have the measurement tools required to analyze accurately the effect of productivity changes on profitability.

Measurement system

On the basis of a productivity-measurement system developed by the American Productivity Center in 1980, Touche Ross & Co. has created a relatively simple methodology for identifying opportunities to increase productivity. It is an extension of commonly used methods for determining aspects of productivity, such as miles per gallon of gas, units of output per labor-hour, the return on funds employed (ROFE) and the yield of finished goods from raw materials.

In each of these measurements, productivity (miles per gallon) is the ratio formed by dividing units of output (miles travelled) by units of input (gallons of fuel used):

$$\text{Productivity} = \frac{\text{Output units}}{\text{Input units}}$$

When productivity is measured in successive periods—for example, each time the car is filled with gas—a trend is shown.

In Touche Ross & Co.'s total productivity measurement system, a company's total productivity is analyzed by dividing the total production units of goods and services by the total resource units required for production. These resources include materials, labor, capital, energy and miscellaneous items such as taxes and insurance.

Productivity formula

To relate such diverse quantitative units of measurement as number of goods and amounts of service, pounds of materials, labor-hours, BTUs and capital equipment, they must be expressed in dollar values. The resulting formula is a profitability ratio that must be greater than one to produce a profit:

$$\text{Total Productivity} = \frac{\text{Total output units goods \& services}}{\text{Total input units of resources}}$$

Expressed as a profitability ratio, the formula would be:

$$\text{Total Profitability} = \frac{\substack{\text{Total Output} \\ \text{\$ Goods \& Services}}}{\substack{\text{Total Input} \\ \text{\$ of Resources}}}$$

Profitability, viewed as a percentage of sales, increases or decreases as a result of:
☐ A change in the unit selling price.
☐ A change in the unit cost of resources.
☐ A change in the quantity of resources used per unit of output.

For the sake of analysis, these changes can be separated into two categories: *price recovery change,* which is a change in the relationship between the unit selling price and the cost per unit of input, and *productivity change,* which is a change in the number of input units required to produce a unit of output.

A crucial problem facing many companies is their inability, because of competition, to recover increases in the cost of materials, labor or other resources by raising prices. Nor are they able to decrease the cost of the resources or substitute others. Therefore, if the profit margin in sales is to be maintained or increased, productivity must be improved.

Value of system

By keeping track of price recovery change and productivity change from period to period, the total productivity measurement system enables a company to study the effects of the changes, identify significant opportunities and focus management resources on appropriate action. It is especially valuable because it pro-

Table 1: Typical Business Total Productivity Measurement System Consolidated Statement (in thousands of dollars)

Resource Description	Current Year Period 2			Base Year Period 1			Change Ratios			Effect on Profits		
	V_2 $ Value	Q_2 Quantity	P_2 Price	V_1 $ Value	Q_1 Quantity	P_1 Price	V_2/V_1	Q_2/Q_1	P_2/P_1	Total	Due To Productivity	Due To Price Recovery
	(A)	(B)	(C)	(D)	(E)	(F)	(G)	(H)	(I)	(J)	(K)	(L)
Output (sales)												
Products and Services	$12,000	1,091	$11.00	$10,000	1,000	$10.00	1.2000	1.0910	1.1000			
Input (cost of sales)												
Total Labor	4,968	432	11.50	4,000	400	10.00	1.2420	1.0800	1.1500	(168)	44	(212)
Total Materials	3,864	690	5.60	3,000	600	5.00	1.2880	1.1500	1.1200	(264)	(177)	(87)
Total Capital	1,417	8,333	.17	1,000	8,333	.12	1.4170	1.0000	1.4167	(217)	91	(308)
Total Energy	528	11,000	.048	500	12,500	.040	1.0560	.8800	1.2000	72	106	(34)
Total Misc.	560	500	1.12	500	500	1.00	1.1200	1.0000	1.1200	40	45	(5)
Total Input	11,337			9,000			1.2597	1.0789*	1.1676*	(537)	109	(646)
Net Profit	663			1,000								
Profit % Sales	5.5%			10%								

Effect of profit % change (.055−.10) × $12,000 = ($540)

— $109 favorable change due to improved productivity
— ($646) unfavorable change due to inadequate price recovery
Column J equals output V_2/V_1 − input V_2/V_1 × base year input V_1.
Column K equals output Q_2/Q_1 − input Q_2/Q_1 × base year input V_1.
Column L equals column J − column K.

*All totals and subtotals require weighted change ratios.

$$Q_2/Q_1 \text{ subtotal} = \frac{Q_2 P_1}{V_1} \text{ for items included in the subtotal.}$$

P_2/P_1 subtotal = Subtotal V_2/V_1 ÷ Subtotal Q_2/Q_1.

---

vides data for each input category. The analysis can be performed for an entire company or an individual product line.

An application of the system is shown in Table 1. Although the figures are hypothetical, they are based on data from client companies.

Column A shows the total dollar sales (output) of products and services and the resources used (input) by cost category in a given period—say, the current year.

Column B shows the number of units sold and units of input during the same period. Units of input are expressed in such common measures as labor-hours, tons of material, capital dollars and BTUs.

Column C gives the price per unit of both the outputs and inputs.

In columns D, E and F the same information is given for the base or a comparison period.

Columns G, H and I show the change, expressed in decimals, that occurred from Period 1 to Period 2.

For example, total dollar sales (value) during Period 2 were 20% higher than in the previous period (that is, 120% of $10 million, or $12 million). The number of units sold (quantity) was a little over 109% of the quantity in Period 1.

Therefore, the price per unit sold was 110% of the base year price—an increase of 10%, from $10 to $11. During the same two periods, the total costs increased by nearly 26%—that is, the cost in Period 2 was 126% (1.2597) of the cost in Period 1.

Indications of ratios

These change ratios indicate a decline in profitability. Although sales rose by 20%, total cost of the resources used in production rose by 26%. Increased productivity did not make up the difference, although there was a slight improvement. A rise of 9% in the quantity of goods and services produced required an increase of nearly 8% in the quantity of resources used.

The ratios also show that price increases failed to recover increases in costs, because the selling price rose only 10% while the average cost of resources went up 17%.

Once the various change ratios are computed, it is possible to determine quite easily the effect of each change on profits:

Step 1: Total Effect on Profits (Column J).—Subtract the value change ratio for each input category from the total output value change ratio, and multiply the result by the category's value in Period 1.

Output V_2/V_1 −
Input V_2/V_1 ×
Input V_1 =
Total effect on profits

Step 2: Effect of Productivity Changes on Profits (Column K)— Subtract the change ratio for each input quantity from the total output change ratio, and multiply the result by the category's value in Period 1.

Table 2: Typical Business Total Productivity Measurement System Performance Ratios

Resource Description	Change Ratios			Performance Ratios*		
	V_2/V_1	Q_2/Q_1	P_2/P_1	Profits	Productivity	Price Recovery
Output						
Products & Services	1.2000	1.0910	1.1000			
Input						
Total Labor	1.2420	1.0800	1.1500	.9662	1.0102	.9565
Total Materials	1.2880	1.1500	1.1200	.9317	.9487	.9821
Total Capital	1.4170	1.0000	1.4167	.8469	1.0910	.7765
Total Energy	1.0560	.8800	1.2000	1.1364	1.2398	.9167
Total Miscellaneous	1.1200	1.0000	1.1200	1.0714	1.0910	.9821
Total Input	1.2597	1.0789	1.1676	.9526	1.0112	.9421

Profits are down approximately (4.7%).
Productivity has improved profits by approximately 1.1%.
Price recovery has decreased profits by approximately (5.8%).

*These ratios computed by dividing the output change ratios by their corresponding input change ratio.

Output Q_2/Q_1 −
Input Q_2/Q_1 ×
Input V_1 =
Productivity effect on profits

Step 3: Price Recovery Changes on Profits (Column L)—Subtract the effect of productivity on profits from the total effect on profits.

Step 4: Performance Ratios—In order to show trends in the relationship between the changes in output and changes in input, divide the output change ratios by the corresponding change ratios for each category of input (see Table 2).

When this is done for total input, it is evident that productivity increased profits by 1.1%, but price recovery decreased profits by about 5.8%, with the result that profits were down by approximately 4.7%.

Conclusions drawn

From these data, certain conclusions can be drawn about the company's profitability, and the conclusions raise specific questions about pricing and productivity which point out avenues of investigation for management.

Conclusion: Materials are the greatest cause of profit deterioration, due to a decline in the yield from materials and a failure to recover increases in the cost of materials by raising prices of goods produced.

□ Why are the yields declining?
□ What types of productivity measurement can be applied to provide ongoing information about yields from materials?
□ Can substitutes be found for low-yield or high-priced materials to reduce costs?

Conclusion: Capital costs are leading to profit deterioration, because an increase in the cost of capital has been offset only partially by gains in productivity.

□ How can the cost of capital be reduced?
□ Where can capital requirements be reduced—by decreasing inventories, write-offs of unused equipment, increased utilization of capacity, some other means?

Conclusion: Increases in labor costs are causing profit deterioration, because wage rates are rising faster than the selling price of goods produced. Improvements in labor productivity, however, are reducing the losses somewhat.

□ Can labor productivity be improved sufficiently to offset wage-rate increases?
□ What types of measurement systems can be applied to provide ongoing information about labor productivity?
□ Can automation be introduced, and what would be the associated capital costs?

Conclusion: In view of the fact that there has been a 12% decrease in energy used, despite a 9% increase in total output, there has been a significant increase in the productivity of energy. But the gain is offset by an increase of 20% in the unit cost of energy, whereas the selling price of goods produced has increased by 10%.

□ Why has the productivity of energy improved, and what can be done to make sure the improvement continues?
□ What can be done to reduce energy consumption further?
□ What alternative energy sources are available?

Conclusion: As a result of productivity gains for miscellaneous items, there has been an improvement in profitability.

□ What are the reasons for the productivity gains?
□ How can the gains be continued and possibly increased?

Similar insights can be obtained at a glance by examining Table 2. Although performance ratios do not show the exact effects of changes on profits, they do indicate specific problems that should be evaluated.

As a result of two key concepts, fully absorbed costs and quantity usage data, the productivity performance measurement system might require some modifications in a company's accounting system. First, all costs related to the production of a specified good or service—that is, the "output" in Table 1—should be identified by input category.

This might require allocations from shared cost centers, and definitely necessitates the blending of direct and indirect costs. For example:

Labor: All labor associated with obtaining the specified output should be included by separate line item, such as the labor that is required directly for the manufacturing process (by major manufacturing centers), indirectly for the manufacturing process, for distribution and for management and other aspects of overhead (by department).

Materials: All materials should be identified, including raw materials, manufacturing supplies and items used for administrative functions.

Capital: Included are both cash and noncash capital costs, such as interest expense, rent, depreciation and lease expense. The only capital cost not included is the return on equity capital that is represented as net profit.

Energy: Among the items should be the costs of all electricity, natural gas, fuel oil, gasoline and alternative energy sources used in direct manufacturing, distribution and overhead cost centers.

Miscellaneous: This is a comprehensive category for all costs not itemized elsewhere, including taxes, insurance and cash payments for hauling and distribution, brokerage expenses and consultation or other contract services.

The second key concept, quantity usage data, involves the use of quantitative data. To compute productivity indices and their effects on profits, the quantities of input items are required. Although information of this sort is often collected by companies, it usually has to be extracted from various documents in the accounting system.

For ongoing productivity measurement, quantitative data should be stored in accounting ledgers, in addition to dollar values. The measurement selected should be based on the units used for pricing input items—for example, labor-hours, pounds or tons of raw material, debt-capital dollars, total dollars on which depreciation is based, square feet of rental space, kilowatt hours, gas therms.

Minor categories

Only major cost categories require so much detail. For minor categories, surrogate quantities can be developed by using a price index. A company might not be interested in details about office supplies, for example, and therefore could record all of them as one line item.

To do this, certain of the major supplies should be selected to represent all the items in the category. Then price data for these representative supplies are obtained for Period 1 and Period 2, and the average change in price between the two periods is calculated.

An index is then established by assigning a surrogate price of 1.00 to the line of items in Period 1. The percentage of change in the average price of the representative supplies is added to the index of 1.00 to obtain a price index for Period 2.

If the average price was increased by 35%, the price index for Period 2 would be 1.35. This amount is divided into the total dollars spent on supplies to obtain a surrogate quantity. Therefore, if $16,000 were spent on supplies in Period 2, the surrogate quantity of supplies would be 11,851 ($16,000/1.35).

Using the index of 1.00 for Period 1 as a base, surrogate quantities can be determined for each succeeding period, and changes in the use of quantities can then be ascertained. The method is similar to that used by the government for the Consumer Price Index and other indices.

With this and the other tools provided by the total productivity measurement system, companies can identify the causes of deterioration or improvement in profitability, determine where attention should be focused to apply available resources with the maximum beneficial effect, project the impact of possible changes in productivity and pricing, track the progress of efforts to increase productivity and take corrective action when necessary.

Adoption of a productivity measurement system is an essential step toward maintaining the vitality of a company in an increasingly competitive world. IE

Gary N. Brayton, P.E., is a senior consultant at Touche Ross & Co. He has 13 years of experience in implementing productivity improvement programs in both the private and public sector. His consulting experience includes operations review and improvement, development of information systems and the development of financial evaluation models. Brayton received his BSIE degree from Arizona State University and his MBA from Golden Gate University. He is a registered manufacturing engineer and a senior IIE member.

III. PRODUCTIVITY IMPROVEMENT TECHNIQUES AND APPLICATIONS

Productivity improvement can be broken down into the following objectives:

1. improving effectiveness (doing right things);
2. improving efficiency (working right);
3. improving quality of work life (creating a more enjoyable working environment).

Productivity improvement can result from the use of many different techniques. A few of these are illustrated in the following section.

*Reprinted from **Industrial Engineering**,*
January 1985.

Strategic Planning: A Crucial Step Toward A Successful Productivity Management Program

By D. Scott Sink, P.E.
Virginia Polytechnic Institute and State
University

Formal productivity management programs are becoming more common in U.S. organizations. Their development is being fueled by the current economic and competitive environment. It is reasonable to suspect that interest in formal productivity management efforts will continue throughout this decade and beyond.

Unless a productivity management program is well planned, its chances of success are low. This article will focus on a participative strategic planning process that will result in a two-to-five-year plan for your productivity effort. The process, if executed correctly, can set the stage for successful development and implementation of your productivity effort.

Example programs

There are a number of sources through which one can investigate case examples of specific productivity management program designs, including the American Productivity Center's Case Study Series; American Productivity Management Association; National Productivity Network; Peters and Waterman, 1982; Buehler and Shetty, eds., 1981; *Working Smarter* (eds. of Fortune); Criner, 1984; Zager and Rosow, 1982; and *National Productivity Review*—see "For further reading." These sources, and others not listed, can provide background, guidance, innovative ideas, approaches, techniques and, perhaps most importantly, gained and shared wisdom.

The successful design, development and execution of a productivity management effort will require a good balance of wisdom, knowledge and skills. As someone said, "Good decision making comes from wisdom judgment and experience. Wisdom, judgment and experience come from

Figure 1: Components of a Productivity Management Program

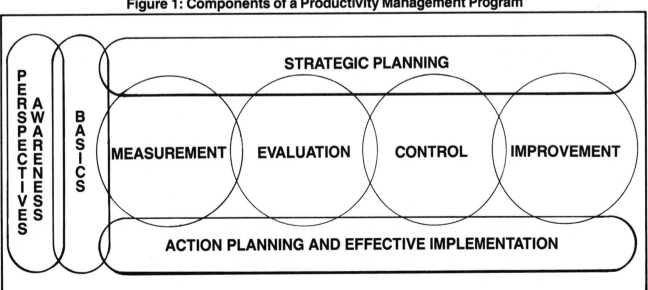

bad decision making."

A few "gems of wisdom"

The following are some conclusions, based on my studies and experience, that I feel make sense relative to developing successful productivity management programs.

Top management support and legitimization. I suggest that the appropriate level of top management in your organization be invited to participate in a strategic planning process (an off-site retreat) for the explicit purpose of developing a two-to-five-year plan for the productivity effort. If you have been given the assignment to develop a productivity program for your organization, you must involve top management in an effective, efficient, gratifying and satisfying planning process for that program.

If your top management refuses or respectfully declines to get involved in this process, I suggest you respectfully decline the assignment and go on to something else. There are too many crusades to lead or be part of in an organization today to be involved with one that is not blessed or that is blessed in name only.

If you get your top management people to the planning retreat and find they can't or won't provide direction for the effort or can't respond to the questions asked in the process, again I suggest you respectfully decline the assignment.

Middle level managers and staff members must wake up to the political reality of life in complex organizations. If a "productivity manager" or a group within the organization given the responsibility for developing a productivity program cannot generate genuine support and legitimization for the effort, at all levels, the program is doomed. You can generate quality support and legitimization by involving all levels of management, particularly upper levels, in the planning for the program.

You will need a structured and effective process in order to accomplish this.

Program champion, effective leadership. Peters and Waterman, in *In Search of Excellence,* point out the importance of product and service "champions." They stress the fact that innovative companies are very effective at responding to changes in their environments. They have a bias for action. "Tools don't substitute for thinking. Intellect doesn't overpower wisdom. Analysis doesn't impede action," they observe. They go on to point out that excellent companies allow their innovative product and service champions a lot of flexibility, "a long tether."

I think we can all readily identify champions in our organizations. Unfortunately, perhaps, many of us can't think of too many who are effective or who champion the right causes. The existence of effective champions or innovators, in the right spot at the right time, is in my opinion a critical but often intangible dimension of success. Most of our organizations stifle innovation and either constrain the champions into submission or drive them out of the system.

We need to understand and be able to recognize the characteristics of potential champions and then place them in critical positions. This is especially important with regard to productivity management programs. Too often, however, top management decides to start a productivity program and then staffs it with a director it couldn't quite find any other position for.

Effective leadership is the most critical factor in the successful development of a productivity effort. When top management places ineffective leaders in positions of power relative to the productivity effort, a signal is being sent that productivity really isn't very important. If a productivity program is going to be any-

thing more than a public relations window dressing attempt to create awareness of the importance of productivity, management must staff the program with effective leaders.

Internal legitimacy, staff-operations linkages. The question of whether to have a productivity center with some degree of centralization or to completely decentralize productivity efforts to the operating units is a critical issue. Legitimate concerns are expressed suggesting that if one starts a productivity program, operating managers will assume they are no longer responsible or accountable for productivity.

The analogy may be drawn between productivity and quality. Quality control departments have, at least psychologically, tended to diffuse responsibility and accountability for quality. They have tended to lead to the mentality that suggests you can inspect quality into a product or service.

It appears that there is a critical and very difficult question associated with the centralization/decentralization of productivity efforts. I suspect the answer to this and related questions is extremely organization-dependent. It is, therefore, impossible to make any general prescriptions. However, I am aware of at least one firm that has explicitly and I think effectively dealt with this issue.

The Honeywell Aerospace and Defense Group has initiated a productivity and quality center. The director of this center reports directly to a staff vice president who is in charge of human resources, management development and the productivity and quality center.

The center focuses upon facilitating innovation in the areas of quality of work, quality of management and quality of work life. It is staffed with professionals from operating groups in the company who are brought in on two-year assignments. After their two years with the center they return to operations groups.

Many of the professionals who come into the center view the assignment as a sort of sabbatical. They can work on special projects they have always wanted to accomplish but have never had time for while in operations. They work on projects that will assist their operations in becoming more productive. They become internal consultants in some cases. They attempt to transfer knowledge and often technology applications. They try to become "change-masters," spark plugs for innovation.

Most center staff, to date, have been highly sought after by operating units following their two-year stints. These professionals often develop new sets of skills while in the center that are very useful during their subsequent careers.

This case example illustrates how one productivity program has attempted, I think quite successfully, to deal with the critical issue of staff-operating unit legitimacy. If a productivity program does not earn the respect of the operating units, its long-term effectiveness, and even its survival, are in question. Productivity programs have to be viewed by the operating units (their customers) as being credible. This requires constant introspection on the part of the productivity program staff relative to mission, goals, objectives, products, services and performance.

Strategically thought through and effectively communicated two-to-five-year plan. Too many productivity efforts in the U.S. have looked like a "random walk" process. Many such efforts over the past ten years began and ended with quality circles and one rather arbitrarily selected productivity measurement technique as their only products and services. Many began and ended with mission statements, goals, objectives and plans developed by one person, usual-

ly the productivity director.

Often, a command from on high comes down to start a productivity effort. A group or individual gets the assignment and immediately rushes off to the first seminar that comes along with productivity in the title. The concepts or techniques presented in that seminar are adopted, a plan is sketched out to begin implementation of "pilot studies," and off they go. Operating and staff groups throughout the company begin getting newsletters from something called a productivity center. Fears, hidden agendas, turf domain problems, uncertainty and ambiguity relative to this productivity effort emerge and heighten. Measurement techniques, control procedures and improvement programs are hastily developed and randomly implemented on target areas in the company.

Resistance to change increases over time as managers having the productivity program "done to them" assess the amount of and legitimacy of power behind the effort. If the influence and power behind the effort are high, operating and staff managers disguise their resistance. If they are low (only apparent lip-service given), the resistance is more open and the productivity effort finds it difficult to find receptive areas to perform its "magic." The effort dies a slow but welcomed death.

This scenario is exaggerated to make some critical points. First, many productivity programs begin and are developed with very naive and unrealistic perceptions regarding how individuals, groups and organizations change. At the heart of any productivity program must be a solid understanding of change and innovation. If the management and staff of a productivity effort do not possess adequate organizational development skills, the program will be less than effective. References in this

area that may be of interest include *The Change Masters,* by Kanter, and Morris's *Implementation Strategies for Industrial Engineers* (see "For further reading").

Change and innovation will be at the heart of all productivity improvement. Management of a productivity effort must be effective at shaping, causing, nurturing, promoting and reinforcing change and innovation.

We can affect the extent to which individuals, groups and organizations are aware of the need for change; willing and able to change; and committed to paying the price of change. However, unless the change is directed, unless there is clarity and consensus as to where we should be headed, there is not likely to be performance improvement.

The strategic planning process can provide consensus direction for attempts to improve productivity. Furthermore, it provides the opportunity to think through how we can systematically tackle longer range goals and objectives by taking smaller steps that move us toward larger objectives.

If the strategic planning process is executed correctly, it can build commitment to the productivity effort and shape internalization of the program's goals and objectives. This will reduce resistance to change and increase cooperation and coordination between the productivity program and the operations and staff units (the customers) in the organization.

My last "gem of wisdom" is, then, that a productivity program must execute a participative strategic planning process throughout the organization that is intended to involve all levels of management in the development of a two-to-five-year plan for the program.

Program components

There are distinct and reasonably discrete components to a productivity

Table 1: Comparison of Typical and Forward-Looking Strategic Planning Processes

Characteristic Strategic Planning in "Type A" Organizations	Characteristic Effective Strategic Planning Processes for the '80s and '90s
• Formal	• Structured but less formal
• Focus on plan	• Focus on plan and process
• Budget driven	• Plan drives budget, not reverse
• Top management and their consultants	• Involvement at all levels of management coordinated over time
• Marketing and financial imbalance	• Balance between marketing (product portfolio), financial and operations issues
• Myopic	• Broadened perspective, more data tapped/considered
• Short-sighted	• Longer term planning horizon
• Limited involvement and participation	• Pervasive discourse on plans
• Detached from many pragmatic operational realities	• Driven/managed top down, data fed forward and backwards

program. Figure 1 depicts the components and their interdependency. I will briefly state the goals for each component in the form of critical questions.

Awareness. Is there a clear, consistent, comprehensive consensus that something needs to change? The first step in change for individuals, groups or organizations is a cognitive and affective awareness that change must happen in order to survive, compete, perform, etc.

Many productivity efforts start and end here. Productivity improvement requires more than awareness. This component may be a necessary condition, but it alone is not sufficient. Slogans, posters, pep rallies, company songs, company shirts and hats, etc., are nice, but limited in their long-term impact.

Basics. Is there a consistent, comprehensive, clear understanding of what the organizational system (department, division, plant and/or firm) is about? Of what performance means? Of what productivity is? Of the performance management process? You cannot manage what you cannot measure. You cannot measure that which you cannot define. The focus in this component is on educating and developing common knowledge about performance, pro-

ductivity, survival and growth in the world of the '80s, '90s and beyond.

Measurement and evaluation. What constitutes performance? Are we measuring what we really want to get out of the system? What measurement and evaluation systems do we currently use? Which work? Which don't? Why? What new measurement and evaluation approaches and techniques are available? Which have potential for our applications? Are we measuring A while hoping for B? Are we rewarding A while hoping for B? The purpose of this component is to systematically improve performance and productivity measurement and evaluation systems.

Control and improvement. What are we doing? What is available? What is appropriate? How can we improve the quality and effectiveness of our efforts in this area? Are our control and improvement efforts directed by objective measurement and evaluation systems?

The purpose of this component is to coordinate, plan for and integrate overall productivity control and improvement efforts. Too many productivity programs have improvement tactics that resemble a random walk process.

Strategic and action planning and effective implementation. Do we

effectively link planning to action? Is our productivity program planning comprehensive, systematic, well thought through? Do our program plans contain all the necessary components?

Stages

In an earlier article (*Industrial Engineering,* October 1982—see "For further reading"), I presented six stages that many productivity programs have evolved or will evolve through. They were:

☐ *Stage 0:* The basics—refocus on fundamental management and engineering principles and practices.

☐ *Stage 1:* Organizational systems performance measurement—focus on performance measurement and evaluation systems at individual, group and organizational levels.

☐ *Stage 2:* Integration of productivity planning with business planning—expand the scope and improve the effectiveness of strategic and operational planning systems.

☐ *Stage 3:* Participation in planning, problem-solving, design, decision-making—design, develop, experiment with participative management processes.

☐ *Stage 4:* Productivity measurement systems refinement—enhance the sophistication of performance and productivity measurement systems. (See Sink, et. al., 1984.)

☐ *Stage 5:* Maintaining excellence—continued and ongoing maintenance and development, sponsor and promote appropriate levels of change and innovation.

I am convinced that, although many of these stages can be worked on simultaneously and therefore do overlap, the average planning horizon required for most U.S. organizations to accomplish these stages is five to 10 years. Without a well thought through strategic plan for your productivity effort, the chances for a successful five-to-10-year program are slim.

Your plan must be developed so that it can survive and maintain consistency through changes in management. However, the plan and the process must be flexible enough to provide for ongoing input and evolution. I suggest the planning process be executed once a year. Planning is analogous to sailing a sailboat versus driving a motor boat.

The references I have provided at the end of this article will reveal many more factors to consider as you develop your productivity program. I feel that the most critical factor has to do with the way in which a two-to-five-year plan for your effort is developed. Recognition on your part that there are critical components and stages of evolution for a productivity program will also affect the success of your efforts.

I will now discuss a strategic planning process that I have found to be very effective at developing a good, solid two-to-five-year plan while simultaneously creating support for the program and setting the stage for effective implementation. I assume that the ultimate goal of all productivity programs is effective implementation of new procedures, policies, techniques, technology, etc., that have a high probability of improving performance of the organization.

Strategic planning process

Strategic planning has traditionally been an activity or function executed by top management. The typical planning cycle and process in most businesses tends to be very lock-step and bureaucratic in character. We simply go through the motions and fill in the necessary forms. It is, most frequently, a very individualistic process and does not effectively tap "group wisdom." It is rarely a group-oriented process, and the various perceptions and knowledge of the medium to long-term picture for the organization are seldom acquired from a large number of organizational participants. Planning has become what top management tells middle management to get the line personnel to do.

I contrast the characteristics of strategic planning in "Type A" organizations (typical American organizational structures and processes—see Ouchi, "For further reading") with those characteristics needed for effective strategic planning processes in successful organizations in the '80s and '90s. Table 1 presents this comparison, and you will note significant differences between the characteristics.

The thesis is that in order for organizations to compete in an increasingly complex, dynamic and competitive technological, social and marketing environment, they will have to sponsor, promote, develop and create proactivity, innovation and change at all levels of the organization. Capital investment in new technologies (hardware, software, and methods) is critical to success. However, we are rapidly learning that you cannot just throw money (technology) at a problem or opportunity and expect it to be remedied or capitalized upon.

We manage *socio*-technical systems (empasis on the former). Technological innovation must be driven by the people in the system. The people in the system tell us where it hurts, and management responds by evaluating the symptoms, causes and potential solutions. Proactive participative problem-solving, goal-setting, decision-making drives technological innovation. It is so deceptively simple.

The Japanese understand this intuitively. American managers listen, nod, but do not hear. We know but do not do. Our actions (theory X behavior) speak so loudly that our subordinates (managers, supervisors, employees) can't hear what we're saying (theory Y philosophies).

Employees today (managers, supervisors, employees, staff) are better informed, have a greater level of knowledge, and most importantly, have strong expectations, if not needs, to participate in problem-solving, planning and decision-making. The problems, plans and decisions we are confronted with are more complex, and the data necessary to deal with them are more dispersed.

The need for acceptance of solutions, plans and decisions is more critical in terms of achieving effective implementation today than it was yesterday. All this leads up to the need to tap more data in more effective ways as we develop strategic plans.

In reality, personnel at all levels of the organization have strategic (long range), tactical (medium range) and operational (shorter range) views of the organizational systems within which they work. Figure 2 depicts this. There are, of course, primary, secondary and tertiary views and perspectives at each level based upon position and focus. However, the point is that all three views exist at all levels and in particular, the tactical and strategic views should be tapped at lower levels in the organization than is now the case.

In order for organizations to be successful in the '80s and '90s (effective and innovative, productive and, in the longer term, profitable) they must develop and implement strategic planning processes that are more flexible, adaptive, effective and, in particular, more participative in character. The trends we see and hear being described in *Megatrends,* by Naisbitt; *Managing in Turbulent Times,* by Drucker; and *The Third Wave,* by Toffler, all support this observation. Case studies that have been written regarding strategic planning process development in excellent companies also support this prescription (see Fombrun and also Barrett, "For further reading"). We

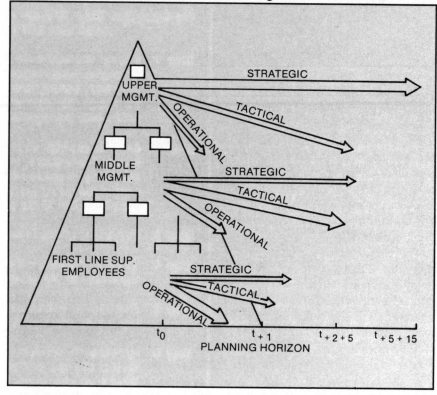

Figure 2: Strategic, Tactical and Operational Views from Perspectives of Various Organizational Levels

simply have to find ways to involve more of the organization in various elements of strategic planning. And we have to find ways to ensure the linkage of strategic planning to action planning and effective implementation.

Eight-step planning process

The eight-step process described here incorporates developments by management consultant Maurice Mascarenhas and work done in the area of structured group processes. The process is a structured, group oriented, participative strategic planning approach. Its goal is to achieve consensus among a group of managers with respect to strategic goals and objectives, specific action programs and implementation steps.

The assumption underlying the process is that effective movement toward goals and objectives evolves from positions of consensus and the ensuing commitment to follow through. This assumption has been tested in a variety of organizations, and with the right amount of leadership intervention at appropriate times we have found the assumption to be very valid.

The eight-step process can be directed toward any number of desired focuses. For instance, the strategic planning process could be directed toward:
☐ Overall business planning.
☐ Productivity management program planning.
☐ Departmental planning.
☐ Quality management program planning.
☐ Cost reduction program planning.

Table 2 presents the basic eight-step process as it might be applied to the development of a plan for a productivity management program.

Note that the first three steps are an attempt to get the persons involved to critically consider factors (both external and internal) that will

or should influence the strategic plan. We simply have the participants in the process silently respond to steps one and two and then, in round-robin fashion, solicit their responses one at a time. The result is a list of factors and planning premises posted on flip chart paper around the walls of the meeting room.

The process of discussing these factors and premises creates a group gestalt for trends, issues, considerations, other viewpoints and perceptions, etc. We have found this process to be an uncomfortable and difficult one for most managers, albeit a very critical and important one. No one has doubted the importance of these first three steps; participants simply discover the difficulty of thinking about these issues.

Step four utilizes the nominal group technique (NGT) (see Delbecq, Van de Ven, and Gustafson, 1975, and Sink, 1983) to generate a prioritized consensus list of strategic goals and objectives for the program or organizational system that is being focused upon. The NGT is a structured group process that is highly effective at shaping group consen-

sus with respect to specific important focal questions. It has been incorporated into the strategic planning process because it is a very effective and efficient mechanism for obtaining group consensus on long-range goals and shorter range action plans for productivity programs. The facilitator for the strategic planning process must be skilled at executing the NGT.

The simplest way to think about step four is that we are attempting to develop a clear perception of where the program or the organizational system is headed during the next planning horizon (often two to five years).

Step five focuses upon developing a list of performance objectives or measures for the program or the organizational system. The focus is on obtaining from the group members their perceptions of how success should be evaluated. How will we know we have succeeded or are succeeding in accomplishing our goals and objectives?

Step six develops consensus and priority action programs that should be budgeted for and worked on in the next year. In other words, what has to occur in the next year in order for us to move successfully toward our priority goals and objectives?

Step seven then provides the opportunity for a team (two to three persons) from the planning group (six to 15 persons) to develop program planning in terms of specific implementation steps and resources to be allocated.

Finally, step eight is executed after the planning session itself and is the evaluation process. Have we carried through with the action program implementation? Are we moving towards our goals and objectives? What deviations/adaptations have occurred?

Executing the process

In order to execute this process, one needs to develop a sort of "plan for the plan." First, and most obviously, the need to develop a two-to-five-year plan for a productivity program must be clear. Second, there needs to be a strong hunch that a participatively developed plan has benefits that outweigh the costs.

I believe the primary benefits you should be considering are: (1) improved quality of the strategic plan itself; (2) the fact that involvement often leads to commitment, which often leads to more effective implementation; (3) that involvement, particularly of the form prescribed by this structured process, is an excellent data collection and auditing device for the organization; (4) improved clarity with respect to expectations on the part of upper management; (5) the team building that usually takes place as a result of this process.

I believe the primary costs are: (1) time (the process takes one and one-half to two days in a retreat type setting); (2) money (a typical session in a retreat type setting with outside facilitators, factoring in lodging, meals, etc., will run between $5,000 and $10,000, depending upon where

Table 2: A Practical Eight-Step Productivity Management Program Planning Process

Pre-Condition/Step: Strategic and tactical planning awareness for the corporation and affected groups.

Step 1: Internal strategic audit (looking within). What internal factors (strengths, weaknesses, conditions, trends, persons, programs, assumptions, etc.) should be considered during the design, development and potential implementation of our productivity management program?

Step 2: External strategic audit (looking around). What external factors (competitors, strengths, weaknesses, trends, conditions, organizations, assumptions, etc.) should be considered during the design, development and potential implementation of our productivity management program?

Step 3: Planning premises, assumptions. Importance—certainty grid.

Step 4: Strategic Planning—2-5 year goals and objectives (desired outcomes) for the productivity program.

Step 5: Prioritization and consensus of performance objectives in key results areas relative to the productivity program.

Step 6: Identification, priorization and consensus for/of strategic, tactical and operational action programs.
What programs, plans, resources, etc., will have to be budgeted for in both the short run (1 year) and longer run (2-5 years) in order for this program to succeed? Simultaneous operational focus (1 year) with a strategic (innovation, growth, continued development and support) focus (2-5 years).
Note: some programs will be doable in one year, others will take 2-5 years or more. Therefore, some sub-programs are to be budgeted for in year 1 but are building for longer term/payoff programs.

Step 7: Program planning and resource allocation.

Step 8: Program evaluation, review, maintenance.

the session is held. An in-house, low budget session, with internal facilitators, could be executed for less than a thousand dollars. Note that neither alternative considers the cost of these managers' time.); (3) preparation costs (it often takes two to four months' lead time, or longer, and two to five days' actual preparation time to pull one of these sessions off. A great deal of attention to detail is required for a session to succeed.); and (4) risk (whenever one involves management and staff in a process such as this, there are various kinds of risks. External facilitators are wise investments, at least initially, because they can anticipate and deal effectively with contingencies that

arise which could compromise the success of the planning effort.)

The third element of the "plan for the plan" is actually part of the preparation for the session itself. You will need to determine how much involvement you will have: who will be involved, how many will be in each session, etc.

The size of the organization for which the productivity plan is being developed will be a major factor in determining the answers to these questions. In small firms of say 300 to 600 persons, there will probably be one top-management planning retreat, one middle-management planning retreat, and perhaps a supervisor planning retreat.

Table 3: Typical Retreat Agenda

DAY 1 (Evening)	Reception
	Dinner
	Introduction
DAY 2 (AM)	STRATEGIC PLANNING
8:00-9:00	Step 1
9:00-9:30	Step 2
	BREAK
9:50-11:00	Step 3
11:00-11:45	Open Discussion
	Briefings on productivity program, productivity basics, etc.
	LUNCH
DAY 2 (PM)	STRATEGIC PLANNING
1:00-2:30	Step 4
	BREAK
3:00	Discussion
	Break for Recreation
	EVENING OPEN
DAY 3 (AM)	MEASUREMENT AND EVALUATION PLANNING, AND ACTION PROGRAM PLANNING
8:30	Step 5
10:00	BREAK
10:30	Step 6
12:00	LUNCH
1:00	Review Step 6
	Step 7
2:30	BREAK
3:00	ACTION TEAM BRIEFINGS
4:00	RECREATION
7:00	Closing Reception
	Dinner
	Wrap-Up, Debrief

If the organization is unionized, the union executive committee should be involved at all levels and in all of the sessions.

The planning retreat sessions should have between 10 and 24 participants. These are ideal sizes and can be violated with some modification to the process itself. If you have 15 or more participants, you will require two facilitators, as in steps 4 and 6 you will need to split the group in half and run two separate NGT sessions.

Participants for planning retreats should be strategically selected. They should be critical decision makers in the organization. They should represent key user groups of the eventual productivity program. They should represent areas of anticipated resistance in the firm. Developing the list of attendees for the planning sessions is a critical activity that merits serious attention.

There are numerous logistics details associated with these sessions. Most will become obvious as you think through setting up the sessions. Little things can have big impact on the ultimate success of the effort. Culture, expectations, informal rules and even gamesmanship play an important part in these planning efforts.

Because each organization is different, it is impossible to prescribe specifics relative to some of these details. However, consider the role that socialization plays in setting the stage for accomplishing the goals of the session.

For example, the evening before the session we usually have a reception, dinner, and then a very brief introductory session. The first day of the retreat we begin at 8:00 AM and break at 3:00 PM so that some recreational activities (golf, tennis, swimming, sailing, fishing, etc.) can be enjoyed.

Planning these types of events into the retreat can "break the ice" and lead to higher overall effectiveness. However, keep in mind that many of these details are culture dependent.

A typical agenda for a strategic planning process is outlined in Table 3.

As we move this process to lower levels in the organization we often shorten the session by skipping steps 1-3. We often simply brief lower levels of management and staff on the results of these steps from other sessions and provide ample time for discussion and expression of other viewpoints. The session can often be executed in one day if this alteration is made.

At the supervisor level, we entertain only steps 6 and 7. This allows us to involve supervisors in the process, but on a reduced basis. Steps 6 and 7 can be executed in a half day.

Results of planning process

The goal of this process is to provide a solid foundation for the design, development and implementation of a productivity management program. If the process is executed properly, this foundation will have been laid in a number of ways.

First, you will have generated an outstanding data base from which to actually formulate the formal plan. Second, you will have generated some ownership of the plan and the program itself on the part of key managers in the organization.

Third, you will have identified key resistance points that will need to be anticipated and planned for as you begin to implement the program. Fourth, you will have a clear sense of direction for the program, and there will be fewer hidden agendas as you proceed to execute the program.

The strategic planning process does not, however, represent a final product. I view these retreats simply as data collection, consensus shaping and commitment building activities. The formal plan will have to be developed once the retreats are completed.

Keep the formal plan simple, concise, clear and non-bureaucratic. The final plan should be extensively presented, reviewed and discussed. It should outline how the productivity program will proceed and focus much less on what it will do.

I prefer a productivity program that coordinates and facilitates change and innovation to a program that attempts to push or drive productivity improvement. I feel the final plan should reflect this orientation. The strategic planning process I have outlined for you will help to accomplish this.

Conclusion

A structured, participative, potentially top-down executed strategic planning process for the design, development and implementation of a productivity management program has been outlined and discussed. Far too many strategic plans are developed in a vacuum.

Changing behaviors of individuals, groups and organizations—innovation—is at the heart of all productivity improvement. The process outlined for you in this article represents an approach to shaping proactive and positive change and innovation in your organization.

Is your productivity program going to stifle innovation and change or promote it?

For further reading:

American Productivity Management Association (an association of productivity managers from U.S. corporations. Periodic meetings are held at locations around the country. Focus is on sharing experiences, ideas, etc. APMA: Skokie, IL.

American Productivity Center Case Study Series. APC: Houston, 1982 to present.

Barrett, E.M. "OST at Texas Instruments" Harvard Business School Case Study, HBS Case Services, HBS: Boston, 1981.

Criner, E.A. *Successful Cost Reduction Programs for Engineers and Managers.* Van Nostrand Reinhold Co.: New York, 1984.

Delbecq, A.L., Van de Ven, A.H. and Gustaf-son, D.H. *Group Techniques for Program Planning: A Guide to Nominal Group and Delphi Processes,* Scott, Foresman and Co.: Glenview, IL, 1975.

Drucker, P.F. *Managing in Turbulent Times,* Harper and Row, New York, 1980.

Fombrun, C. "Conversation with Reginald H. Jones and Frank Doyle," *Organizational Dynamics,* Winter 1982.

Kanter, R.M. "Dilemmas of Managing Participation," *Organizational Dynamics,* Summer 1982.

Kanter, R.M. "Shaping Corporate Change: Key to 1990's—Innovation and Risk-Taking" (an interview published by the American Productivity Center), *Productivity Brief,* No. 35, April 1984. Houston.

Naisbitt, John. *Megatrends,* Warner Books Inc., New York, 1982.

Kanter, R.M. *The Changemasters: Innovation for Productivity in the American Corporation.* Simon and Schuster: New York, 1983.

Morris, W.T. *Implementation Strategies for Industrial Engineers,* Grid Corp. Columbus 1979.

National Productivity Network (NPN) (an association/network of regional productivity centers in the U.S. Centers exist, primarily at major universities, in Maryland, Georgia, Florida, Pennsylvania, Virginia, New York, Illinois, Oklahoma, Arkansas, Texas, Utah, Oregon, California and Arizona. The network meets semi-annually. Managers from organizations are welcome.)

National Productivity Review: New York, NY (A leading productivity journal; each issue usually contains an article on productivity programs. Author highly recommends this journal.).

Ouchi, W.G. *Theory Z: How American Business Can Meet the Japanese Challenge.* Addison-Wesley Publishing Co.: Reading, MA, 1981.

Peters, T.J. and Waterman, R.H. *In Search of Excellence: Lessons from American's Best-Run Companies.* Harper and Row Publishers: New York, 1982.

Productivity Improvement: Cast Studies of Proven Practice. Vernon M. Buehler and Y. Krishna Shetty (Eds.) AMACOM: New York, 1981.

Sink D.S. "The ABC's of Theories X, Y, and Z," *1982 Fall Annual Institute of Industrial Engineers Conference Proceedings,* IIE: Norcross, GA, November 1982.

Sink, D.S. "Building A Program for Productivity Management: A Strategy for IEs," *Industrial Engineering,* Oct. 1982, p. 42.

Sink, D.S. *Productivity Management: Planning, Measurement and Evaluation, Control and Improvement.* John Wiley & Sons: New York, 1985 (to be released in early spring).

Sink, D.S. "Using the Nominal Group Technique Effectively" *National Productivity Review,* Spring 1983.

Sink, D.S. and Mize, J.H. "The Role of Planning and Its Linkage to Action in Productivity Management" *1981 Spring Annual Institute of Industrial Engineers Conference Proceedings,* IIE: Norcross, GA, May 1981.

Sink, D.S., Tuttle, T.C., and DeVries, S.J. "Productivity Measurement and Evaluation: What is Available?" *National Productivity Review,* Summer 1984.

The Innovative Organization: Productivity Programs in Action. Robert Zager and Michael P. Rosow (Eds.) Pergamon Press: New York 1982.

Toffler, A. *The Third Wave,* Bantom Books Inc., New York, New York 1981.

Working Smarter. (by the Editors of *Fortune*). The Viking Press: New York, 1982. Ⅲ

D. Scott Sink, P.E., is an associate professor in the department of industrial engineering and operations research at Virginia Polytechnic Institute and State University and director of the Virginia Productivity Center at VPI. He received his BSIE, MSIE and PhD degrees from Ohio State University. Sink is a senior member of IIE and currently director of the management division. He is a recipient of the Haliburton Award of Excellence, the Dow Outstanding Young Faculty Award and the Oklahoma Society of Professional Engineers Outstanding Young Engineer of the Year award. He is a registered professional engineer in the state of Oklahoma. His book titled, *Productivity Management: Planning, Measurement Evaluation, Control and Improvement,* will be released by John Wiley and Sons early this year. He is actively involved in productivity management program planning, development and execution with a number of major U.S. firms as well as with many smaller firms and plants, and these experiences form much of the basis for this article.

*Reprinted from **Industrial Engineering**, December 1984.*

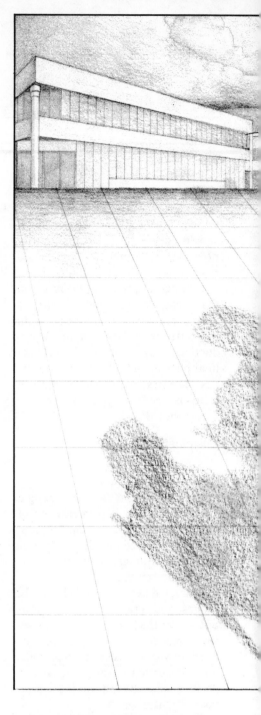

Worker Participation In Office Space Planning And Design Process Pays Large Dividends

**By Ed Frazelle
and Clarence L. Smith, Jr.**
North Carolina State University

The evolution of work duties and responsibilities in today's office has closely paralleled that in the manufacturing sector. Just as we divided the work to transform raw materials into finished products through some manufacturing process, we have specialized office job positions and created complex organization hierarchies to create, process, store and retrieve the product of the office of today—information.

The autocratic, dictatorial management style seen in most offices during the past 80 years enabled information to be passed from one level to another in some fashion. Structured reporting responsibilities looked good on paper but, as we have all experienced, simply did not ensure good communications. Usually, more structure meant more meetings and/or memoranda with little, if any, benefit to the quality of our office communications.

Needs have changed

Since the exchange of knowledge and information is the primary activity in most offices and is a function of the people and the tasks they perform, top management must be aware of and respond to the changing needs of today's office work force.

Alternative work styles, cafeteria benefit packages, office automation, an enhanced worker environment and participative management practices have all made significant contributions to improving our office productivity and quality of information. It has been shown time and again that increased office worker satisfaction and improved productivity and quality are complementary elements.

The traditional information resources for the average office worker have also changed in ways that have placed greater emphasis on sound office space planning and design. No longer do office workers rely solely on pens, paper, typewriters, telephones and mail. The desktop computer, printers, monitors, fireproof files, twisting and tilting terminal trays, and high-tech furniture with ergonomically designed chairs and footstools are the new resources which make the office of today look like the office of the future.

In this area of office space planning and design (OSPD), worker input can pay tremendous dividends for workers and offices alike.

People planning

Allowing worker involvement in planning and designing office space simply makes good sense. Who is in a better position to discuss and design any office work environment than the persons performing the particular jobs? However, past history tells us that these persons are often neglected in the planning process. The autocratic, dictatorial management styles of the past did not allow for it.

Consequently, it does not come as a surprise that the National Bureau of Standards (NBS) recently noted the lack of appropriate planning as the primary reason for failures in the

Illustration by Bill Adams

automated office. The study indicated that problems which often arise in the visual, acoustic and thermal environments could be significantly reduced if more effective planning addressed these issues. Worker participation in the planning process is critical.

Productivity growth

A 1979 survey by Lou Harris/ Steelcase confirmed that the design of office space and facilities to a large extent governs office worker productivity.

In the study, a number of workers were asked to identify the environmental factors which were most important in their job performance. The top five they listed were:

1.) Ability to concentrate without noise and other distractions.
2.) Heat, air conditioning and ventilation.
3.) Access to needed tools, equipment and materials.
4.) Conversational privacy.
5.) Lighting.

These factors can, and should, be addressed as part of the OSPD function. They can either hinder or enhance productivity.

Office facility planning should be an ongoing activity; i.e., a process. More often than not, however, office facility planning is reactive—a procedure enacted in the face of a need that has unexpectedly arisen and caught management's attention. Typically, this is a need for more space, more efficient space, better looking space or a better location. The key point is that an ongoing, participative office facility planning process would transform those unanticipated needs into anticipated and planned-for occurrences.

Just like manufacturing facility planning, office facility planning should be a systematic process which encourages employee participation and promotes innovation. The recommended process is described below.

1.) Identify the objectives of the overall office operation. The strategies and objectives developed for any office activity are the basis for the tasks required and resources needed to accomplish the overall mission of the organization. Determining the objectives of overall office operations defines specific structure and standards.

2.) Identify the specific office functions required to accomplish the office operation's objectives. Typical office functions include data processing, accounting, personnel, etc.

3.) Define the space requirements for each office function. Space standards can and should be utilized in this step; however, only for initial estimates. There are frustratingly many exceptions.

For example, consider a purchasing agent who needs extra space for the vendors who are in and out of his or her office, or the space planner who requires extra desk space to work with bulky layout drawings. Also consider the non-smoker who refuses to share an office with a smoker.

The only way to identify these

exceptions is to receive input from the affected employees, by either interview or questionnaire. The summation of these space requirements, including provision for anticipated short-term growth, represents the total office facility space requirement.

4.) Define the interrelationships among office functions. The participation of the entire workforce is crucial here. Communications surveys, management interviews and relationship charts are all effective means of compiling the interrelationship data.

Many organizations use all of these data collection tools. They then permit the decision process to turn into a battle of political clout between upper level managers. This negligence renders worthless the valuable employee input which cost so much to obtain. Strive for participative, not political, decision making.

5.) Determine the office facility location. Typically at issue is the urban versus suburban location decision. An urban location offers creative and vibrant work surroundings. On the other hand, a suburban location, typically an office park complex, offers the pleasantries of a controlled landscape. Other issues include traffic problems, access to public transportation, and parking space. The most seriously affected party, the workforce, should be consulted on each of these issues.

The result will be an answer to the question, "What would be the loss in workforce and impact on the remaining workforce for each office facility location alternative?" The answer should be weighed along with such considerations as rental and building costs, corporate image, and distance from suppliers and customers, among others, before determining the office facility location.

6.) Generate alternative office space plans. Alternatives should be considered at each level of the office space planning hierarchy depicted in Figure 1. Alternative office locations include suburban and urban sites. Alternative structural concepts include the spine concept advocated by Frazelle (see "For further reading"), the core concept and the conventional concept. Alternative layout designs include open, closed, a combination of the two, and landscape.

An example alternative office plan would be a spine structure with open offices in a suburban location.

OSPD activity can occur in any or all sections of the hierarchy. OSPD activity in the office location and structural design sections is infrequent and usually represents a major effort. In contrast, most office space planning activity involves layout redesign.

For major layout redesign efforts, nominal group and/or brainstorming sessions involving several office space planners are helpful. Computer-aided layout techniques should also be utilized, since these techniques offer tremendous benefits to the planning process.

7.) Evaluate alternative office space plans. The evaluation process begins with the identification and weighting/ranking of the important criteria of the alternative plans. Again, input from all affected parties should be utilized.

Once the relevant criteria have been identified and weighted/ranked, each alternative plan should be evaluated based on those criteria. The same parties should be consulted during the evaluation process, so that they can offer foresight from their individual perspectives. Based on the weighting/ranking and evaluation information obtained, a final weighted score or ranking of the alternatives can be derived.

8.) Select an office space plan. Based on the information obtained in step 7, select a plan. Once the selection is made, do not look back. Proceed with confidence. Confidence in the selection will help ensure a successfully implemented plan.

9.) Implement the selected office space plan. Implementation involves supervising the installation of the layout, managing resistance to change and debugging the new office layout. Each of these phases of implementation is challenging enough.

However, if employees have not participated in the office space planning process up to this point, even the best plan may prove impossible to implement. The attitudes toward change normally exhibited by the workforce—such as loss of control, fear of the unknown, and a lack of understanding—will be magnified to the extent that the potential benefits of the office space plan are jeopardized.

10.) Maintain and adapt the office space plan. Office operations are changing at a more rapid rate than ever before. The office space plan should be continually updated to reflect these changes. Since the workforce is the most knowledgeable about, and first to know of, such change mechanisms as organizational redesigns, new office equipment and the assortment of day-to-day operating problems, worker input is invaluable.

11.) Return to step 1. Remember, office space planning is an ongoing interactive process!

A success story

Some organizations have already begun to merge the office space planning process with participative management techniques. A Vancouver, BC, office employee involvement/ space planning effort resulted in reduced video display glare effects, adjustable work stations and instructions on how to use them, enhanced lighting and acoustics and "high touch" fabrics, shapes, colors and textures for an enhanced environ-

Figure 1: Office Facility Planning Hierarchy

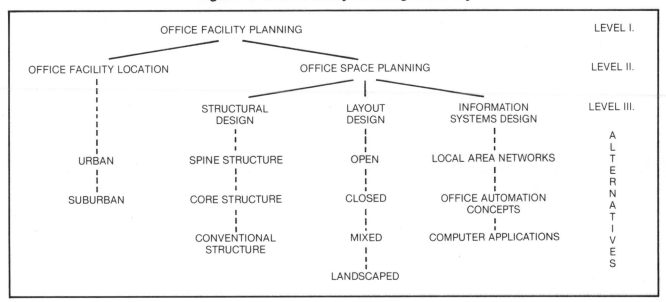

ment. An approximate 15% gain in productivity occurred.

Future considerations

The only certain characteristic describing the future of office operations is uncertainty. Robert Propst, author of *The Office: A Facility Based on Change,* tells us that "a great many of our irritations (in office operations) stem from services and facilities that respond too slowly, or not at all, to our new objectives and values." In that light, mobility must become the primary objective of office facility design.

Another consideration will be the use of the computer as a more and more important resource for office facility planners. In July, 1983, the International Facility Management Association co-sponsored a survey of 252 office facility managers. Forty-one percent of the respondents reported that they currently used computers, 33% planned to do so in the next two years, and another 17% were searching for the appropriate computer system.

These computer systems can be and are being used to help generate alternative layouts, project space requirements, maintain space and furniture inventory, manage space planning projects and evaluate alternative layouts.

However, no matter what the future holds, effective office space planning will continue to be charac-

terized by the integration of a participative management approach and a systematic planning and design process.

For further reading:

Cullinane, Thomas P. and Tompkins, James A., "Facility Layout in the '80's: The Changing Considerations," *Industrial Engineering,* September 1980, p. 35-39.

Frazelle, Ed and Smith, Clarence, "Two Dimensional Office Mobility: Planning for Productivity," *The Office,* to appear December 1984.

Frazelle, Ed and Smith, Clarence, "Soko Circles: The Gateway to Improved Warehouse Productivity," *Handling and Shipping Management,* September 1984.

Frazelle, Ed, "The Spine Concept in Office Space Planning: Will it Work?" *Proceedings of the 1984 Annual International Industrial Engineering Conference,* May 6-10, 1984, pp. 213-217.

Smith, Clarence, "Awareness, Analysis and Improvement Are the Vital Components of Productivity," *Industrial Engineering,* January 1984, pp. 82-86.

Smith, Clarence, "Ergocircles—Extending Quality Circles with Ergonomics," *Proceedings,* 1984 IAQC Conference, Cincinnati, OH, 1984.

IE

Ed Frazelle works as a consultant to industry and government specializing in facilities planning and designing. In addition, he is the managing editor of *IIE Transactions.* A senior member of IIE, he holds office at the region and chapter levels. He is a recipient of the Gilbreth Memorial Fellowship, among other Institute honors. Frazelle is active in several international professional and honor societies. He earned his BSIE and MSIE from North Carolina State University.

Clarence L. Smith, Jr., is undergraduate administrator in the department of industrial engineering at North Carolina State University. He teaches, consults and conducts research in work analysis and design, productivity improvement strategies and participative management, with particular emphasis on office productivity. A senior member of IIE, Smith is a past Raleigh chapter president and currently serves as Region III membership chairman. He is the founding president of the Central Carolina chapter of the International Association of Quality Circles. A member of Tau Beta Pi and Alpha Pi Mu, Smith received his BSIE and MSIE degrees from North Carolina State University.

Reprinted from 1984 Fall Industrial Engineering Conference Proceedings

PARTICIPATIVE MANAGEMENT IN THE UNION SHOP

A. Adnan Aswad

School of Engineering, The University of Michigan-Dearborn
Dearborn, Michigan 48128

Kathryn Stillings
Diablo Systems/Xerox
Fremont, California 94537

ABSTRACT

The literature is reviewed, and through a series of interviews of management and union personnel, the key criteria for the success of participative management in unionized organizations are examined.

INTRODUCTION

American management has responded to the worldwide competitive challenge to increase productivity with concerted efforts to modernize production equipment and management techniques, and it has adopted Participative Management as the main approach for producing quality goods at a minimum cost.

Participative Management (PM) techniques have been introduced under various labels such as "Employee Involvement," "Quality of Work Life," "Workplace Democracy," "Consultive Management," and many others. However, the primary premise in all, as Ouchi [5] states it, is that "workers are the key to productivity." Thus, PM is to be achieved through a sharing of power and authority previously the sole province of management, pushing decision-making down to the lowest practical levels, and a respect for the knowledge and abilities of the workers at all levels of the organization. It is a clear statement of goals and objectives and using all of the available resources to achieve them.

The anticipated outcomes are more creative decisions and more effective implementation through the participation of those impacted. It is a means of developing a more meaningful work climate for the workers, providing "ownership" of the problems and of the solutions.

The participative approach has been successfully applied and documented in many non-unionized companies, notably Hewlett-Packard, Tandem Computers, and Intel. However, approximately 60 percent of American manufacturing labor is represented by some form of a union. The primary question addressed here is whether Participative Management can be effectively implemented in a unionized environment, whether the concept of sharing power with the workers is compatible with the parallel hierarchy of union management.

Through a review of the literature and a series of personal interviews of management and union personnel, the key criteria of **Trust, Honesty, Openness, Mutual Goals, Search for Equitable Solutions, and Sharing of Power** for the success of PM in the union shop are examined.

Union-management relationships have been traditionally adversarial, reinforcing a "we-they" position in negotiations and discussions. Unions have been organized along a rigid centralized pattern, in sharp contrast to the decentralized approach fostered by participation. Unions have traditionally worked for rights and benefits; through participation, workers are being given a sense of partnership in the operation and welfare of the business.

The question is, then, can these differences be resolved? Is the union a superfluous player in the game, a limiter to the success of Participative Management? Are the workers receptive to the changes generated by Participative Management? Can management follow through on the concepts, to really share the power of planning and implementation with the workers?

AN HISTORICAL PERSPECTIVE OF PARTICIPATIVE MANAGEMENT

At the beginning of the industrial revolution in America, master-slave relationships gave rise to the necessity of the union for the protection of the worker. Management and owners behaved in arbitrary ways, and there was a clear need to develop curbs for the excesses in areas of worker rights, expectations, working conditions and wages. The union formation movement was legislatively supported through the Wagner Act of 1935, which provided recognition to union rights and powers. The Taft-Hartley Act of 1947 provided strengthened union legal status through powers of enforcement of union rights.

Standards of fairness evolved systematically as limitations on management prerogatives. Collective bargaining provided a system of government, formalizing the nature of relationships and structuring the rights and responsibilities of both parties. Thus unions developed for worker protection, to equalize the relationship between management and workers, and to provide for a system of orderly operation. A full review of the historical perspective may be found in Cole [1], Robinson [6], and Selznick [7].

The union-management relationship, however, evoked differing perceptions from the parties involved. One manager stated that: "the union gets all the rights, management gets all the responsibilities." As management viewed union behavior as that of irresponsible, greedy children, and the union saw the management as arbitrary and unreasonable, it is no suprise that a strong adversarial relationship evolved.

Such perceptions gave rise to a progression of management theories describing how people behave, what they want and how they should best be managed. Simmons and Mare [9] review the literature of Theories "X" and "Y" by D. McGregor, Theory Z by W. Ouchi [5], and F. Taylor's concept of Scientific Management and its antidote advocating a human relations approach exemplified by the Hawthorne experiments in 1924. In the 1930's, Joseph Scanlon developed the concept of the "Scanlon Plan," a means of sharing the gains of productivity with the workers, rewarding suggestions, committee input and group performance. Variations of gains-sharing plans have surfaced since then and to many experts form a key element in the concept of participative management. The contention is, it is not enough to welcome the input of the workers, but they must also realize a share in the output of their efforts.

As resistance to Taylorism produced a revolutionary increase in union power, Management Rights clauses became a key contractual component in the 1940's. These clauses re-affirmed management's exclusive right to manage and limited the union's authority to participate in management prerogatives. The unions' demands for "more" instilled a management fear of losing control of the business.

Since the turn of the century, labor relations underwent all of these changes, each of which was perhaps right for the times and the prevailing economic climate. In the mid-1970's, however, there was a realization that the old ways were not working, that America was losing ground competitively. The world market was expanding, and America was not maintaining what it viewed as its fair share and appropriate role. And the concept of "Participative Management" emerged, a combination of theories of previous eras as well as heuristic contributions from its world-wide competitors.

PARTICIPATIVE MANAGEMENT: THEORY AND PRACTICE

Participative management requires fundamental changes in labor-management relationships, a shift from the autocratic approach engendered under Taylorism, a reduction of the suspicion and mistrust, to develop a democracy. All of these terms of change are a massive undertaking under the best of circumstances, but the question being addressed here is, with the union as a player in the game, can that level of change be realized? Can the formal, structured adversarial relationships be modified to work together with management and labor on the same team?

According to Ouchi [5] and Simmons and Mare [9], as the concept of Participative Management was introduced to managers and labor leaders, a number of perceptions evolved on the negative side. It would be instructive to examine these conceptual roadblocks to implementation of a successful participatory approach:

a) Management Myths

A number of myths emerged on the management side, representing a reluctance to share power with labor, such as:

1) Participation is no more than "being nice" to workers.

2) Meetings lead to less time to work.

3) Workers do not want to think about their jobs.

4) Workers only care about money and doing as little as possible under the best conditions.

5) Efficiency and democracy are incompatible.

6) Group decision making is too time-consuming.

7) Workers cannot handle important decisions.

8) Americans are too individualistic and self-serving to work in groups.

Within these comments is a clear reflection of the influence of Theory X and Taylorism, the management practices that have been the primary training and practice for years.

b) Union Concerns

Similarly, the union has voiced concerns about the move to Participative Management, including:

1) It is merely window dressing for speed-up, a union-busting tactic, a way to keep the "natives" distracted working on trivial things while management pulls shenanigans and automates them out of a job.

2) Participative Management is just another means of management control of the workers.

After long years of behaving in an adversarial way, neither the management nor the union was open to free acceptance of the concept of Participative Management. The union was suspicious of the motivations and intents, and management found difficulty accepting that workers actually would want the role or that it was appropriate for efficency.

It would be instructive at this point to examine two programs of Participative Management in terms of the key criteria of **Trust, Honesty, Openness, Mutual Goals, Search for Equitable Solutions, and Sharing of Power;** their perceived effectiveness; and the level of support and commitment by the impacted parties.

A large automotive company introduced the program of "Employee Involvement" (EI) during the mid-1970's. According to a survey conducted in June and July 1982 at seven company locations, each of which had a project underway at least 12 months:

"The results of the survey confirm what local union and company leaders have known for some time: Employees believe EI is having a beneficial effect on their jobs and the work environment, and the EI process is working well where there is a high level of local management and local union commitment. . . . The survey revealed major changes in employee opinions now versus before EI in all categories, including problem-solving, supervisory practices, job satisfaction, product quality, and cost consciousness. . . . The survey also revealed that EI is functioning well as a process. Very positive employee attitudes are evident in the areas of management support for problem-solving efforts, effectiveness of EI groups and the overall level of union and management commitment to the process. Significantly, 91 percent of the EI participants and 80 percent of the non-participants believe the Union was 'right' in getting involved in EI."

There was substantial evidence of change, such as a wide awareness to improve, with people talking openly about their problems. More disputes were being resolved through discussion, without resorting to the formal grievance procedure for remedy. For example, one plant stated that they generated 1300 grievances for standards and violation of work rules in 1981, but only 120 in 1982. Both management and union attributed much of this reduction to the success of the EI efforts.

In contrast to the positive tone about the EI program at the above automotive company, discussions with workers, union leaders, and managers at a small unionized electronics company in California yielded a picture of a very negative program and detrimental results.

While management instituted a Quality Circle program and a Workmanship problem-solving program, there was almost universal disappointment with the results of the efforts.

The union official called it "a paternalistic quick-fix for their own poor management abilities," saying that they could not seem to follow through on any program and kept starting new things all the time. The union was not consulted in the planning, but was 'given' participation. She claimed a lack of respect for contractual obligations and an unwillingness to listen and respond to workers. She viewed the Labor-Management meetings held monthly as another forum for management manipulation. As an example, the union official cited the recent meeting where a video tape was shown to union leaders that illustrated a Japanese printer company operating with only three direct labor workers in evidence

and the balance of work being performed by robots. The ensuing commentary by management appeared to be thinly veiled threats to the union to limit upcoming contract negotiation expectations.

As a further point, the union representative asked rhetorically why workers should increase their contributions and participate in doing management's job, particularly when the company had focussed its efforts on reducing the skill level required and forcing grade and pay reductions and implementing arbitrary labor cutbacks. She viewed the management as greedy, demanding, and dishonest, with the union role necessary to protect the workers in the daily confrontation and contract violations.

In the same company, the Manager in Charge of Operations expressed a disappointment in the participative efforts, saying that it was not working as expected. He identified the primary problems as lack of continuity; frequent management personnel changes, especially the drivers of the participative program; but particularly a lack of priority. As difficult economic and competitive circumstances forced changes, "management of things got the attention, with people programs set aside--the program got stalled out due to expediency."

His perception was that the company invoked a lot of changes for participation, but the quality level was about the same, there were disappointing productivity results, the union evidenced increasing contractual rigidity and hostility, the low level of trust was unaltered, and management was in the position of imposing more restrictive controls due to worker behavior such as absenteeism.

He candidly admitted that management had a habit of giving out half-truths, which encouraged the union/worker suspicions. All of the planning was generated on a top-down basis, with a resultant problem of communicating downward to the shop floor. The biggest problem he saw, the greatest resistance, was getting all the salary workers on board to support participation.

Despite the admitted lack of success with an approach that had been in effect philosophically for about 18 months, he remained optimistically confident that participation will become a normal behavior pattern in the time frame of five to seven years. He observed that, for the most part, workers and the union remained unsold on any benefits achievable through participation, but he still contended that it was the direction the company must head.

In discussions with the manager who designed and implemented the participative approach for the company, there was a statement that despite the lofty objectives, the company's plan degenerated. Problem-solving groups were a misnomer; they served primarily as a one-way information channel for management instruction of the workers. Issues dealt with were immediate quality problems and process changes, with solutions given by support staff. As he phrased it, "what has really come out of the effort is a better way to support the line, more product and assembly support . . . with people uncomfortable with the participation approach, they went back to the traditional behaviors once the drivers left."

The irony is that though representatives of all levels of management and labor denied that the participative approach was effective for their organization, when asked, managers insisted that they were a participatory company.

To summarize, the large automotive company, in contrast to the electronics company, is well on its way toward achieving its goals using PM by having implemented:

1) Increased Communications to share knowledge, identify success patterns, highlight problems, and promote trust.

2) Joint Councils of Management and Union Leadership in order to formulate mutual goals in an open and honest manner and to invite involvement in proper decision making at the lowest practicable level.

3) Negotiated Contract Language to define without ambiguity the expectations and goals of both parties.

4) Revised Work Rules which are more flexible and adaptable to changing technology and are more conducive to generating equitable solutions for increasing productivity.

5) Training Programs and Incentives to create a better educated and skilled workforce both in the management and union ranks and thus improve the quality of their decisions.

However, both of the above companies still have one major unsolved problem ahead of them, namely, **The Sharing of Power**, which we will address next.

SHARING OF POWER

To accept Participative Management implies that responsibility and commensurate authority are spread throughout the organization. Unless mechanisms for the sharing of power and authority are instituted in the organization, the inherent risks for persons at each level of the management and union hierarchies will inhibit participation in decision making.

Specifically, the sharing of power with the workers and the union runs counter to trends of previous management practices, which strived to limit and control the authority of the unions. It requires a major transformation in the nature of the issues and the contracts to be negotiated.

On the other hand, in developing an approach to sharing the power, we should keep in mind that the union has a self-interest in maintaining its own viability. Many of the changes resultant from the participatory approach can be perceived as threats to union interest and can evoke a resistance movement. As one union leader indicated, if participation ignores the union and provides what was traditionally provided by the union, workers will start to question the necessity of belonging. It was his contention, though, that the union still has the same basic functions, to mediate for the workers with the management. He saw no perceptible difference of the union roles now versus previously, with the exception perhaps that the union was expanding into new areas. Thus, in terms of sharing of power, it must be remembered that the union in its fight for survival has an interest in limiting what powers should be shared and in restricting the changes as not to impinge on rights previously obtained through union efforts. Moreover, the union wants to ensure that the workers remain aware of the necessity of the union structure to protect their interests.

One UAW leader at Solidarity House stressed the importance of separating contractual issues from participative issues. It was his contention that the function of the union is to represent the workers and to provide mechanisms such as collective bargaining and grievance channels for disputes. And the issues subject to participative discussion must be clearly segregated from the negotiated issues. He felt that the two processes must be and were remaining separate and do not affect each other.

Clearly, the union role is being redefined through the introduction of Participative Management. One suggested model [9] is that the union serve as a structured conduit for communication and information flow and as an aid to cooperative efforts with management. This requires a move from the role of protector of workers against management to that of a contributor aligned with management in goals and means. Several managers noted that there has been some change in tenor in the collective bargaining process toward the above-mentioned model. This might perhaps explain the growing tendency for workers to reject union-negotiated settlements as not being in their interest and to "overthrow" their present leadership. Such trends will not cease until a cooperative sharing of power between management and union is realized.

CONCLUSIONS AND COMMENTS

Significant strides have been made in the implementation of Participative Management in some union shop companies. Mechanisms for nurturing **Trust, Honesty, Openness, Mutual Goals, and Search for Equitable Solutions** have been successfully instituted in the large automotive company we studied. Similar and appropriate mechanisms can readily be emulated by other organizations wishing to adopt PM techniques. However, in the union shop the proper **Sharing of Power** between management and union remains a major problem needing a solution. There must be a clear definition of the power to be shared. Roles of workers, union and management must be sharpened, with clear delineation of expectations. It has been said that too much democracy leads to anarchy--with no person or persons in charge of the store. A possible approach might be development of a joint management and union hierarchy for decision-making, with the understanding by all levels of their responsibilities and authorities.

A caution must be noted, that Participative Management could conceivably become an end in itself, rather than the means to manage and operate effectively. It is the results that are important rather than the mechanisms of management.

Introduction of change and the evolution of new management concepts are time-consuming processes; this study and evaluation was done at a midpoint in the process, with concepts and approaches being developed and revised as a result of corporate experiences.

One senior manager made the comment that "given two companies at similar points in their life cycles, both with typical top-down management, it will take at least twice as long for a union shop to implement a participative approach than a non-union shop." The union has maintained a centralized organization which operates counter to the participative move for decentralization, and the collective bargaining process presents an artificial barrier to change, maintaining the "we-they" relationship as a limiter to mutual cooperation.

It would be instructive to revisit the application of Participative Management in about five years, giving the process time to mature and settle into its level of acceptance within the corporate structure. This would provide the opportunity to test the theoretical bases of the management approach, as well as follow up the results presented. It would also provide further clarification as to whether the union can fit into the process, whether the apparent conflicts of management-union-worker participation are operating or resolved.

ACKNOWLEDGEMENT

The authors gratefully acknowledge the contributions of all those who were interviewed in the course of this research in the large and small companies and in the UAW. Their names shall remain anonymous until such a day when the fruits of their labor ripen and Participative Management becomes a full reality.

BIBLIOGRAPHY

[1] Cole, David L., The Quest for Industrial Peace, Greenwood Press, Westport, Connecticut, 1963.

[2] Cole, Robert, Work, Mobility and Participation: A Comparative Study of American and Japanese Industry, University of California Press, Berkeley, 1979.

[3] Cole, Robert, "Learning from the Japanese: Prospects and Pitfalls," AIIE Proceedings, 1981.

[4] Drucker, Peter F., Concept of the Corporation, The John Day Company, New York, 1972.

[5] Ouchi, William, Theory Z, Addison-Wesley Publishing Company, Reading, Massachusetts, 1981.

[6] Robinson, James W., Introduction to Labor, Prentice-Hall, New Jersey, 1975.

[7] Selznick, Phillip, Law, Society, and Industrial Justice, Transaction Books, New Brunswick, New Jersey, 1969.

[8] Serrin, William, The Company and The Union, Alfred A. Knopf, New York, 1973.

[9] Simmons, John and William Mares, Working Together, Alfred A. Knopf, New York, 1983.

BIOGRAPHICAL SKETCHES

Dr. Adnan Aswad is a Professor of Industrial and Systems Engineering and Associate Dean of the School of Engineering at the University of Michigan-Dearborn, where he has taught since 1965. He received his B.S. in Mechanical Engineering in 1955 from Robert College. He later earned M.S.M.E., M.S.I.E., and Ph.D. degrees, all from the University of Michigan-Ann Arbor. During the period 1955 to 1963, Mr. Aswad worked as a Design engineer overseas and as a Stress Analysis engineer at the International Harvester Company. He is a senior member of the Institute of Industrial Engineers and past president of the Detroit Chapter (IIE) and is a member of the American Society for Engineering Education (ASEE), The Institute of Management Science (TIMS), and the Society of Manufacturing Engineers (SME).

Ms. Kathryn Stillings is presently Lead Industrial Engineer at Diablo/Xerox in Fremont, California. Previously, she served as Lead Industrial Engineer at Ford Motor Company in Dearborn, Michigan. She holds a B.A. in Industrial Sociology from the University of Washington in Seattle and a B.S. in Industrial Engineering from the University of Michigan-Dearborn. She is a Senior Member of IIE and a member of the Society of Value Engineering, SAVE.

*Reprinted from 1984 Fall Industrial
Engineering Conference Proceedings*

LABOR AND MANAGEMENT'S COOPERATIVE EFFORTS
TO INCREASE PRODUCTIVITY

Clifford Sellie, Chairman/CEO

STANDARDS, INTERNATIONAL INC.

7511 N. Waukegan Road

Chicago (Niles), Illinois 60648

Do Labor-Management Cooperative Efforts really increase productivity? In all cases? Of course not. Just why do some programs succeed and others fail?

From the manager's viewpoint, what advantages can they gain? And what can they lose? Different types of managers have different needs and interests. These need to be identified.

Likewise, there are many different facets from Labor's viewpoint. That viewpoint also needs to be taken into consideration.

There are a variety of programs and of steps that can be taken to improve the chances of success. The results are well worth the effort, to everyone.

CAN COOPERATIVE EFFORTS INCREASE PRODUCTIVITY?

There have been dramatic accomplishments from cooperative efforts by Labor and Management --- and dramatic flops.

Successful cooperative programs have provided a climate where: products are excellent and keep improving every year; employees and management mutually respect each other; productivity and effectiveness are high; customers and suppliers are enthusiastic supporters and sources of profitable ideas, enjoy doing business with their contacts at the company; the company does not get fat and complacent during prosperity, does not go bankrupt during hard times.

In contrast, there are many professed cooperative groups that miss the mark by a country mile. In some cases, the programs are a joke, with lip service to the concepts and autocracy the practice. In other cases, management is weak-kneed and spineless, dominated by employees --- individuals, groups or unions. Instead of mutual respect, employees and management doubt the others are capable of "walking and chewing gum at the same time."

What are some of the reasons for the successes? What are some of the reasons for the failures?

COMMON SENSE PROGRAMS: DOING WHAT COMES NATURALLY

Programs of cooperation are just plain common sense where there are common mutual objectives. This is true in all human activities --- industrial, academic, sports, government, military; you name it. But it is important to identify common objectives when establishing cooperative programs.

Common Goals and Conflicting Goals

First of all, there must be no conflict of goals between the employees and management in the areas selected for cooperation. For example, the employees cannot be delegated the task of determining their own wage. They can deal with questions regarding how to maximize the pie, but not how the pie is to be divided. In this area, the division of the pie, there is a conflict of interest between employees and the company.

As far as the size of the pie is concerned I believe that there is no conflict and that maximizing the size of the pie is in everyone's long term interest.

The conflict as to wages can be diminished some by using clear Job Evaluation programs. In this case, the employee can determine the skills necessary to advance up the pay scale. That still leaves the conflict once a year, or every 2 or 3 or 5 years, as to what wages should be for the different pay scales.

Likewise, the employees cannot be delegated the task of determining their own time standards without great potential for conflict. However, if the company uses methods-based Standard Data, the conflict can be diminished some. The employee can be shown how the methods determine the time. This is far less emotional than personal judgments, "rubber yardsticks," as to the correct time standard for a job.

I believe communication, information, and facts are as important in areas of conflict as of common interest. You want understanding in both cases. It is just that you don't expect or get the same cooperation and support where there is a conflict of goals. You can lessen the hostility, but not necessarily the opposition, by good facts and good communication. I remember one time in particular where we had tough negotiations with a Machinist's Union before we were able to get agreement on a methods-based Standard Data program. The Chief Steward, after it was over, surprised me greatly. Instead of swearing at me, as he had been doing, he used his strongest vocabulary to tell me: "I'm blankety-blank glad you're fixing up those blankety-blank standards. I'm blankety-blank tired of spending half my blankety-blank day sitting in that blankety-blank washroom."

We did not get cooperation while reviewing an idea and starting the ball rolling, due to conflicting goals. It did verify the possibility of getting acceptance, despite conflicting goals, where there is factual data to support an idea.

Partial Reduction of Conflict in Areas of Conflict

Henry Seroka, well-known I.E. for the Allied Industrial Workers', AFL-CIO, and I prepared an outline labeled "Basics of Union-Management Success" for a Productivity Seminar. At times, we have found this outline helpful in solving problems in areas of conflict. (Copies available on request.) We have also found Job Evaluation and Standard Data helpful in the two most common areas of conflict, namely Wages and Time Standards.

More Areas of Mutual Interest

There are more areas of common objectives than of conflict between labor and management. However, it does require (1) common sense and (2) management knowledge to identify these areas correctly. The good manager will be practical and factual. He should be able to focus the plans and procedures and structure for cooperation on the areas of mutual interest.

At the same time, there should be mutual recognition that there are areas of conflicting interest. There is no benefit to anyone, on an on-going basis, from confusing issues.

Common sense! We all know the gag: It's called common sense, because it's so uncommon. How about management knowledge? Good management knowledge is an acquired skill. Managers are made, not born. Some have more latent managerial talent or interest. Some have more conducive environment. But all tend to benefit from study and practice of good management techniques. And it is human nature to indulge in some bad management practices. Accordingly, it is essential that anyone in a managerial position study the basics of cooperation, of successful cooperative programs and of failed cooperative programs. Theory is wonderful where it is supported by facts. Idealistic theory, without concern for reality, is as bad as cynical theory.

We have found where common sense and good leadership knowledge exist that cooperative labor-management efforts to increase productivity are outstanding, year after year.

Which Program Should You Use?

There are many categories and names for these cooperative programs, such as:

- Labor Management Teams
- Worker Councils
- Employee Involvement
- Quality Circles
- Quality of Work Life
- Work Simplification
- Job Enrichment
- Suggestion Systems
- Productivity Teams.

They all work. Which program should you use and what name should you call it? I don't care about the name you use. I do care about the type(s) of program you select. They should be based on your company's objectives.

Company Needs and Objectives
What Are They?

Some of the common reasons given by current advocates for installing a cooperative program are:

1. To improve quality of work life.
2. To share responsibility and power.
3. To develop enthusiasm, loyalty and pride.
4. To improve productivity and maintain competitiveness.
5. To meet and/or beat competition and provide job security for all company employees.

How well do any or all of these reasons meet the typical company's needs? Do any meet your company's needs?

Prime Objectives

Prime objectives in most organizations are to provide products and/or services,

1. that meet customers' needs,
2. on time,
3. at a profit.

Well-run organizations do a superior job of meeting these three prime objectives. When they meet these objectives, there can be many desirable by-products, including: Quality of Work Life, Job Enrichment, Enthusiasm, Loyalty, Pride, etc.

Where other objectives are placed ahead of these three prime objectives, the net results may be zero or negative. Why? Let's look at what may happen when other objectives come first.

Other Objectives

1. How about using Quality of Work Life as a prime objective? People are important, aren't they? Isn't it true that "a happy team is a productive team"? Not always. A productive team is usually a happy team. But a happy team may merely be happy.

Where employees have achieved full Quality of Work Life --- where management, supervision, staff and union all are fully satisfied --- shouldn't that company be approaching Nirvana, perfect bliss? Maybe. Maybe, instead, they are approaching bankruptcy. Their costs may be sinking them. Or their customers may be leaving them.

2. What about Sharing Responsibility and Power as an objective? Let's take the "heat off" management, give everyone a piece of the action. Sounds appealing.

Sharing responsibility, more correctly stated as "placing responsibility with the person doing the work" is a well-known principle of Quality Assurance and Industrial Engineering. There is a Quality Assurance maxim: "The person doing the work builds in the quality, not the inspector." There is an Industrial Engineering maxim: "The person doing the work is the person who knows most about the job." Sharing responsibility has tremendous advantages, when done in accordance with good Quality Assurance and good Industrial Engineering principles.

How about sharing power? Too often this becomes an excuse for management's abdication of its responsibilities. If it is true that the person doing the job has the greatest knowledge of the job, that applies to management personnel also.

Executives and supervisors must take responsibility and must have the power and courage to do their jobs.

It is common in autocratic firms for management's functions to be too broad and to include the details of work done by others. Under Labor-Management Cooperative Programs, management may be tempted to abdicate the responsibility and power for work that only they can do and should do. The dividing line is broad and not easy to recognize. But it must be identified to prevent chaos.

3. How about the objectives of developing Enthusiasm, Loyalty and Pride? Where these are placed as the major objectives, the company may forget about the customers. They may forget about costs; they may forget about staying in business.

4. How about Labor-Management Cooperation to Improve Productivity and Maintain Competitiveness? Or,

5. How about Meeting and Beating Competition to Provide Job Security?

"The winning team is the team that is jumping with joy." Where the company is meeting and beating competition, we have found more real, not faked, Enthusiasm, Loyalty and Pride.

These last two objectives are the most compatible with the three prime objectives.

WILL COOPERATIVE PROGRAMS REALLY WORK? YES! WHY?

Meets Basic Human Needs

Labor participation meets basic human needs. We all have group and individual desires to be involved, to be needed, to contribute, to be important in what we do.

Follows Good Management Practice

Labor-Management cooperation meets the advice to management - historically and currently; from Sloan to Drucker:

- Delegate,
- Decentralize,
- "Decision making on-site,"
  where the work is done.

Knowledge of the work to be done usually fits the following pattern:

1. greatest by the worker,
2. next by the immediate supervisor,
3. then by interested staff or manager.

WHO WILL BENEFIT?
AND WHO WON'T?

Poor Managers

Labor-Management Cooperation is not recommended for weak company Managers. Weak
Managers may see it as a threat. They
will refuse to use Labor-Management Cooperation even where formally installed or
authoritatively dictated. If I don't turn
the key, the car won't run.

Weak Managers may see it as an excuse.
They will not manage. They will abdicate.
Bill Gomberg, former AFL-CIO staff headquarters Industrial Engineer, made a very
pertinent comment. "He (Paul Hoffman,
President, the Studebaker Corporation)
went in for union-management cooperation
in a tremendous way. In fact, the union
was running the plant until it ran it
right into the ground . . . . Not because
the union was running it. It ran into the
ground because management had vacated its
function. And the union's running it was
a symptom of management's abandonment of
its function."

The Studebaker Corporation was an early
warning signal to American industry in
general, and to our automobile manufacturers in particular, of the dangers of
soft, complacent management, remote from
the operations. Fortunately --- not from
our country's viewpoint --- but fortunately from the viewpoint of foreign competition --- that disaster signal was generally ignored.

Good Managers

Labor-Management Cooperation is recommended for good Managers. A strong
Manager who is self-confident will not see
it as a threat. "NIH~, is a common
organization illness; more fully identified as the "Not Invented Here" sickness.
"NIH" is not a slogan of a good Manager.
He will benefit from the added information
and understanding and will continue to
take final responsibility and authority.

Weak Union Leaders

Labor-Management Cooperation is not recommended for weak Union leaders. Weak Union
leaders may see it as a threat. They may
be afraid for their control; may lack
self-confidence; be afraid of working with
company management. They may be afraid
that management will take advantage of
their cooperation and may see their function as being the continual responsibility to stir up trouble. They may be
indoctrinated into the philosophy of Union
management hostility.

Uninformed Labor Union leaders may mis-
understand the importance to the Union of
company success. They may have forgotten
the axiom: "The Company can exist without
the Union, but the Union cannot exist
without the Company." They may not understand the desires of their own people, the
need of their people to feel a part of and
proud of their company.

Uninformed Labor Union leaders may be
short-sighted, focused on temporary gains
instead of long-term gains. They may
focus only on getting a bigger share of
the pie rather than having a bigger pie to
split up, even though the latter may mean
more income and better security for their
Union members.

Good Union Leaders

Participative Management is recommended
for strong Labor leaders. A good Labor
leader who is self-confident will recognize the benefits, short-run and long-run,
to the employee from participating. He
will understand both the psychic and the
job security benefits to the employees
from being part of a strong competitive
firm. An informed Labor leader will not
see Participative Management as an ego
threat. He will be glad to have all members of the company team --- employees,
supervisors, staff and management ---
working together, instead of fighting.

A good Labor leader will continue to take
responsibility to represent the employees
against unfair or negative action, at the
same time supporting good positive action,
regardless of the source. "NIH" is not a
slogan of a good Labor leader.

He will help develop employees as a strong
positive group, helping the company and
themselves; instead of trying to develop
an antagonistic negative group, hurting
the company and themselves.

WHAT ARE THE SECRETS FOR SUCCESS?

Communication

Communication is vital in a cooperative
program. The concept of informative communication in this context could be defined as an explanation of the conditions
and actions which affect and enrich cooperation between the individual employee
and the company as an organization. Information must be distributed in all directions within the company. Any employee
should be free to ask a question of anyone else on any subject. It is not
certain that he or she will receive a
reply which the employee considers to be
complete, but the employee has the right
to receive a motivated answer and to be
referred, if necessary, to someone who can
give a more detailed reply.

It is not possible for us to inform everyone in writing about everything, as the cost would be excessive and the information would not in any case be read by everyone. It is more important that our employees have the desire to be informed and the ability to acquire information, and that their initiatives are under no circumstances rejected by the person to whom the question is addressed.

- -

An excellent proof of the need for all communication to be clear is available in the following item found in "Queries, Quandaries and Quips":

DEAR MR. HARRIS: Some time in the 1950's, I believe Sen. Pepper of Florida was defeated for re-election largely because of a leaflet that was circulated throughout the state which turned voters against him. Do you know the approximate contents of that leaflet? - J.A.R. Akron, Ohio . . .

It went as follows: "Are you aware that Claude Pepper is known all over Washington as a shameless extrovert? He is also reliably reported to practice nepotism with his sister-in-law and he has a sister who was once a thespian in Greenwich Village. He has a brother who is a practicing homo sapiens, and he went to a college where the men and women openly matriculated together. It is an established fact that Mr. Pepper before his marriage practiced celibacy. Worse than that, he has admitted to being a lifelong autodidact."

Expressions of Interest

These expressions of labor interest and management interest can be made to individuals or groups. Both direct communication to individuals and group communication are recommended.

Expressions of Appreciation

Prompt personal thanks plus bulletin boards, awards, house bulletins, etc., help keep the programs alive and well.

Do Not Ignore Any Source of Cooperation

Excellent cooperative resources are:

1. Employees; hourly workers, supervisors, staff, management.
2. Suppliers; they are excellent resources for production design and cost control.
3. Customers; they are excellent resources for market research and product ideas.

Properly approached and truly recognized, all three are bountiful suppliers of

successful idea.

Only the fearful use organization structure as a shield -- to avoid giving or receiving information or to dodge facts.

Follow Through

Suggestions with potential need to be highlighted. They need (a) logging, (b) investigation, (c) reporting, (d) implementation, (e) reporting, (f) follow-up, (g) reporting. This reporting, at the different stages of completion, should be both to the suggester and to management.

Rejections of suggestions also need reporting, tactfully and with encouragement.

WHAT ARE THE STEPS TO SUCCESS?

Training & Retraining in Basics

Understanding the basics of the following subjects is recommended for everyone interested in improving themselves, whether in a cooperative or an autocratic organization: Quality Circle concepts; Job Instruction techniques; Work Sampling principles; Disciplining techniques; Scheduling techniques; Listening concepts; Motion Economy principles; Work Simplification principles; Performance Measurement techniques; Performance Evaluation principles; Production Costs, where directly affected; Industry Economics; Calculation of costs, for costs they directly affect; Calculation of Profit Margins, where they have profit margin responsibility; Incentive Rules, where appropriate; Union Rules, where appropriate; Company Policies.

Ideas, suggestions, implementation -- the tools of cooperation --- are more effective when based on factual knowledge and good technical skills. We have found training employees, supervisors, staff, and management in these basics well worth the effort --- for immediate results and for long term results.

We have found it useful when explaining policies and rules, to explain the purpose as well as the rule. Otherwise, we may be merely encouraging the confusion encountered by one Chicago "Scofflaw." (Scofflaw is a Chicago term for a driver who accumulates many parking tickets, without paying for them.) This particular driver was finally arrested and brought before the judge. When the judge showed him all the parking tickets he had accumulated for parking in the street in front of his office, this fellow vehemently protested his innocence. He indignantly exclaimed: "I don't deserve this! I should not have gotten any tickets for parking there!" The judge asked, "Didn't you see the sign?" The driver, even more indignantly than before, said, "Yes,! The sign said: Fine for parking. So, I parked there."

Recognition, Recognition, Recognition

Cooperative Labor-Management Programs are human needs programs. Planning, listening, understanding and recognition are essential for success.

There is an added benefit from recognition. It is a key indicator of whether there really is a cooperative program in fact or just in theory. There is an old Norwegian joke about a very pleasant guy who hated to be rude to his neighbors even though they imposed on him. So he started to say, "Yes, I'll do that as soon as I get this sieve filled with water." Occasionally, when he would get an indignant, "You can't fill a sieve with water," he would explain: "I know. But that's as good an excuse as any when I don't want to do something."

In our Clients' programs for cooperation, we provide forms and procedures for identifying and recognizing idea sources and idea implementation. When we find lack of action, we often find many "sieve and water filling" excuses. The real reasons are apt to be insecure managers with large inferiority complexes. You can be 99% positive that's the real reason when a manager's reports and conversations are filled with "I did this," "I thought of that," etc. What a shame! What a beautiful way to handicap one's managerial growth!

Is Financial Pay for Cooperation Helpful or Unhelpful?

Caution is recommended on using financial motivation to gain Labor-Management Cooperation. The long term deleterious results of financial rewards for participation (such as gain-sharing, suggestion sharing, suggestion ownership, etc.) are apt to be counter-productive.

Productivity Incentive Plans May Be Helpful

On the other hand, properly selected incentive programs, aimed at productivity improvement, can be beneficial and helpful. Effective productivity incentive programs can benefit everyone --- company, employee, customer. Ineffective incentive programs can damage everyone --- company, employee, customer.

Despite propaganda to the contrary, an appropriate productivity incentive program is not damaging to Labor Relations nor Labor-Management Cooperation. Some of the best run companies with good Cooperative Labor-Management programs, also have strong incentive programs. "A well-designed and properly applied wage incentive system can offer a sound basis for industrial peace and management labor cooperation." (Mitchell Lokiec, Director, Management Engineering Department, International Ladies Garment Union)

Profit Sharing - Good Ideas for Wrong Reasons

Profit Sharing can help the company; and the employees; and cooperation! It seldom is needed to obtain cooperation. Often is needed as an excuse, by weak Labor leaders, for encouraging employees to do what they are proud to do --- cooperate. It often is needed by Companies with weak Managers, so they don't give away the store when times are prosperous.

WHO SHOULD PARTICIPATE?

EVERYONE: From Janitor to Chairman of the Board.

- -

BIOGRAPHICAL SKETCH

Clifford Sellie is Chairman and Founder of STANDARDS, INTERNATIONAL INC., a Chicago based Management Consulting and Engineering Research firm serving over 1000 Clients world-wide. The firm's Clients include many of the largest conglomerates plus a broad range of progressive small and medium-sized companies. STANDARDS, INTERNATIONAL has continued to serve a number of their Clients on a repeat basis, some for over 30 years.

Mr. Sellie is an internationally recognized authority on Work Management and Work Motivation. He and his firm have actively participated in numerous company and association research programs, particularly on Pre-Determined Time Techniques (MTM, WF, DMT, MTA, etc.) and Incentives (financial and non-financial). Author of many articles on work measurement, work simplification, wage incentives, supervisor bonus programs, union relations, cost reduction and management controls; he is a frequent speaker at Universities, Engineering and Trade Associations.

Former faculty member of the University of Minnesota. He holds a BA, St. Olaf College; MA, University of Minnesota. Member of Institute of Industrial Engineers, American Management Association, Society for Advancement of Management, MTM Association for Standards and Research, Past President, Industrial Management Society.

Registered Professional Engineer, State of Illinois. Listed in Who's Who in Commerce and Industry and World Who's Who in Industry.

Reprinted from 1984 Fall Industrial Engineering Conference Proceedings

LOCOMOTIVE MAINTENANCE PRODUCTIVITY MEASUREMENT SYSTEM

Janyce L. Tomcak
Industrial Engineer
Missouri Pacific Railroad
Union Pacific System
St. Louis, Missouri

ABSTRACT

In order to obtain maximum utilization of the Missouri Pacific Railroad locomotive fleet and the labor force responsible for locomotive maintenance, it was necessary to develop a computer based productivity measurement system. Objectives: (1) To develop a practical solution to a complicated locomotive repair productivity problem through the use of existing computer reportings, minor additions to existing reportings, and equitable and reasonable standard time values for the various locomotive crafts. (2) Publish and distribute results periodically to management, providing a means of evaluating the performance at the locomotive facilities.

INTRODUCTION

The basic concept of the Missouri Pacific Railroad's computer based locomotive maintenance productivity measurement system involves the periodic application of the following procedures:

1. Calculate the number of locomotives placed at or released from the locomotive facilities.
2. Determine the type of locomotive and the type of inspection or repair performed.
3. Utilize the appropriate standard time value to compute the standard man-hour values for each type of work accomplished.
4. Identify the payroll man-hours expended for each type of work reported.
5. Compute and evaluate the performance.
6. Distribute the performance results.

BACKGROUND

Familiarization

There are five primary locomotive maintenance facilities located throughout the Missouri Pacific Railroad system. The primary shops are listed below with the MoPac abbreviations:

1. Ft. Worth (FTWR)
2. Houston (HOUS)
3. Kansas City (KCMO)
4. North Little Rock (NLRK)
5. St. Louis (SLMO)

Besides the switch engines (SENG), there are four categories of locomotives. Using the builder and the number of axles as identification, the locomotive categories are defined below:

1. EMD4
2. GE4
3. EMD6
4. GE6

The following scheduled company inspections and unscheduled repairs can be performed at the primary locomotive facilities and are shown with the appropriate abbreviation:

1. Trip inspection (TRIP)
2. 45-day inspection (MM)
3. Periodic inspection (QM)
4. Semi-annual inspection (SM)
5. Annual inspection (AM)
6. Two-year inspection (TM)
7. Unscheduled maintenance (OT)
8. Accident damage (AD)

The crafts shown below are responsible for the maintenance of the locomotives:

1. Machinist
2. Electrician
3. Sheet metal worker/Pipefitter
4. Blacksmith/Boilermaker
5. Carpenter
6. Laborer

History

Missouri Pacific Railroad's Vice-President of Operations authorized the formation of a Locomotive Repair Task Force to study locomotive maintenance shop operations including, but not limited to:

1. The distribution of shop forces by day and shift
2. The standardization of repair procedures
3. Material requirements
4. The establishment of work performance standards

The original Task Force which was appointed in 1979 consisted of representatives from the Mechanical, Materials, and Industrial Engineering Departments. With a concentration on the operations at the five major shops, the initial efforts of the Task Force were directed toward (1) the development of a comprehensive materials order list of the parts required for each type of locomotive and each type of inspection and (2) the establishment of a simple manual performance system based on estimates of the man-hours required to perform locomotive servicing and inspections.

MEASUREMENT OF SHOP PRODUCTIVITY

Work Performance Standards

The Master Mechanics at the five major shop locations were provided with survey forms to collect actual inspection and changeout times for each craft. The data was validated by Task Force observations. Unscheduled repair times were collected for locomotives placed in the shops for failure or scheduled inspections. It was determined that all standard time values should include a five percent allowance for personal time and a four percent allowance for meals. The trip inspection standard time values include an additional ten percent allowance for unavoidable delays at the service track. The remainder of the standard time values include a five percent allowance for unavoidable delays and a five percent allowance for fatigue. The five percent fatigue allowance is excluded from the trip inspection standard time values as the work pace is not continuous at the service track.

The scheduled maintenance standard time values then became:

((STANDARD INSPECTION TIME) +

(STANDARD REPAIR TIME) +

(STANDARD TRIP INSPECTION TIME)) x ALLOWANCES

The unscheduled repair standard time values were adjusted as follows.

(STANDARD UNSCHEDULED REPAIR TIME) x ALLOWANCES

Figure 1 (See Appendix) is an illustration of the table of standard time values developed for the performance system.

Manual Performance System

A manual performance system was the predecessor of the computer based system. The manual system required that each major shop compile the quantity of work accomplished for each type of maintenance and the number of man-hours expended by each craft on each type of maintenance. A daily shop report of the locomotive maintenance data was sent to the general office in St. Louis where Industrial Engineering compared the standard man-hours to the reported man-hours and obtained a productivity measure.

The standard man-hour values for each type of maintenance were derived from the following

summation of locomotive type standard man-hour values.

$$\sum_{i=1}^{5} (\text{STANDARD TIME VALUE})_i \times (\text{NUMBER OF UNITS})_i$$

The equation below demonstrates the calculation of performance.

$$\text{PERFORMANCE} = \frac{\text{STANDARD MAN-HOURS}}{\text{REPORTED MAN-HOURS}}$$

The advantage of the manual system was that it could be implemented quickly. The supervisory and clerical time required to compile the data and the vulnerability of this type of system to errors or falsification promoted the search for a computer based performance system.

COMPUTER BASED PERFORMANCE SYSTEM

Existing Data

The existing computer programs for generating locomotive inspection work orders were enhanced to produce a data base containing the locomotive inspections and repairs performed at the locomotive maintenance shops. A locomotive unit for which the work is not completed in the same time period is considered one-half of a unit.

A list and sample of the data retrieved from the MoPac Transportation Control System (TCS) reportings are shown below:

1. Location code
2. Type of reporting (ie. placement, release, etc.)
3. Locomotive unit owner and number (ENG for MoPac system locomotives)
4. Locomotive unit assignment location code
5. The two letter designation of the work performed for each locomotive placed at or released from a shop
6. Date (Julian)
7. Time (24 hour clock)

```
13418   PA   ENG   1669   13418   MM    840922350
13418   RA   ENG   1669   13418   MM    840930100

16337   PA   ENG   1671   13418   QIQM  841062130
16337   RA   ENG   1671   13418   QIQM  841071500
```

It was not possible to capture a trip inspection count through the TCS reportings. Manual trip inspection compilations from the shops were required to complete the necessary inspection and repair data base for the performance system. All of the information necessary to determine the standard man-hours for a shop was available from this data base and the application of the predetermined standard time values.

Additional Payroll Data

The Mechanical Department payroll records already contained the craft man-hours expended on different locomotive models at each location. To accommodate the calculation of locomotive maintenance performance at the five major shops by comparison of standard man-hours to actual man-hours, the Task Force developed codes to be recorded on the Mechanical Department OCR daily time card (See Appendix Figure 2) under the "Special" column of the "Function Code" section. The current coding instructions for the special function codes are listed below:

Locomotive Maintenance Special Function Codes

1 - Trip Inspections - all covered locomotive craft employees who work on trip inspections, lubrications, and inspection write-ups on the service track.

2 - 45-day Inspections - all covered locomotive craft employees who inspect or work on scheduled 45-day MoPac maintenance.

3 - Periodic Inspections - all covered locomotive craft employees who inspect or work on scheduled quarterly MoPac maintenance.

4 - Semi-Annual Inspections - all covered locomotive craft employees who inspect or work on scheduled semi-annual Federal maintenance or scheduled semi-annual MoPac maintenance.

5 - Annual Inspections - all covered locomotive craft employees who inspect or work on scheduled annual Federal maintenance or scheduled annual MoPac maintenance.

6 - Two year Inspections - all covered locomotive craft employees who inspect or work on scheduled two year Federal maintenance or scheduled two year MoPac maintenance.

7 - Unscheduled Maintenance - all covered locomotive craft employees who work on unscheduled maintenance.

8 - Accident Damage - all covered locomotive craft employees who work on units shopped for accident damage.

9 - Component Rebuilding - all covered locomotive craft employees who work on locomotive component rebuilding or maintenance and repair in the back shop.

10 - Non-locomotive Maintenance - all covered locomotive craft employees who work in diesel shop support functions, equipment repair and maintenance, buidling maintenance, crane operation, crew hauling, fueling, sanding, and any other support job not directly involved in locomotive maintenance.

The Task Force is in the process of redefining the special function codes in order to support the development of a more intricate performance system.

Refinement of Standards

A high degree of accuracy is preferred in the development of a measurement system and remains a goal of the Task Force in the refinement of the Locomotive Maintenance Productivity Measurement System. In order to accommodate a wider variety of work performed at each shop, a group of major unscheduled maintenance types were identified and the following codes were designed to be incorporated in the TCS reportings for locomotive shoppings. The original unscheduled maintenance codes and the proposed modifications are listed below:

Unscheduled Maintenance Codes

AB - Air Brake
AC - Air Compressor Changeout
AD - Accident Damage
AG - Auxiliary Generator Changeout
AL - Alternator Changeout
BA - Battery Changeout
CF - Cooling Fan Changeout
*CH - Compressor Head Changeout
CP - Coupler Changeout
DB - Dynamic Brakes
DG - Draft Gear Changeout
DT - Double Truck Changeout
EB - Engine Blower Changeout
EC - Engine Changeout
EL - Electrical Work
EP - Engine Protector Changeout
EW - Engine Work
EX - Exhaust System
FD - Fire Damage
FI - Fuel Injector/Nozzle Changeout
FP - Fuel Pump/Filters Changeout
*FS - From Storage
FZ - Freeze Damage
GR - Ground Relay
GT - Gear Train Changeout
GV - Governor Changeout
*IS - Into Storage
LT - Load Test
MB - Main Bearing Changeout
*MD - Modification
MG - Main Generator Changeout
*MT - Miscellaneous Truck
OC - Oil Cooler Changeout
**OT - Other Things
PA - Power Assembly Changeout
PT - Paint Locomotive
*PU - Paint Underframe
RD - Radiator Changeout
RV - Reverser Changeout
SJ - Strip Job
SN - Sanders
SP - Snow Plow Changeout
SR - Speed Recorder Changeout
ST - Single Truck Changeout
TC - Turbocharger Changeout
TR - Traction Motor Changeout
TW - Turn Wheels
*WA - Wash Locomotive
WC - Wheel Changeout
WL - Water Leak
WP - Water Pump Changeout
WS - Wheel Slip

*Proposed addition
**Proposed deletion

ADMINISTRATION OF SHOP PERFORMANCE

Monthly Reports

Summary type and detail type computer generated reports are available for management to evaluate the performance of each shop and the major shops of the system as a group. Listed below are the reports presently produced:

149

Productivity Measurement Reports

1. Locomotive maintenance and performance report - for individual major diesel repair facilities
2. Unscheduled maintenance performance report- for individual major diesel repair facilities
3. Summary of the MoPac system standard hours, actual hours and performance
4. Summary of the MoPac system locomotive maintenance based on locomotive assignments
5. Summary of inspection and repair work assigned to one shop and completed by another
6. MoPac system craft standards - for each locomotive type and type of maintenance
7. Craft hours and costs report - for individual major diesel repair facilities
8. Craft performance report - for individual major diesel repair facilities
9. MoPac system craft performance report - for each type of maintenance and type of locomotive
10. MoPac system craft performance report based on equivalent man-months
11. Performance graphs - for individual major diesel repair facilities and the MoPac system

The Kansas City, April, 1984, locomotive maintenance and performance report is reproduced in Figure 3 (See Appendix). The report includes the total number of locomotives receiving a type of maintenance for each locomotive assignment shop and for each type of locomotive. The standard man-hours, actual payroll hours, and a performance rating are presented for the time period of the report, the previous month, the average for the past three months, and the average for the past six months.

A sample of the Kansas City, April,1984, unscheduled maintenance performance report is illustrated in Figure 4 (See Appendix). The report exhibits the unscheduled maintenance standard time value of each locomotive type and the number of locomotives receiving unscheduled maintenance for each type of locomotive. The individual unscheduled maintenance codes retrieved from the TCS system and the appropriate standard time values appear in this report and are utilized in the calculation of the standard man-hours shown as "OT Standard Hours". Actual payroll hours recorded as special function code seven are represented as "OT Payroll Hours". The last computation is the unscheduled maintenance performance displayed as "OT Performance".

The MoPac system performance graphs and the individual performance graphs demonstrate the service track, diesel shop, and overall performance percentages which are presented in the Kansas City performance graph in Figure 5 (See Appendix).

Distribution of Reports

All of the reports are distributed to the Chief Mechanical Officer, the Superintendent of Motive Power, the System Locomotive Inspector, and the Task Force members. The Mechanical Superintendents receive all of the reports except the graphs of shops located outside their district. The Master Mechanics receive only the reports concerning their shop and the system summaries. A special function code time distribution report which monitors the labor forces is distributed to the Master Mechanics, the System Locomotive Inspector, and the Task Force members.

Advantages of the System

1. Increases management awareness of productivity
2. Provides a basic comparison of work performance and type of work performed at the locomotive shops
3. Provides an informational guide for management through a method of system-wide standardization and comparison
4. Assists in the identification of problem areas
5. Furnishes an analysis of operations
6. Supplies a gauge for planning and operation changes
7. Is more accurate and less time consuming for Mechanical supervisory and clerical employees than the previous manual system

Disadvantages of the System

1. Complex programming to be maintained
2. Standard time values to be developed as maintenance requirements are revised
3. Accuracy of computer input to be controlled
4. Reports to be prepared as more conclusive results for management decisions are required
5. Time constraints to be observed as new reports are implemented

FUTURE DEVELOPMENTS

Reevaluation of the Standard Time Values

With the initiation of the unscheduled maintenance codes, the scheduled maintenance standard time values became antiquated. The Task Force is in the process of updating and reevaluating the scheduled maintenance standard time values. As the performance system progresses into more sophisticated stages, it will be necessary to make a complete update of the standard time values. This should be done on a regular basis as time and personnel permit.

Definitive Locomotive Component Input

MoPac TCS has completed an advanced locomotive component reporting system which the Task Force intends to use as a data base for an accurate component changeout count. The unscheduled maintenance codes defined as component changeout which cannot be verified with the new TCS component system will be credited as repairs for which standard time values have been developed.

Heavy Repair and Component Rebuild Input

The MoPac is centralizing the heavy repair and component rebuilding areas of locomotive maintenance. The Task Force has been requested to study the inclusion of this facet of locomotive maintenance into the productivity measurement system.

Compatibility with Union Pacific Railroad Systems

The recent merger between the Missouri Pacific Corporation and the Union Pacific Corporation has offered the opportunity to advance the system. Compatibility between existing reports and measurement systems is currently under consideration by all interested parties.

Goals

Extensive programming is necessary to maintain and update the computer based productivity measurement system. The ultimate goals of the Task Force encompass the continuing improvement of the measurement system on all requested productivity measurements with the minimum of upkeep.

BIOGRAPHICAL SKETCH

Janyce L. Tomcak received a Bachelor of Science degree in Applied Mathematics from the Georgia Institute of Technology in 1975. Ms. Tomcak is currently an Industrial Engineer in St. Louis, Missouri with the Missouri Pacific Railroad where she has been employed since 1976. Ms. Tomcak has been a member of the Locomotive Repair Task Force since 1981. She is an officer on the governing board of the American Council of Railroad Women, a member of the Southwestern Railway Club, and a member of the National Association of Railway Business Women.

Appreciation is expressed to J. S. Hollerorth, J. R. D'Agostino, M. T. Manninger, G. D. Byrd, and the Mechanical Department.

AVERAGE SERVICING, INSPECTION, AND REPAIR STANDARD TIME VALUES *

MAINTENANCE CRAFT	TRIP HOURS	MM HOURS	QM HOURS	SM HOURS	AM HOURS	TM HOURS	OT HOURS	AD HOURS
MACHINIST	1.19	2.85	8.06	12.72	16.86	25.01	7.61	3.17
ELECTRICIAN	0.38	1.40	6.28	8.64	11.88	14.80	2.88	1.22
SHEET METAL WKR	0.26	1.50	1.23	2.33	3.71	5.56	2.36	1.36
BLKSM-BOILERMKR	0.01	1.05	1.14	1.64	2.12	2.26	1.37	6.97
CARPENTER	0.06	1.06	1.22	1.64	1.99	2.39	1.30	0.78
LABORER	1.69	1.75	1.65	2.43	2.30	2.53	1.16	0.03
TOTAL	3.59	9.61	19.58	29.40	38.86	52.55	16.68	13.54

SWITCH ENGINE SERVICING, INSPECTION, AND REPAIR STANDARD TIME VALUES *

MAINTENANCE CRAFT	TRIP HOURS	MM HOURS	QM HOURS	SM HOURS	AM HOURS	TM HOURS	OT HOURS	AD HOURS
MACHINIST	1.12	1.04	7.60	12.36	16.06	24.16	7.61	3.17
ELECTRICIAN	0.36	0.56	5.92	8.39	11.32	14.29	2.88	1.22
SHEET METAL WKR	0.24	1.30	1.16	2.26	3.53	5.37	2.36	1.36
BLKSM-BOILERMKR	0.01	0.02	1.11	1.62	2.07	2.21	1.37	6.97
CARPENTER	0.06	0.02	1.21	1.62	1.97	2.34	1.30	0.78
LABORER	1.65	1.73	1.62	2.36	2.19	2.45	1.16	0.03
TOTAL	3.44	4.67	18.62	28.62	37.14	50.82	16.68	13.54

EMD 4 AXLE SERVICING, INSPECTION, AND REPAIR STANDARD TIME VALUES *

MAINTENANCE CRAFT	TRIP HOURS	MM HOURS	QM HOURS	SM HOURS	AM HOURS	TM HOURS	OT HOURS	AD HOURS
MACHINIST	1.15	1.15	8.14	12.44	16.54	23.38	7.61	3.17
ELECTRICIAN	0.37	0.64	6.35	8.45	11.66	13.84	2.88	1.22
SHEET METAL WKR	0.25	1.32	1.25	2.28	3.64	5.20	2.36	1.36
BLKSM-BOILERMKR	0.01	0.02	1.14	1.63	2.10	2.18	1.37	6.97
CARPENTER	0.06	0.02	1.23	1.63	2.10	2.30	1.30	0.78
LABORER	1.66	1.74	1.66	2.38	2.26	2.37	1.16	0.03
TOTAL	3.50	4.89	19.77	28.81	38.30	49.27	16.68	13.54

GE 4 AXLE SERVICING, INSPECTION, AND REPAIR STANDARD TIME VALUES *

MAINTENANCE CRAFT	TRIP HOURS	MM HOURS	QM HOURS	SM HOURS	AM HOURS	TM HOURS	OT HOURS	AD HOURS
MACHINIST	1.15	6.75	7.83	13.35	18.62	38.21	7.61	3.17
ELECTRICIAN	0.37	4.90	6.10	9.07	13.13	22.61	2.88	1.22
SHEET METAL WKR	0.25	1.81	1.20	2.45	4.09	8.50	2.36	1.36
BLKSM-BOILERMKR	0.01	1.10	1.13	1.67	2.24	2.92	1.37	6.97
CARPENTER	0.06	1.12	1.22	1.68	2.24	3.12	1.30	0.78
LABORER	1.66	1.81	1.63	2.55	2.54	3.87	1.16	0.03
TOTAL	3.50	17.49	19.11	30.77	42.86	79.23	16.68	13.54

EMD 6 AXLE SERVICING, INSPECTION, AND REPAIR STANDARD TIME VALUES *

MAINTENANCE CRAFT	TRIP HOURS	MM HOURS	QM HOURS	SM HOURS	AM HOURS	TM HOURS	OT HOURS	AD HOURS
MACHINIST	1.33	4.97	8.49	13.60	17.69	23.94	7.61	3.17
ELECTRICIAN	0.42	1.42	6.61	9.30	12.47	14.17	2.88	1.22
SHEET METAL WKR	0.29	1.78	1.30	2.49	3.89	5.33	2.36	1.36
BLKSM-BOILERMKR	0.01	1.09	1.14	1.68	2.17	2.20	1.37	6.97
CARPENTER	0.07	1.12	1.23	1.69	2.17	2.33	1.30	0.78
LABORER	1.78	1.77	1.69	2.60	2.41	2.43	1.16	0.03
TOTAL	3.90	12.15	20.46	31.36	40.80	50.40	16.68	13.54

GE 6 AXLE SERVICING, INSPECTION, AND REPAIR STANDARD TIME VALUES *

MAINTENANCE CRAFT	TRIP HOURS	MM HOURS	QM HOURS	SM HOURS	AM HOURS	TM HOURS	OT HOURS	AD HOURS
MACHINIST	1.19	7.25	8.43	13.88	20.23	43.56	7.61	3.17
ELECTRICIAN	0.42	5.27	6.57	9.43	14.26	25.77	2.88	1.22
SHEET METAL WKR	0.29	1.87	1.29	2.54	4.44	9.69	2.36	1.36
BLKSM-BOILERMKR	0.01	1.11	1.14	1.70	2.34	3.18	1.37	6.97
CARPENTER	0.07	1.13	1.23	1.70	2.34	3.41	1.30	0.78
LABORER	1.78	1.87	1.68	2.65	2.76	4.41	1.16	0.03
TOTAL	3.76	18.50	20.34	31.90	46.37	90.02	16.68	13.54

FIGURE 1.

* (Fictitious Standard Time Values)

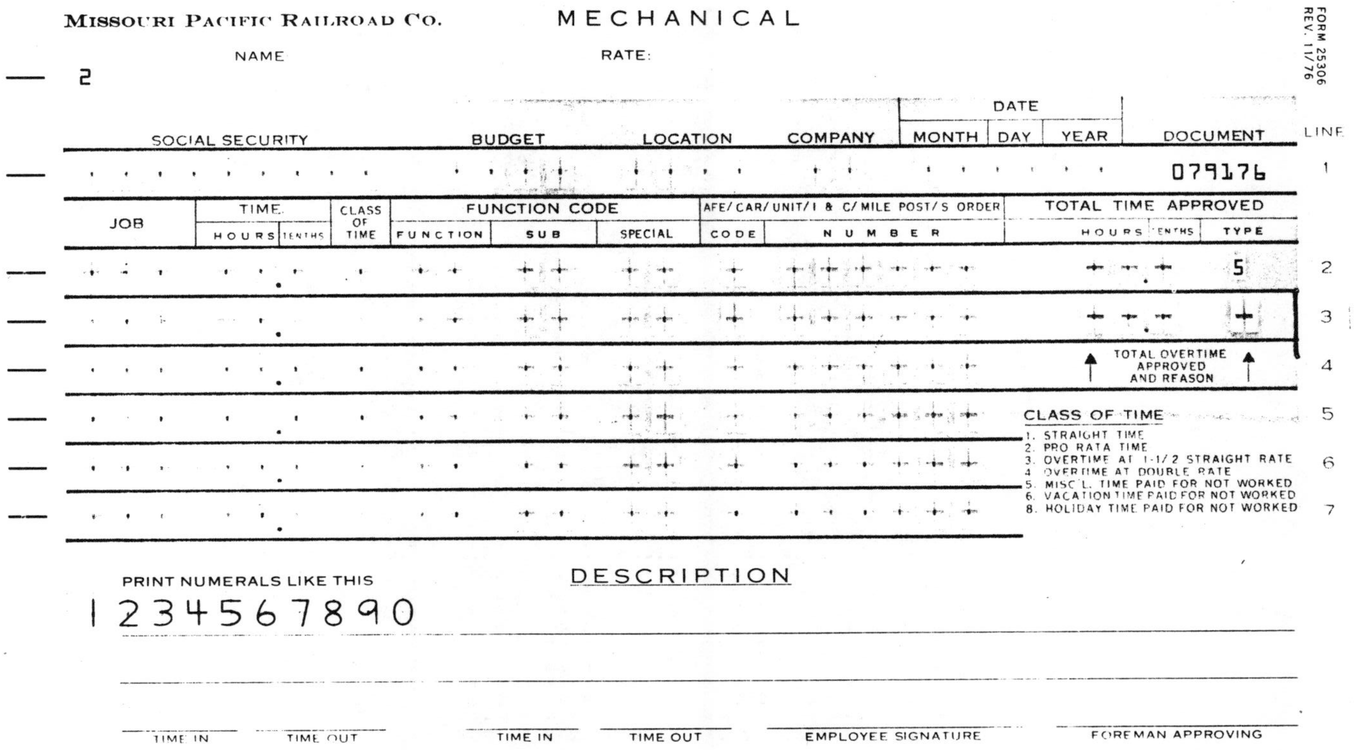

MISSOURI PACIFIC RAILROAD CO.   MECHANICAL

NAME   RATE:

2

SOCIAL SECURITY   BUDGET   LOCATION   COMPANY   MONTH | DAY | YEAR   DOCUMENT

DATE

079176

TOTAL TIME APPROVED

JOB   TIME   CLASS   FUNCTION CODE   AFE/CAR/UNIT/1 & C/MILE POST/S ORDER   TOTAL TIME APPROVED
        HOURS | TENTHS OF   HOURS | TENTHS | TYPE
                        TIME FUNCTION | SUB | SPECIAL | CODE   NUMBER
                                                                                              LINE

1

2   5

3   1

4   TOTAL OVERTIME
                                                                                APPROVED
                                                                                AND REASON

5

6

7

CLASS OF TIME

1. STRAIGHT TIME
2. PRO RATA TIME
3. OVERTIME AT 1 1/2 STRAIGHT RATE
4. OVERTIME AT DOUBLE RATE
5. MISC. TIME PAID FOR NOT WORKED
6. VACATION TIME PAID FOR NOT WORKED
8. HOLIDAY TIME PAID FOR NOT WORKED

PRINT NUMERALS LIKE THIS

1234567890

DESCRIPTION

_____ _____   _____ _____
TIME IN   TIME OUT   EMPLOYEE SIGNATURE

_____ _____   _____ _____
TIME IN   TIME OUT   FOREMAN APPROVING

FIGURE 2.

153

APR 1-30,1984 LOCOMOTIVE MAINTENANCE AND PERFORMANCE REPORT- KCMO

245 ASSIGNED UNITS
194 ROAD UNITS 51 SWITCH UNITS

ASSIGNED TO

	FTWR	HOUS	KCMO	NLRK	SLMO	SYS&FORN	TOTAL
TRIP INSP	0.0	0.0	0.0	0.0	0.0	1911.0	1911.0
45 DAY	11.0	1.0	31.5	14.0	8.5	5.5	71.5
QUARTERLY	4.0	1.0	23.0	7.5	5.5	1.0	42.0
SEMI-ANN	4.0	1.0	19.5	3.0	2.0	0.0	29.5
ANNUAL	0.0	0.0	6.5	1.5	0.0	0.0	8.5
TWO YEAR	0.0	0.0	9.0	1.0	1.0	1.5	11.5
O T MAIN	31.0	2.5	78.0	34.0	14.0	35.0	194.5
A DAMAGE	0.5	0.0	5.0	4.0	0.0	2.0	11.5
TOTAL	50.5	5.5	172.5	64.0	31.0	1956.0	2279.5

LOCOMOTIVE CLASS

	SENG	EMD4	GE4	EMD6	GE6	TOTAL
TRIP INSP	0.0	1274.0	0.0	637.0	0.0	1911.0
45 DAY	12.5	30.5	4.0	23.0	1.5	71.5
QUARTERLY	4.5	25.5	2.5	9.5	0.0	42.0
SEMI-ANN	6.5	15.0	1.0	7.0	0.0	29.5
ANNUAL	1.0	3.0	0.0	4.0	0.0	8.0
TWO YEAR	2.0	8.0	0.0	1.5	0.0	11.5
O T MAIN	24.0	107.5	6.0	57.0	0.0	194.5
A DAMAGE	2.0	4.0	1.0	4.5	0.0	11.5
TOTAL	52.5	1467.5	14.5	743.5	1.5	2279.5

	SERVICING	INSP-OT	TOTALS
STANDARD MAN-HOURS	4917.64	4431.25	9348.84
ACTUAL PAYROLL HOURS	7157.39	6645.10	13802.48
PERFORMANCE RATING	68.71%	66.68%	67.73%

AVG LAST MONTH

	SERVICING	INSP-OT	TOTALS
STANDARD MAN-HOURS	4740.08	4467.61	9207.69
ACTUAL PAYROLL HOURS	7392.48	6738.00	14130.48
PERFORMANCE RATING	64.12%	66.30%	65.16%

3 MONTH MOVING AVG

	SERVICING	INSP-OT	TOTALS
STANDARD MAN-HOURS	4788.97	4459.71	9248.68
ACTUAL PAYROLL HOURS	7289.52	6612.73	13902.25
PERFORMANCE RATING	65.70%	67.44%	66.53%

6 MONTH MOVING AVG

	SERVICING	INSP-OT	TOTALS
STANDARD MAN-HOURS	4887.16	4328.40	9215.56
ACTUAL PAYROLL HOURS	7654.67	6268.30	13922.97
PERFORMANCE RATING	63.85%	69.05%	66.19%

FIGURE 3.

APR 1-30,1984 UNSCHEDULED MAINTENANCE - KCMO *

		SENG	EMD4	GE4	EMD6	GE6		SENG	EMD4	GE4	EMD6	GE6		SENG	EMD4	GE4	EMD6	GE6
STD	AB	4.75	4.75	4.75	4.75	4.75	AC	13.01	13.01	19.01	13.01	19.01	AG	21.01	21.01	21.01	21.01	21.01
UNITS	AB	0.0	4.5	0.5	1.0	0.0	AC	0.0	1.0	0.0	0.0	0.0	AG	0.0	0.0	0.0	0.0	0.0
STD	AL	161.01	161.01	161.01	161.01	161.01	BA	3.01	3.01	3.01	3.01	3.01	CF	13.01	13.01	13.01	13.01	13.01
UNITS	AL	0.0	0.0	0.0	0.0	0.0	BA	0.0	1.0	0.0	0.0	0.0	CF	0.0	0.0	0.0	0.0	0.0
STD	CH	7.14	7.14	7.14	7.14	7.14	CP	1.34	1.34	1.34	1.34	1.34	DB	5.01	5.01	5.01	5.01	5.01
UNITS	CH	0.0	0.0	0.0	0.0	0.0	CP	1.0	1.0	0.0	0.0	0.0	DB	0.0	0.0	0.0	0.0	0.0
STD	DG	4.75	4.75	4.75	4.75	4.75	DT	8.01	8.01	8.01	8.01	8.01	EB	9.01	9.01	9.01	9.01	9.01
UNITS	DG	0.0	0.0	0.0	1.0	0.0	DT	0.0	1.0	0.0	0.0	0.0	EB	0.0	0.0	0.0	0.0	0.0
STD	EC	153.01	153.01	153.01	153.01	153.01	EL	4.07	4.07	4.07	4.07	4.07	EP	2.01	2.01	2.01	2.01	2.01
UNITS	EC	0.0	0.0	0.0	0.0	0.0	EL	8.5	33.0	2.5	17.5	0.0	EP	0.0	0.0	0.0	0.0	0.0
STD	EW	5.39	5.39	5.39	5.39	5.39	EX	17.01	17.01	17.01	17.01	17.01	FD	12.49	12.49	12.49	12.49	12.49
UNITS	EW	13.5	74.0	5.5	43.0	0.0	EX	0.0	2.0	0.0	0.0	0.0	FD	0.0	0.0	0.0	0.0	0.0
STD	FI	2.02	2.02	2.02	2.02	2.02	FP	2.01	2.01	2.01	2.01	2.01	FS	25.01	25.01	25.01	25.01	25.01
UNITS	FI	0.0	12.0	0.5	3.5	0.0	FP	0.0	1.0	0.0	0.0	0.0	FS	0.0	0.0	0.0	0.0	0.0
STD	FZ	12.49	12.49	12.49	12.49	12.49	GR	5.01	5.01	5.01	5.01	5.01	GT	153.01	153.01	153.01	153.01	153.01
UNITS	FZ	0.0	0.0	0.0	0.0	0.0	GR	4.0	1.0	1.0	4.0	0.0	GT	0.0	0.0	0.0	0.0	0.0
STD	GV	3.01	3.01	3.01	3.01	3.01	IS	17.01	17.01	17.01	17.01	17.01	LT	5.64	5.64	5.64	5.64	5.64
UNITS	GV	0.0	0.5	0.0	0.0	0.0	IS	0.0	0.0	0.0	0.0	0.0	LT	0.0	7.5	0.0	6.5	0.0
STD	MB	11.01	13.01	11.01	13.01	19.01	MD	8.00	8.00	8.00	8.00	8.00	MG	107.01	107.01	107.01	161.01	161.01
UNITS	MB	0.0	0.0	0.0	0.0	0.0	MD	0.0	0.0	0.0	0.0	0.0	MG	0.0	0.0	0.0	0.0	0.0
STD	MT	5.90	5.90	5.90	5.90	5.90	OC	6.01	6.01	6.01	6.01	6.01	OT	8.00	8.00	8.00	8.00	8.00
UNITS	MT	0.0	0.0	0.0	0.0	0.0	OC	0.0	0.0	0.0	0.0	0.0	OT	2.0	1.0	0.5	0.5	0.0
STD	PA	9.01	9.01	9.01	9.01	9.01	PT	105.01	105.01	105.01	105.01	105.01	PU	5.01	5.01	5.01	5.01	5.01
UNITS	PA	0.0	7.0	0.0	4.5	0.0	PT	0.0	0.0	0.0	0.5	0.0	PU	0.0	0.0	0.0	0.0	0.0
STD	RD	7.01	7.01	7.01	7.01	7.01	RV	2.01	2.01	2.01	2.01	2.01	SJ	251.01	251.01	251.01	251.01	251.01
UNITS	RD	0.0	1.0	0.0	1.5	0.0	RV	2.0	0.0	0.0	0.0	0.0	SJ	0.0	0.0	0.0	0.0	0.0
STD	SN	1.51	1.51	1.51	1.51	1.51	SP	3.01	3.01	3.01	3.01	3.01	SR	2.01	2.01	2.01	2.01	2.01
UNITS	SN	0.0	0.0	0.0	0.0	0.0	SP	1.0	0.0	0.0	0.5	0.0	SR	0.0	5.0	0.0	1.0	0.0
STD	ST	4.51	4.51	4.51	4.51	4.51	TC	25.93	25.93	25.93	25.93	25.93	TR	5.01	5.01	5.01	5.01	5.01
UNITS	ST	0.0	1.0	0.0	0.0	0.0	TC	0.0	0.0	0.0	0.0	0.0	TR	0.0	2.0	1.0	0.5	0.0
STD	TW	2.80	2.80	2.80	2.80	2.80	WA	9.01	9.01	9.01	9.01	9.01	WC	5.01	5.01	5.01	5.01	5.01
UNITS	TW	1.0	0.0	0.0	0.0	0.0	WA	5.0	19.5	0.0	4.5	0.0	WC	0.0	2.0	0.0	1.0	0.0
STD	WL	5.39	5.39	5.39	5.39	5.39	WP	3.01	3.01	3.01	3.01	3.01	WS	4.07	4.07	4.07	4.07	4.07
UNITS	WL	2.0	9.0	0.0	13.0	0.0	WP	0.0	1.0	0.0	4.0	0.0	WS	0.0	1.0	0.0	0.0	0.0

```
OT STANDARD HOURS   1563.13
OT PAYROLL HOURS    2457.20
OT PERFORMANCE      63.61%
```

FIGURE 4.

* (Fictitious Standard Time Values)

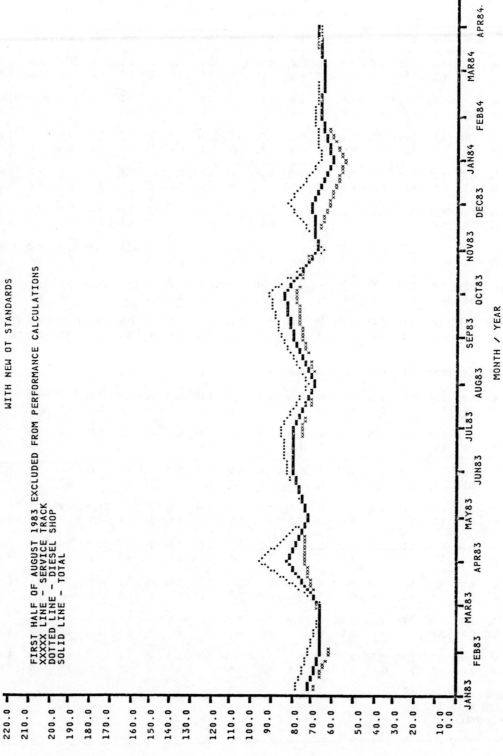

FIGURE 5.

Reprinted from 1984 Annual International Industrial Engineering Conference Proceedings.

COST OF QUALITY AND PRODUCTIVITY IMPROVEMENT

Michael P. Quinn, Assistant Vice President
Egbert F. Bhatty, Assistant Manager

Manufacturers Hanover Trust Company

ABSTRACT

Cost of Quality is a methodology which can be applied by banks and other financial institutions to reduce their total costs while, at the same time, increasing the productivity of the operation. The methodology has been successfully applied by the Management Consulting Services department of Manufacturers Hanover Trust Company in both the domestic and international divisions.

This paper presents the theory behind the Cost of Quality program, and then describes in step-by-step fashion how the concepts of the Cost of Quality program can be applied to any clerical operation.

INTRODUCTION

The banking industry, in recent years, has become increasingly aware of the strategic importance of quality in the delivery of its products.

Heightened competition between banks has led to the use of quality as the key to differentiation of relatively standardized financial products. In addition, consumers' expectations of quality have been raised to the point where they exhibit a low tolerance for mistakes in their financial transactions. And finally, there is the historical requirement that bank transactions be processed quickly and accurately.

These three factors have led to the current emphasis on introduction of quality control as a formal discipline in the banking business.

However, many managers who have formalized the quality control process in their banks are not as yet fully aware of the impact that an effective quality control program can have on productivity as well as on operating costs.

This paper will show how a Cost of Quality program increases productivity, and reduces costs, while, at the same time, improving the quality levels of the operation.

COST OF QUALITY THEORY

Cost of Quality theory recognizes quality as an economic consideration -- it costs money to build quality into a product.

Our experience at Manufacturers Hanover Trust Company (MHT), as well as the reported experience of other financial institutions, shows that costs associated with quality control can range from 15% to 40% of a department's total operating expense.

The magnitude of these costs, and the drain that these costs represent on a bank's profits, necessitate that these costs be identified and controlled.

At MHT we have developed a Cost of Quality program that seeks to (1) identify, measure, and reduce these quality-related costs, while (2) improving quality levels, and (3) increasing the productivity of the operation.

Quality-related costs can be grouped into four categories:

(1) Preventions costs,

(2) Appraisal costs,

(3) Internal Failure costs,

(4) External Failure costs.

Prevention costs are costs incurred to reduce, eliminate, and prevent defects. Such costs typically include the costs of job training, quality report preparation, quality planning and analysis, computer/manual systems design (when geared to preclude defective output), developing and presenting quality-related seminars, and user-testing of quality control systems.

Appraisal costs are costs incurred to detect errors, and to evaluate the quality of the work done. Such costs typically include the costs for proofing, verifying, inspecting, checking, signing, balancing, and recapping.

Failure costs are costs incurred whenever a product -- whether it be a simple deposit transaction, or a complex foreign exchange deal -- fails to meet quality standards. Failure costs can be divided into two groups: internal failure costs, and external failure costs.

Internal Failure costs are costs incurred in correcting the errors (caught at appraisal) before delivery of the product to the customer. In other words, internal failure costs are the costs of re-doing work a second time. Such costs typically include the labor cost of reworking items that were incorrectly processed, the cost of scrapped forms, and the cost of wasted computer and other equipment time.

External Failure costs are costs incurred in correcting errors (not caught by the appraisal process) after delivery of the product to the customer. Such costs typically include the cost of compensation paid, penalties, difference write-offs, interest payment on late deliveries, as well as the cost of investigations.

The costs associated with Prevention, Appraisal, Internal Failure, and External Failure can, as stated earlier, be substantial -- ranging from between 15% to 40% of a department's total operating expense. However, it is the distribution of these costs that is more interesting.

Our own experience, combined with other financial industry data, shows the distribution of quality costs to be approximately as shown in Graphic 1.

Quality Category	Percent of Total Cost of Quality
Prevention costs	5%
Appraisal costs	50%
Internal Failure costs	15%
External Failure costs	30%

Graphic 1: Distribution of quality costs in financial services industry

Clearly, as the figures above show, in the financial services industry, we have all the time to do the work over, but never enough

time to do it right the first time!

These figures also exemplify the typical approach to improving the quality of work -- check the work, and then recheck again!

Does this increase in Appraisal costs -- the expense incurred in checking and rechecking the work -- lead to an increase in quality?

This relationship is explored in Graphic 2.

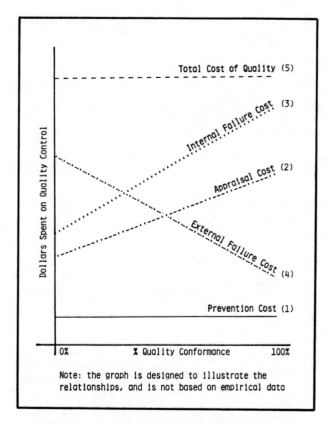

Graphic 2: Effect of increase in Appraisal costs on quality

As more Appraisal work is done, increasing Appraisal costs (2), more defects are caught internally, increasing Internal Failure (i.e. rework) costs (3).

As more defects are caught internally (i.e. before delivery), fewer are caught externally, thus decreasing External Failure costs (4).

While the Total Cost of Quality (5) may be impacted, it often remains nearly unchanged because the decrease in External Failure costs (4) is offset by the increases in

Appraisal costs (2) and Internal Failure costs (3).

Prevention costs (1) remain unaffected.

So the answer to our question -- does an increase in Appraisal costs lead to an increase in quality? -- is "Yes", but without any noticeable reduction in the Total Cost of Quality. All that has happened is a shifting of quality dollars from one area (External Failure) to another (Appraisal).

The Total Cost of Quality, however, is impacted by the Cost of Quality approach, which, in contrast to the typical approach, emphasizes the prevention of errors rather than their correction.

What happens when Prevention costs are increased?

This relationship is explored in Graphic 3.

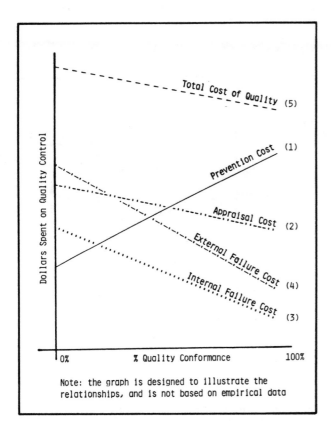

Graphic 3: Effect of increase in Prevention costs on quality

As Prevention efforts increase (1), both Internal failures (3) and External failures (4) decrease.

Appraisal costs (2) may be reduced also -- due to the overall improvement in quality.

The net result is an improvement in quality conformance, and a reduction in the Total Cost of Quality (5).

Lower quality costs translate into an increase in productivity.

This, then, is the essence of the Cost of Quality program: that dollars spent on Prevention produce a much more significant return than dollars spent on Appraisal. In other words, Prevention dollars buy more than Appraisal dollars!

COST OF QUALITY PRACTICE

At MHT we have applied the Cost of Quality methodology to clerical operations in both our domestic and international divisions.

Our results to date have been as follows:

-- 37% reduction in the Total Cost of Quality,

-- 10% reduction in Total Cost of Quality as % of Department Operating Expense,

-- 12% increase in employee Productivity.

Our methodology comprises a 12-step program. The department manager/supervisors will spend 10% to 15% of their time on project activity over the investigation/installation period.

Step 1: Identify the quality problem

The analyst and the department manager together develop a PIF -- a project initiation form -- that defines the quality problem clearly and precisely. The PIF should also state in clear and unambiguous terms the solution required, and how the results will be quantified.

Step 2: Develop and analyze the distribution of quality costs

The distribution of quality costs for the department (Graphic 4) shows how much is being spent on each of the 4 different quality categories. The distribution of quality costs should be prepared with great care, because its analysis can yield clues to the nature of the quality problem.

The distribution of quality costs is prepared as follows:

Step 2a: Draw up the department's Personnel Worksheet

The analyst, in conjunction with the department manager, identifies

(1) the different functions performed in the department -- for example, data input, verification, etc.,

(2) the number of employees who perform each function, and

(3) the mid-point salary per hour for each function.

The data on employees, functions, and salaries is used in conjunction with the Activity Worksheet (see 2b below) in computing the cost of the quality-related activities of the department.

Step 2b: Draw up the department's Activity Worksheet

The analyst, in conjunction with the department manager and supervisors, develops

(1) the department's activity (task/operations) list,

(2) establishes how long it takes to complete each activity,

(3) matches activities (for example, check customer name) with functions (verifier),

(4) identifies quality-related activities (e.g. verify amount), and

(5) assigns each quality-related activity to its appropriate quality category -- Prevention, Appraisal, Internal Failure, External Failure.

Having developed data on the different activities performed in the department, the time taken to perform each activity, how much it costs per hour to perform each activity, and the quality category of each activity, the analyst is now in a position to develop the Cost of Quality Summary Worksheet.

Step 2c: Develop the Cost of Quality Summary Worksheet

The Cost of Quality Summary Worksheet is used to calculate the monthly quality cost for each activity in each quality category. These are then summated into the monthly Total Prevention Cost, Total Appraisal Cost, Total Internal Failure Cost, and Total External Failure Cost.

The analyst should ensure that the monthly quality cost figures reflect not just the salary costs, but should also include the fringe benefits paid by the company.

Once the total costs by quality category (Total Prevention Cost, etc.) are known, the analyst is in a position to prepare the Cost of Quality Report.

Step 2d: Develop the Cost of Quality Report

The total costs for each quality category (Total Prevention Cost, Total Appraisal Cost, etc.) are summated into the Total Cost of Quality. This is the amount that the department spends each month on quality-related activities.

Equipment expenses, Occupancy expenses, and Other Operating expenses are excluded from the calculation of Total Cost of Quality because (1) they constitute a small percent of the Total Cost of Quality, and also because (2) the major portion of these costs cannot be affected in the short run.

The analyst also calculates the total cost of each quality category (Total Prevention Cost, etc.) as a % of the Total Cost of Quality.

This is the distribution of quality costs (Graphic 4), and should be represented in both absolute dollar amounts, as well as percentage terms.

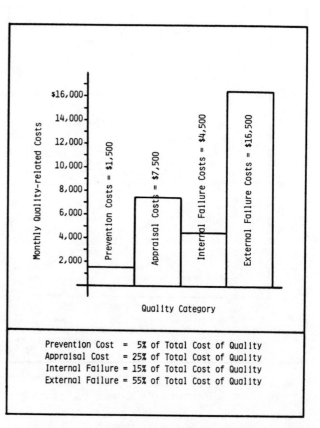

Graphic 4: Distribution of quality costs in a banking department

As can be seen from Graphic 4, the Total Cost of Quality comes to ₡30,000 per month. If Department Operating Expense is assumed at ₡100,000 per month (for purposes of illustration), then Total Cost of Quality as % of Department Operating Expense = 30%. That is, nearly one-third of the monthly operating expense of the department is spent on quality-related activities. And yet, despite this high level of expenditure, there are quality problems in the department!

What is wrong? Where is the quality control system breaking down? It is to answer questions like these that the analyst analyzes the Report and the distribution.

Step 2e: <u>Analyze the Cost of Quality Report and the distribution of quality costs</u>

What does Graphic 4 show? What conclusions can we draw from the distribution of quality costs shown?

For one, it shows that the department is heavily involved in correcting errors <u>after</u> delivery to the customer -- some ₡16,500 being spent monthly (55% of the Total Cost of Quality) on correcting errors pointed out by customers!

Secondly, the department spends ₡7,500 each month on checking and verifying the work (Appraisal cost), catches and corrects some errors (Internal Failure costs), but somehow lets a large number of errors to get through to the customers (External Failure cost). Clearly, the checking and verification function is sloppy!

Thirdly, there is a tremendous emphasis on correcting errors (Internal Failure cost + External Failure cost = ₡21,000 per month), and almost no effort is put into preventing errors (Prevention cost = ₡1,500 per month).

Conclusions that an analyst can draw at this stage

(1) the department needs to shift quality dollars away from the management of failure, and into the prevention of errors, and

(2) the checking, verification, and inspection procedures of the department need to be tightened considerably.

Step 3: <u>Determine the different error types</u>

The analyst, in conjunction with the manager and supervisors, draws up a list of errors being made by the department, and designs a form to record current-week errors. The same form can be used to collect data on errors from the files of immediate past weeks.

The size of the sample required can be determined by using the formula to calculate the sample size given in a statistics textbook.

The objective of this step is

(1) to determine the total number of errors per day,

(2) to determine the different types of errors made, and

(3) to determine the frequency of the different error types.

Next, Pareto Analysis is done to identify those few errors which are responsible for generating the major portion of the Total Cost of Quality.

The analyst should also examine the large list of remaining errors, and pick out any error types which <u>can</u> lead to unexpectedly large External and Internal Failure costs.

The remaining errors can be left for the department manager to handle later.

Step 4: <u>Determine how the errors are generated</u>

Having determined the different error types, the next step is to determine <u>how</u> each error is generated -- that is, to determine whether the error is internally-generated or externally-generated.

Internally-generated errors are (1) errors made by the department itself -- for example, input error, or verifier error, and (2) errors made by other departments within the company which lead to incorrect input by the department where the Cost of Quality study is being conducted -- for example, the cable department can stamp the wrong account number on a cable that has come in for the letters of credit department.

Externally-generated errors are errors made by the customer, or another bank, or another company -- for example, a corporate customer may give the wrong settlement instructions in a foreign exchange deal.

It is important to classify all errors into internally-generated errors and externally-generated errors, because the internally-generated errors are almost entirely controllable. They can be either (1) prevented from arising altogether, or (2) reduced to the barest minimum, thus reducing the Total Cost of Quality.

Externally-generated errors are more difficult to control, but measures have to be designed to effect a reduction in these also.

Step 5: <u>Determine which function(s) are responsible for/contribute to internally-generated errors</u>

This step fixes the responsibility for the error. It seeks to determine which function within the department (data entry, verification, etc.) is primarily responsible for making the error. It also seeks to identify all other functions within and outside the department which could contribute to the making of the error.

The function primarily responsible for making the error is, in most cases, located within the department itself -- for example, a wrong credit is processed, and in most cases it is the data entry function that has made the error.

In other cases, the function primarily responsible for making the error is located in another department within the company -- for example, a foreign exchange deal is input and verified exactly as specified by the Trader on the deal ticket. However, it is eventually found that there was an error in the deal -- the Trading function having specified the wrong receiving bank.

Today, with the increasing use of computers in clerical operations processing, it behooves the analyst to check whether the function primarily responsible for making the error is not the computer system itself. For example, the computer system installed in a letters of credit operation may not be able to uniquely identify an l/c number issued by a bank in one country from the same l/c number issued by a different bank in another country.

Examples such as the above occur, and the quality control analyst must be especially thorough in checking out the computer systems installed in the department: more and more clerical errors are the result of poor systems design and/or faulty computer programming.

At this stage of the analysis we have found it useful to draw up a matrix as shown in Graphic 5.

Frequency-rating of the function(s) responsible for making the errors allows the analyst to rank the functions by weakness -- that is, which is the weakest function, which is the next weakest, and so on -- and directs attention to the point(s) of greatest weakness in the department's manual and computerized workflows.

In addition to allowing the analyst to rank the functions by weakness, the matrix also allows the analyst to draw some broad and general conclusions which, in effect, is a grand strategy for reduction of errors.

INTERNAL INQUIRIES -- Type, Frequency, and Function Responsible		
Type of Inquiry	Frequency	Function Responsible for Error
Charges	22 per day	DEC VER
Other dept errors	33 per day	OD
Wrong credit	16 per day	DEC VER SYS
Duplicate payment	15 per day	SYS
Wrong debit	14 per day	NAC DEC VER SYS
etc., etc....		

Graphic 5: Matrix showing function(s) responsible for errors

For example, an analyst confronted with the situation shown in Graphic 5 would arrive at three quick conclusions

(1) there is a general failure of both the manual and the computerized systems within the department -- that is, the control mechanisms in both the manual and the computer systems have broken down,

(2) there is a general lack of understanding among employees of the technical aspects of their jobs, and

(3) there is a lot of inaccurate work flowing into the department from other departments within the organization.

Step 6: <u>Establish SPECIFIC reasons for errors made by the different functions</u>

This is the most delicate step in the entire methodology.

It involves sitting down with

(1) the managers, supervisors, and, occasionally, even the workers of <u>other</u> departments,

(2) the project managers, systems analysts, and programmers in the computer department, and

(3) the manager, supervisors, <u>and</u> workers of the department itself,

and establishing the specific reasons why each function is making the kinds of errors that it does -- in other words, why is each error made? what causes it?

Why is the verifier making errors? Some of the reasons that an analyst may establish as the cause of the errors are

(1) the verifier is a part-time worker,

(2) work does not come through in a smooth, even flow, but arrives in big, unexpected chunks,

(3) the computer system does not allow for "blind" verification.

Why does the data entry clerk make errors? Again, the analyst may establish the following reasons

(1) the account number provided by the other department is wrong,

(2) the input format has not been standardized,

(3) the data entry clerk is required to pick-up and also microfilm the work, and this cuts into production time.

Similarly, reasons are also established for errors made by (1) the other departments, and (2) the computer system.

Step 7: <u>Formulate solutions for elimination and/or reduction of errors</u>

At the end of step 6 the analyst develops a matrix that lists

> (1) the function making the error -- for example, data entry,
>
> (2) the error itself -- say, wrong account number,
>
> (3) the specific reason(s) for the error -- and these may be
>
>> (a) other department provides wrong account number,
>>
>> (b) input format not standardized,
>>
>> (c) pressure to deliver quantity,
>>
>> (d) other duties cut into production time,
>>
>> (e) low emphasis on quality,
>>
>> (f) computer system does not provide unique identifiers for 1/c numbers,
>
> etc., etc.

In Step 7, the analyst's task is to fashion a solution for each of the specific reasons why the error was made. This, again, is done in conjunction with the relevant managers and supervisors.

The objective, of course, is to devise such solutions as will <u>prevent</u> the errors from being made.

In the case of the data entry clerk, some solutions that an analyst might formulate

<u>In the department itself</u>

(1) installation of a system for prescreening the input,

(2) greater emphasis on quality, not quantity,

(3) more training in the technical aspects of the position,

(4) reorganization of workflows in the department.

<u>In other departments</u>

(1) stricter quality control of work going out to other departments,

(2) better understanding of the interface between the inputs and the outputs of the two departments,

(3) better training of workers.

<u>In the computer department</u>

(1) improved design for data entry system,

(2) system-generated rejection of incorrect input,

(3) understanding of data entry clerk's input problems.

Step 8: <u>Establish SPECIFIC reasons for externally-generated errors</u>: as was done for internally-generated errors in Step 6.

Step 9: <u>Formulate solutions to eliminate/ reduce externally-generated errors</u>: as was done for internally-generated errors in Step 7.

These will range from customer education by personal contact or letters through to charging fees -- for errors originating in the customer's organization -- on a sliding scale.

Step 10: <u>Develop a timetable for implementation of solutions</u>

The timetable will list the broad range of solutions -- from training employees to reorganization of workflows through modification of computer systems to educating customers -- and specify start times and completion times for the implementation of the solutions.

Some of the solutions can be scheduled and completed quickly, in a matter of 1/2 weeks. Others will take time: 6/8 weeks to make major changes in a computer system. Yet others may have to be scheduled continuously: for example, the training, re-training, and updating of employees in the changing technical aspects of their jobs.

The next step is the establishment of improvement targets.

| | Errors Per Day | | | | | |
| | Present | TARGETS | | | | |
	Dec 1983	Mar 1984	June 1984	Sept 1984	Dec 1984	Mar 1985
Internally-generated errors						
Wrong bank account number	25	20	18	12	etc.	etc.
Wrong charges	21	21	19	16		
Wrong credit	17	15	10	7		
Duplicate payment	15	14	12	9		
etc., etc.						
Total	78	70	59	44	20	12
% reduction		10%	24%	44%	74%	
Externally-generated errors						
Claim made in error	33	32	etc.	etc.		
Wrong instructions	28	27				
Debit wrong account	25	24				
Credit wrong account	16	15				
Cancel payment	11	10				
etc., etc.						
Total	113	108	etc.	etc.	68	
% reduction		4%			40%	

Graphic 6: Cost of Quality targets

Step 11: Establish improvement targets

Given the schedule for implementation of the solutions designed to eliminate/reduce the internally-generated and the externally-generated errors, the analyst's next step, in conjunction with the department manager and supervisors, is to establish improvement targets. Graphic 6 shows the format we use to track targets and accomplishments.

A few things should be kept in mind during the period of implementation

(1) externally-generated errors -- that is, errors made by customers -- will decrease very, very slowly,

(2) internally-generated errors will decline slowly at first, and then very rapidly as the cumulative effect of all the solutions designed to eliminate/reduce errors is felt,

(3) neither internally-generated errors nor externally-generated errors can be reduced to zero.

However, unless zero defects is aimed at, the reduction in errors will not nearly be as much as the Cost of Quality program can deliver.

Step 12: Monitor the working of the program

The analyst should monitor each and every aspect of the program during the first two weeks of implementation. Despite all the pilot testing, unexpected situations will crop up which only the analyst can handle.

Once the new work habits and new work procedures are institutionalized, the analyst should let the manager take over.

As part of the monitoring process, the analyst should make himself available to the manager over the first quarter of implementation, directing his attention to the sequence of events that flow from the implementation of the Cost of Quality program

(1) Quality dollars are shifted from Appraisal to Prevention,

(2) As efforts to prevent errors take

effect, quality improves -- that is,
errors decline,

(3) As both internally-generated errors and
externally-generated errors decline

    (a) Internal Failure Costs decline,

    (b) External Failure Costs decline, and

    (c) because of the improvement in
       quality, even Appraisal Costs
       decline.

(4) As a result

    (a) the number of inspectors, verifiers,
       and checkers can be reduced, and

    (b) the number of investigators,
       adjusters, compensation clerks
       can be reduced.

(5) This leads to a reduction in the Total
Cost of Quality, while Productivity
increases because the same amount of
work is being done with fewer people.

(6) When the full effect of the prevention
efforts is felt, a second stage of
productivity increases takes place:
because of better training of workers
in the department, smoother manual and
computerized workflows, and cleaner
input from other departments, more
volume can be handled by the
department in the same amount of time.
Or, in other words, fewer people are
required to process the same volume.

The analyst-manager meetings over the first
quarter of implementation are important
because they help the manager assimilate
and internalize the mechanism of the Cost
of Quality program

      how a decrease in quality costs
      leads to an increase in quality
      as well as a two-stage increase
         in productivity.

The analyst should, at the end of the first
quarter of implementation, assist the
manager in

(1) drawing up the new distribution of
quality costs. It will show a
moderate increase in Prevention
Costs, dramatic declines in both
Internal Failure and External
Failure Costs, and some decline in
Appraisal Costs,

(2) calculating the new, lower Total Cost
of Quality as a % of the Department
Operating Expense,

(3) graphing the decline in both the
internally-generated and externally-
generated errors,

(4) establishing the increase in
Productivity, and

(5) quantifying the direct savings
achieved as a result of

    (a) reduction in quality costs,

    (b) reduction in the workforce, and

    (c) reduction in compensation
       payments.

CONCLUSION

This paper has outlined a simple, easy-to-
learn and easy-to-apply methodology called
Cost of Quality.

It is a methodology which quality control
analysts, internal as well as external
management consultants, and managers of
clerical operations can use -- and effect
a reduction in the total cost of quality,
while at the same time increasing the
productivity of their operation.

Although the Cost of Quality methodology
has been tested and successfully applied
in the domestic and international
operations of Manufacturers Hanover Trust
Company, the methodology is essentially
generic in nature, and can be easily
adapted to achieve similar results in any
paperwork operation -- in manufacturing
industries as well as financial and other
service organizations.

BIBLIOGRAPHY

Duncan, Acheson J., Quality Control and
Industrial Statistics, Irwin, 1974

Juran, Joseph M., Quality Control Handbook,
McGraw-Hill, 1974

Crosby, Philip, Quality is Free,
McGraw-Hill, 1979

White, Bruce, "Quality Fitness Test",
Quality, March 1981

Aubrey II, Charles A. and Eldridge,
Lawrence A., "Banking on High Quality",
Quality Progress, December 1981

Orsini, Joyce, "Three Tips for Quality
Control", Bank Administration, April 1982

Eldridge, Lawrence A., and Aubrey II,
Charles A., "Stressing Quality -- The Path
to Productivity", Bank Administration,
June, 1983

BIOGRAPHICAL SKETCHES

Michael P. Quinn is an Assistant Vice President at Manufacturers Hanover Trust Company, and a program manager in the Management Consulting Services department. In addition to having responsibility for Quality Control, he has been instrumental in developing a Productivity program for MHT's Operations Division. He has been a lecturer for the American Institute for Banking, the Bank Administration Institute, the American Society of Quality Control, and the American Management Association.

Egbert F. Bhatty is an Assistant Manager at Manufacturers Hanover Trust Company in the Management Consulting Services department. He specializes in the development and application of Quality-Productivity programs. Mr. Bhatty's previous consulting experience includes work at a British manufacturing multinational, and in American insurance and savings & loan companies.

*Reprinted from **Industrial Engineering**, October 1983.*

Rethinking Productivity:

Models Yield Keys To Productivity Problems, Solutions

By A. Alan B. Pritsker

A. Alan B. Pritsker is president of Pritsker & Associates, a consulting and software corporation which distributes the SLAM, MAP and Micro-NET programs for analyzing and improving systems. After spending seven years at Battelle Memorial Institute, Pritsker was a professor of industrial engineering for 20 years. He has developed several languages for use in building models. An IIE fellow, he has received the AIIE-H.B. Maynard Innovative Achievement Award and the Institute's Operations Research Division Award and Distinguished Research Award.

Information is a key to uncovering ways of increasing productivity. Information gleaned from ongoing operations provides the basis for:
□ Understanding current operations.
□ Understanding and evaluating current productivity.
□ Initiating future designs.
□ Formulating meaningful arguments for operational changes that can lead to productivity improvements.

Industrial engineers use models as information amplifiers and information generators. We solve problems by determining the structural aspects of a system, collecting data from the system, analyzing the data and transforming them into useful form by means of a model which focuses on the aspect of the system that is under study.

In this way, the model is used to support decision-making. Decisions are made by comparing alternative designs, purchasing new equipment, establishing new procedures, implementing organizational changes and developing an infrastructure for the business enterprise.

The focus of this article is the industrial engineer's reliance on modeling to bring about productivity improvements.

Recently my wife Anne presented me with a quiz called "Second Guessing" from *Games* magazine. The quiz involves using intuition and experience to arrive at a guesstimate of an answer for each of several questions. A portion of the quiz is given below.

Each question is to be answered by choosing the most suitable unit of time from among the following: *seconds, minutes, hours, days, weeks, months, years, decades* and *centuries.*

Only ten seconds are allowed for each answer, and calculations using pencil and paper are forbidden.

Here are five of the questions from the quiz:
1. With your rubber flippers on, how long would it take you to swim around the equator?
2. You have a loud voice. In fact, your voice is so loud that when you yell "hello" from New York City, a friend in Los Angeles can hear you. After you yell "hello," how long is it before your friend hears your voice?
3. What is the average life span of an ordinary housefly?
4. You're sunning yourself on the roof of a building about the height of Mount Olympus and you drop your stopwatch after hitting the start button. When the stopwatch hits the

street, it is broken. How long a time is indicated on the stopwatch?

5. You own a square mile of land. If one-tenth inch of rain falls on your land and you catch all the water before it hits the ground, how long will it take you to drink all the water?

The above exercise illustrates at least one point that runs counter to standard industrial engineering practice. With only ten seconds to solve each problem, we don't have sufficient time to do a proper analysis but must rely heavily on our experience and judgment. Industrial engineers do not like to do this, although managers don't mind. Industrial engineers do not like to be wrong; managers care more about making a decision than about being right all the time.

Although this may sound odd at first glance, it may not be a bad way to think. As a manager and decision-maker, I'd say conservatively that at least 50% of the decisions I make have no effect on productivity (or maybe anything else). When I was an academician, I had to be right on all my decisions, and they were all important.

An interesting aspect of the quiz is that the form that the answers should take is given. This simplifies problem solving and also specifies the level of detail required in the analysis. Only a ballpark answer is required.

Scoring answers is a big problem in industrial problem solving, since we often don't know what the correct answer is. For this quiz, a good measure of accuracy would be the distance from the correct answer. For instance, if you answered hours and the correct answer is weeks, you are two units of time off. For those big on performance measures, I would accept a quadratic function for assessing your second-guessing capabilities.

Let's look at the questions individ-

ually and relate them to the IE problem-solving approach and to productivity enhancement possibilities.

Question 1: With your rubber flippers on, how long would it take you to swim around the equator?

Let's discuss first the thought processes used in answering this question. At first, you probably thought "Is this for real?" "Is Pritsker pulling my leg?" or "This is probably a trick question and the time it takes to swim a circle *at* the equator is two seconds."

Interestingly enough, that approach is not atypical. How many times have you thought your boss was half-crocked when he asked you to investigate something like the productivity of the night watchman?

Coming back to the quiz, you remember you are reading *IE* magazine and, therefore, the question must be a serious one. You answer decades because you don't know how to swim and you know centuries isn't the right answer because you aren't going to live that long.

Let's consider how you would solve the problem using an industrial engineering approach. First, you would build a model. Possibly the model would be an equation of the form total time equals the distance to be traversed divided by the swimming rate, plus an allowance time.

If you had one minute to do the problem rather than the ten seconds, estimates could be made for the various factors in the above model, and you would have an answer. You would probably estimate the distance at 25,000 miles and a swimming rate approximately equal to a walking rate of four miles per hour, and decide that estimates of allowances are not really required. Therefore, your answer would be months.

If you decided to spend a day on this problem, you would probably go to the library to get an estimate of

(Here) we don't have time to do a proper analysis, but must rely heavily on our experience and judgment. Industrial engineers do not like to do this, although managers don't mind.

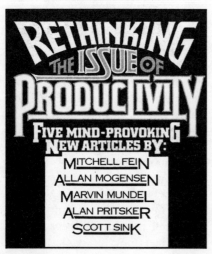

RETHINKING THE ISSUE OF PRODUCTIVITY

FIVE MIND-PROVOKING NEW ARTICLES BY:

MITCHELL FEIN
ALLAN MOGENSEN
MARVIN MUNDEL
ALAN PRITSKER
SCOTT SINK

the distance around the equator, go to the swimming pool and have someone time you with your rubber flippers on and still assume that there was no need to incorporate allowances into the problem. If the form of the answer had not been specified, your answer would be 8.7 months.

Next, there would be some question as to whether we should make the swimming rate a random variable, especially if the question had been raised in a probability course. If it had been given in a time study or work methods course, we would have incorporated allowances in the form of breaks, time spent in the hospital, fatigue and learning factors into the model.

The point is that we would have broken the problem down into its components, made estimates for the components and combined the estimates through a model to obtain an answer to the original question.

In building the model and making the estimates, we would have learned more about the situation under study, and we would be prepared to offer alternative ways of performing the job. Through the process of building models and making estimates, we develop new scenarios and new alternatives that lead to productivity improvements.

Question 2: You have a loud voice. In fact, your voice is so loud that when you yell "hello" from New York City, a friend in Los Angeles can hear you. After you yell "hello" how long is it before your friend hears your voice?

For this question a model requires the distance between New York City and Los Angeles and the speed at which sound travels. You might remember from your high school physics course (no one ever remembers anything from their college physics course) what the numerical value for the speed of sound is, or you

might refer to the *Handbook of Industrial Engineering* to find out that it is 1,125 ft per second. Assuming New York is 2,600 miles from Los Angeles, we find the correct answer to be hours.

If you didn't know the speed of sound or where to locate it, or if it were a quantity that was not known, analytical or experimental procedures could be used to estimate the answer. Again, this is part of the engineering process.

To improve productivity here— that is, reduce the transmission time—we would have to either change the distance or increase the rate of transmission. Since changing the distance does not seem plausible, we would limit our attention to changing the transmission rate.

We know that signals travel faster than sound. We discover that making a conversion from sound to a signal, transmitting the signal and then decoding the signal back to a sound would speed up the transmission considerably.

This example illustrates how the decomposition of a problem into its elements can lead to a method of solving a problem and also to procedures for seeking alternative solutions.

Question 3: What is the average life span of an ordinary house-fly?

Suppose the question were what is the average life of an ordinary employee? Or a computer? Or a machine tool? Any of these would raise many additional issues that would have a bearing on productivity. We would no longer concern ourselves strictly with the geriatric aspects of the question, but would consider the elements that go into making up a life span or job span as well. We might even be bold enough to think about definitions for productive life and the changes in productivity over time.

"How many times have you thought your boss was half-crocked when he asked you to investigate something like the productivity of the night watchman?"

RETHINKING THE ISSUE OF PRODUCTIVITY

FIVE MIND-PROVOKING NEW ARTICLES BY:

MITCHELL FEIN
ALLAN MOGENSEN
MARVIN MUNDEL
ALAN PRITSKER
SCOTT SINK

Questions of accounting, such as depreciation or appreciation of salary, might enter into our analysis. The impact of preventive maintenance, hospital plans, new advances in machinery, robots and continuing education would all be important.

A better question might have been how do we increase the life span of the ordinary employee? The good industrial engineer can obtain productivity improvements by changing the form of the question.

By the way, the average life span of an ordinary housefly is about four weeks.

Question 4: You're sunning yourself on the roof of a building about the height of Mount Olympus and you drop your stopwatch after hitting the start button. When the stopwatch hits the street, it is broken. How long a time is indicated on the stopwatch?

In this situation we would use a model from physics to estimate the answer. The equation is

$$t = \sqrt{2d/g}$$

where t equals time, d equals distance and g equals gravitational acceleration. Substituting estimates into the above equation leads to an answer in seconds.

It is interesting to note that this question uses the most advanced engineering analysis of all the questions posed. Yet in terms of the type of answer desired, it is clearly the easiest. No matter how wrong you were in your estimate of the height of the mountain, your answer would have been seconds based on experience and judgment. In this instance there is no need to build a model, and clearly those who thought about relativity concepts should be put in the R&D department.

Question 5: You own a square mile of land. If one-tenth inch of rain falls on your land and *you catch all the water before it hits the ground, how long will it take you to drink all the water?*

The amount of rainfall is over 400 million cubic inches, which is over a million gallons. Assuming we could drink a couple of gallons a day, the answer is centuries. (If it rained beer, I think it would still be centuries.)

Most of the problems that we deal with in manufacturing situations have a dimensional complexity way beyond that presented by this question. Combining jobs with routes through stations, input and output queues, transporters needed for material handling and resources in terms of tools, fixtures, personnel, etc., requires advanced models and explicit decomposition procedures. Productivity in such situations cannot be increased through judgment alone, unless the questions posed are straightforward.

Obtaining productivity improvements requires the gleaning of additional information from the data that we have at hand, combining it through the building of models and use of the models in an intelligent and meaningful way. The models must be built to solve specific problems and provide information that will increase our decision making capability.

Modeling and analysis can be applied to the complete gamut of problems of interest to the industrial engineer. This includes those encountered in manufacturing operations, banking operations, hospital and medical facilities, and communication and computer systems, among others.

The following discussion refers to manufacturing, but the comments are equally valid for the other areas of industrial engineering use.

Figure 1 displays an aggregate view of a manufacturing plant. The elements of the plant shown in this

view are as follows:

□ Parts.
□ People.
□ Activities.
□ Information.
□ Machines.
□ Messages.
□ Computers.

Each of these elements needs to be understood and characterized as to both function and information content to provide a description of manufacturing operations. Much of the work of industrial engineers is devoted to this definitional type of activity.

When modeling and analysis are used in seeking productivity improvements, these elements need to be both defined and characterized. The characterization need be only as detailed as the problem at hand requires. Second guessing is permissible and in some situations desirable.

The procedures used to manage the elements listed above, and the relations among the elements from the particular point of view of the problem, need to be understood. Languages and techniques for facilitating these modeling functions have been developed. The advent of such languages has simplified the descriptive process associated with building a model of a complex situation such as a factory and, in turn, has increased the productivity of the industrial engineer.

After a model is built and the elements are characterized, the outputs obtained from the model are analyzed. The model can be driven directly from the data collected from the manufacturing plant, or it can be driven through the characterization of the elements. Structural information on the relationships and procedures associated with the elements is used to obtain outputs that resemble the system being studied.

The model is then used to evaluate alternative ways of operating the

Figure 1: Aggregate View of Manufacturing Plant

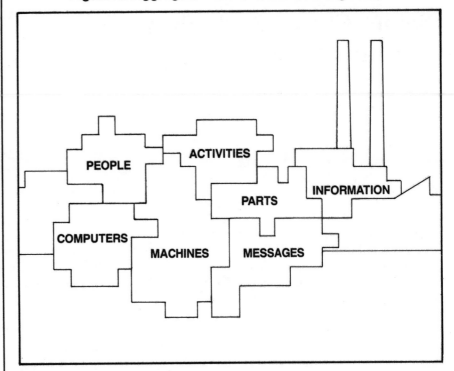

plant, new equipment purchases and plant expansion plans. The proposed changes are analyzed and their productivity improvement potential assessed.

Graphic outputs which show the operation of the plant as well as productive capacities over time for the various designs and changes are then presented to the manufacturing decision-maker.

An iterative process is used for modeling and analysis so that the modeler, analyst and decision-maker will be comfortable with all aspects of any proposed changes for productivity improvement. The project team must be oriented toward effecting changes in the plant that will help achieve the desired goal of productivity improvement.

This implementation step should involve all members of the project team, and plant management typically asks for additional information and further design evaluations. Considerable planning is necessary for successful implementation of a project. Edicts normally do not result in real productivity improvements.

Following implementation of a change, the cycle is repeated, either with a new problem or for additional analysis of how the changes made have affected productivity. In addition, an iterative approach for each step—modeling, analysis and implementation—within the overall procedure is used to achieve the long-term goals of productivity improvements. Because of this iterative approach, the use of modeling languages can greatly facilitate the process.

Furthermore, it is advantageous to use databases for storing both plant information and outputs of models. This facilitates comparisons and analysis and aids in the presentation of graphs, plots and animated traces of operations.

A query language for accessing, combining and displaying plant data and model results is also useful.

One of the most important problems facing the United States in the '80s is that of developing methods for increasing productivity. Faced with a scarce money supply, we will have to adapt current production plants to meet new demands or build new facilities with limited expenditures. Therefore, we must make wise decisions to obtain effective use of both our capital and our technical resources in providing goods and services.

Effective use of computer modeling and analysis will enable decision-makers to make wise choices. The effective use of industrial engineering tools requires an understanding of the pro' 'ems and concerns related to the decision to be made. This includes a definition of the elements and factors that affect the decision, a listing of the measures by which a good decision will be judged, and the building of a model that forecasts the measures of performance for each decision option.

Modeling is the key to the use of computer-based tools for problem resolution. It has four steps:

1. A system is decomposed into its significant elements.

2. The elements are analyzed and described, using second guessing if necessary.

3. The elements are integrated in a model of the system.

4. Performance is predicted through the analysis or simulation of the model.

The modeling phase is followed by analysis and implementation of results, both of which require sound industrial engineering programs and techniques. We have the tools; let's get on with the task at hand.

For further reading:

Miner, R.J. and L. Rolston, *MAP/1 User's Manual*, Pritsker & Associates Inc., 1983.

Pritsker, A.A.B., "Applications of SLAM," *Transactions of the IIE*, March 1982.

Pritsker, A.A.B. and C.D. Pegden, *Introduction to Simulation & SLAM*, Halsted Press and Systems Publishing Co., 1979.

Pritsker, A.A.B. and C. E. Sigal, *Management Decision Making: A Network Simulation Approach*, Prentice-Hall Inc., 1982.

Salvendy, Gavriel, ed., *Handbook of Industrial Engineering*, John Wiley & Sons, 1982.

Sklov, M. and B. Spitzer, "Second Guessing," *Games* Magazine, September 1983.

Standridge, C.R. and S.A. Walker, "Simulation Support Software: Concepts, Current Capabilities, and Applications," *Proceedings of the 1983 Annual Industrial Engineering Conference and Exposition*, Louisville, KY, pp. 552-561.

Talavage, J. and W. Lilegdon, *MicroNET User's Manual*, Pritsker & Associates Inc., 1983.

Yancey, D.P., M.W. Sale, J.P. Whitford, J.J. O'Reilly and W.R. Lilegdon, *The IDSS Prototype (2.0)—Version 2 User's Reference Manual*, Pritsker & Associates Inc., 1983.

*Reprinted from **Industrial Management**, September-October 1983.*

Raising Clerical Productivity with Most® Clerical Computer Systems

Ronald A. Soncini

Charles Dickens' statement that, Quote, "It was the best of times, it was the worst of times," End Quote, certainly can be applied to white-collar-intensive industries today. There has been a substantial rise of white collar workers in the last ten years. According to the Bureau of Labor Statistics, from 1968-79, the white collar population in the U.S. has risen nearly three times as fast as blue collar workers. Of the 47.2 million white collar employees in the U.S. today, nearly one-third are categorized as clerical. On the negative side, white collar labor costs have been racing upward also.

To manage more effectively, supervisors of clerical personnel are trying to get the most work from their staff and hold down costs as much as possible at the same time. If these sound like conflicting objectives, they can be. However, the manager of clerical employees can introduce techniques designed to reach both goals.

The key to lowering costs while improving performance is to increase the productivity of the current workforce. And the most effective method of obtaining more output per worker is to improve the method and to measure the effort and set standards for performance.

With these tools, managers can control the flow of paperwork, the primary clerical output, and reduce the per-piece cost of processing.

The obvious question is how to measure. A number of methods have been used through the years to evaluate clerical work. Probably the most common approach is the historical estimate. It is based on past performance and for this very reason is not a good tool. It only tells you what has been done, not what should be done. Work sampling is another technique, a little more accurate, but still based on the status quo. Time studies involve stopwatches, observing the worker performing the task. One of the biggest disadvantages of time studies is that an analyst rating an operator performing work, must make efficiency judgments which are subject to error. Also, clerks being watched and timed will not work at a realistic pace because clerical personnel resent the "Big Brother" approach to efficiency. Finally, there are predetermined time systems, which I will discuss further in a few minutes.

All of these methods have been applied sporadically to clerical operations and with mixed results. The reason for the apparent lack of interest in measuring and controlling clerical work can be found in an analogy with the petrochemical industry. When gasoline was selling for 30 cents a gallon, there was little incentive to search for new sources of oil because the cost of exploration was too high in relation to the profits that could be expected. In the same way, clerical costs have been low for many years.

There was no justification for managers to institute a program to improve productivity because the return on this investment would be relatively small. Today, however, we are looking at a much different picture.

Another reason office managers have failed to implement work measurement programs is the complexity and length of application time involved in measurement systems, even some of the most accurate predetermined method/time techniques.

A majority of predetermined method time systems are better suited for repetitive work — the assembly line operation with short interval tasks. While there is a certain amount of short-cycle repetitive work in clerical operations, there is also much nonrepetitive work that is more difficult to measure.

When we began to consider adapting MOST® Systems to the clerical field several years ago, we made a careful survey of the needs of clerical staffs in the area of measurement and controls. From our industrial experience, we already knew that we had a predetermined method/time system that was fast, accurate and consistent. It was also well suited to measuring both repetitive and nonrepetitive work. But there were differences between industrial and clerical environments which caused us to modify our basic approach.

RONALD A. SONCINI is Coordinator, MOST® Clerical Systems, for H. B. Maynard and Company, Inc. Based at corporate headquarters in Pittsburgh, Pennsylvania, Mr. Soncini is responsible for the translation of MOST® Work Measurement Systems, the manual technique, into its clerical version and the transfer of this information into a computer software program. He is a graduate of North Texas State University in Denton with a degree in business and holds a master's degree in management from the University of Pittsburgh.

One of the first facts we discovered about clerical operations is that users prefer a computerized work measurement system. We have MOST® Clerical systems in the manual mode, of course. And part of the learning process for users requires a thorough knowledge of the basic technique. But in application, MOST® Clerical Computer Systems is the primary method. There are several reasons for this trend.

First, organizations that have a large number of clerical operations, such as banks, insurance companies, hospitals and utilities, are familiar with computerized programs. To them, having a manual work measurement system would be regressing.

Second, the computer is ideal for an organization with multi-office locations, such as a bank with statewide operations. It provides a uniform application for all units and instant interchangeability of computer-stored data.

Third, it yields accurate, consistent, and well-documented time standards and method description. If the operator inputs the correct data, the computer will not make errors.

And fourth, it attracts highly qualified personnel for standards development work as well as maintaining the present staff.

The second major difference between clerical and industrial versions of MOST® Systems is the basic structure. MOST® Computer Systems for industrial applications is modular, that is, clients can lease any one of a number of software modules according to the type of work measurement programs they want to implement. There are five basic modules and five supplementary ones at the present time. MOST® Clerical Computer Systems, by the nature of the tasks it is designed to perform, is not modular. It incorporates the same five basic modules as MOST® Computer Systems, but all of these modules are necessary to produce the results clerical staff managers are interested in. These are essentially three pieces of data; staffing requirements; performance reporting; and unit costing.

The "Big Three" which form the core of MOST® Clerical Computer Systems are now the primary focus of our attention.

Probably the first measurement management will want to make is performance, both of individuals and groups. This information can be used as basic data for the other two modules. Input will include the employee names, time standards, quantities, non-standard time and waiting time.

Output is generated in the form of two reports. First is the summary report, by individual or section. This program can handle individual standards or overall daily performance on all standards. Standard times for each job are developed using MOST Systems predetermined method/time systems techniques. Then standard times can be used to calculate the actual earned hours for each job to determine a performance percentage. At a glance, a manager can tell how well each employee is performing assigned tasks. If one employee is consistently low on a certain job, it may indicate a lack of skill which calls for additional training or reassignment. If all employees are having difficulty with a certain portion of the work, it may require a methods study to determine what changes need to be made in the work.

Other output generated by the performance report include coverage, productivity and delay, all expressed as percentages of total hours available.

The second document provided by performance reporting is the comparative report. This creates a gross performance rating by work center, section or group. It measures overall group performance, rather than how the individual performs each part of the job. It is more valuable in assessing how the work unit satisfies the expectations of the supervisors and higher management. It also evaluates the ability of supervisors to manage the section.

Comparative reporting also indicates the payback on engineered labor standards. The average efficiency of a clerical operation is about 55 percent. With standards installed and properly maintained, managers should be able to observe an increase in performance up to approximately 90-100 percent in some cases. This percent performance, I might add, is not generated solely by new standards, but also is dependent on some work simplification and methods improvements, This is about the maximum efficiency that can be attained in a clerical operation without some incentive package, but it does represent a sizeable savings for the average organization.

For example, the Bureau of Labor Statistics 1980 report on wages in the banking industry gives a figure of $4.46 an hour for nonsupervisory workers. If these employees perform at 55 percent efficiency, the productivity value is $2.45 per hour. At the 90 percent level, that value increases to $4.01 per hour, a savings of $1.56 per hour per person.

Carrying the calculation one step further, the American Bankers Association reports that there are 225,360 clerical and office employees in the top 50 full-service banks in the U.S. That is an average of 4,506 clerical workers per bank. If we assume 30% of these workers can be placed on production control reports, then $1.56 per hour times 1,352 would be a savings of $2,109 per hour or about $16,874 per day per bank. This amounts to just over $4 million per year per bank by using effective work measurement, work simplification and methods improvement techniques. This is the maximum savings that could be attained if all clerical work could be measured. While this is not possible in all cases, savings of 60% of this figure are still achievable in most cases.

Results of performance reporting can be generated daily, weekly, or monthly on the computer. Using this information, managers can create meaningful staff reports, the second major module of MOST® Clerical Computer Systems.

While performance reports will indicate how productive the workers are, staffing indicate the need for additional staff and can even help predict whether permanent or temporary help will be more economical.

The input is similar to the basic data for performance reporting. In fact, the total quantity portion is derived from the performance reporting sheets.

Output, shown in one case for a staff allowance (which is the same as the productivity of performance level) of 85 percent is 6.3 persons. How this should be interpreted depends on present staffing levels. If there are currently six people in this work center, it means that the present workload could be handled by the available staff by use of overtime or by addition of part-time help. If present staff is five, the manager has several options to consider. It may mean hiring an additional full-time person. Other options are overtime, part-time work, on-call help or using the services of a manpower assistance firm. If present staff is seven, the manager has other courses of action to consider. There is the possibility of transferring one worker, or layoff if there are no job openings in other departments. The manager might even elect to eliminate the extra job by attrition. This is where unit costing enters the picture and we'll talk about the third module of MOST® Clerical shortly.

Staffing reports are not needed as frequently as performance reports. But with the data filed in computer memory, it is easy to maintain and update staffing on a continuous basis and produce answers for a staff review request on short notice.

With the computer, managers can "play games" with staffing problems. Actually, these so-called games are quite serious simulations of staffing problems. The computer has the capability to answer the questions such as, "What if I need to increase workload?" "How much additional staffing will it take and what will it cost?" "Is it more economical to increase staff overtime to do the work?" "Can we justify hiring additional personnel?"

Unit cost is the third clerical module. It is designed to look at the labor cost of processing one unit, such as a check or bank loan, through its paperwork cycle in various departments. It also uses information from the performance reporting module to determine how much time is spent on each piece of paper and how much the total process costs.

In one example, the computer prints out all the steps involved in processing personal "on us" checks. At the end of the summary, it gives a total cost broken out by direct labor, overhead and materials. The last two items can be furnished by company accountants or the computer can be programmed to calculate these figures based on a weighted average for each department or area.

Productivity can now be balanced against expense of operation. A clerical worker might be performing a task manually which could be done more inexpensively and faster by machine. Or machine costs might be high in comparison with work done manually. It gives the clerical manager another tool to use in determining how to lower paperwork costs.

Throughout this description of the three major data modules of MOST® Clerical Computer Systems, we have mentioned development of engineered labor standards as the foundation on which the whole system rests.

MOST® Clerical Systems is a predetermined method/time system. It is based on the principle that all work can be described as a series of motions that take place in a predetermined sequence. These sequence models, as they are called, can be combined to form a method description,

which together with data on the workplace form a MOST® Systems calculation.

In the method description, key words such as "Place," "Turn," "Push," and "Pick Up" tell the computer what values to assign to each move. When combined with data on the workplace such as relative location of workplaces and operators, tools, objects and equipment, the computer automatically calculates the time required for each job. For example, to copy an original on Xerox at mailroom requires 1030 TMUs, which translates into approximately 37 seconds of work, not including process time for the duplicating machine. While the TMU or time measurement unit is the standard yardstick for the industrial engineer, the computer can easily convert these numbers into minutes and seconds if desired.

Calculating the time to do one task provides one bit of data, called a suboperation. This must now be filed in a data bank so it can be retrieved quickly when needed to combine with other bits to form a labor standard. In a manual operation, this would probably be assigned a code or number and placed in an appropriate file drawer. We could have assigned numbers to suboperation data for MOST® Clerical Computer Systems, but it would still have been too cumbersome, and difficult for the applicator to remember long series of numbers that related to a specific task.

What we did do was construct a title from key words that describe the operation. There are seven major categories in a title and a suboperation can be filed by any one or all of these subdivisions. It is a much better mnemonic than numbering because the applicator only as to remember one word about the time standard. For example, if he or she knew that the suboperation required setting of a standard related to a copying operation, the word "copy" could be entered into the computer and all the suboperations that contained the word "copy" would be displayed. The user then could search this list for the appropriate title and retrieve it from the computer memory.

To prepare a time standard, which is the next step in the standard-setting process, the applicator picks and chooses suboperations data from appropriate title sheet organization lists.

This data is then filed in the time standards data base under a number of categories chosen by the applicator or manager. The most common headings under which clerical time standards are filed in computer memory does not preclude any other category which the user feels is desirable for identifying that particular operation.

The end product of MOST® Clerical Computer Systems is a complete documentation of the work area and the method at a certain point in time. It includes technical data provided on manual methods, work conditions, machines, allowances, how standards were developed, staffing conditions and the distribution of work.

All of this information is stored on computer memory where it is readily accessible for updating standards, creating new ones or reviewing present methods.

MOST® Clerical Systems can cut costs and reduce

paperwork in analyzing and improving clerical operations. It is easy to learn and apply because it eliminates much of the detail necessary in other predetermined method/time systems. Yet at the same time, it is an accurate technique for determining the productivity of clerical employees.

Clerical work has been around for a long time but there has been little attention paid to this work. One of the earliest known forms of writing was cuneiform inscribed on clay tablets by the Sumerians 5,200 years ago. These records were inventory lists and business transactions probably prepared by clerks. Today in clerical operations we are faced with the same age-old problems and questions the Sumerians probably asked. How many people should it take to do the work? How long should it take to complete a job? How much does it cost?

MOST® Clerical Computer Systems is providing the answers to all these questions and doing it efficiently by using the latest electronic technology. It offers an alternative to guesswork and estimating that has dominated the clerical work process for years. By measuring the work, it can help managers increase clerical productivity systematically and reduce their costs dramatically.

IV. COMPANY PRODUCTIVITY PROGRAM APPROACHES

Many organizations now have formal productivity improvement programs. In fact, in the last five years many types of productivity programs have become almost as commonplace as safety programs, energy programs, and so on. Some progams have been quite successful, while others have failed miserably. Approaches by thirteen companies are presented in the following section.

*Reprinted from **Industrial Engineering**, December 1984.*

American Express Implements Office Automation To Aid Customer Service

By Jay W. Spechler, P.E.
American Express

The mention of office automation creates many different images in the minds of those who are just beginning the process of understanding this relatively new technology. The application of office automation ranges from memory typewriters and centralized secretarial services to completely automated and integrated customer service functions.

One of the more significant forward steps towards the latter type of office automation took place in the airline industry beginning in the early 1970s. Prior to that time major airline reservation offices were decentralized, with locations in cities of heaviest traffic. Reservations agents recorded flight information on index cards which were transported via conveyor belts to centralized clerical stations for filing.

Manual summary listings of all flights were maintained to prevent overbooking. Communications between reservation offices were usually by telex and very cumbersome. With the introduction of office automation, reservation offices were centralized. Most paper disappeared as customer flight information data bases were created, and reservation agents entered all relevant flight information into computer terminals.

Expansion needed

At the beginning of this decade, the need for further expansion of the application of office automation technology became acute in companies performing major customer service functions (e.g., insurance companies and financial services institutions) having labor intensive clerical functions. In these firms, increasing complexity of transactions, expansion of product lines (new services) and the need for frequent *two way* communications with customers all combined to place great strain on conventional clerical systems.

This was particularly the case at American Express for measurement of and attention to customer service.

An extensive application of office automation at American Express is partially illustrated in Figure 1. In addition to creating data bases on all customers' accounts, there is a capability to communicate with customers through an automated letter generation facility.

One application of this facility is the automatic release of dunning letters to cardmembers whose accounts are past due. Hundreds of different letters on various subjects are contained within the system, virtually eliminating the need for manual, typewritten correspondence.

Five-year implementation

The introduction of nationally integrated advanced office automation systems at American Express has spanned a period of more than five years from concept development to full implementation. This has required the direct participation of hundreds of systems and operating personnel and a carefully planned, multi-phased approach.

Industrial engineering, or performance engineering as it is termed at American Express, has performed a key role in the introduction of office automation technologies. The purpose of this article is to outline the key integrative and facilitative, as well as systems design and development, efforts that the IE function has had responsibility for. The result of a team approach, in which the IE plays a critical role, has been a highly successful office automation system.

Preparing a statement

One cannot overstate the importance of front-end program definition as well as economic and service justification efforts as a preliminary step towards implementing office automation. This is the "outlining of the territory" phase.

Establishing the objective and scope of the automation effort is critical in obtaining appropriate agreements and commitments in the following areas:

☐ Management support.

Figure 1: Model of Office Automation at American Express

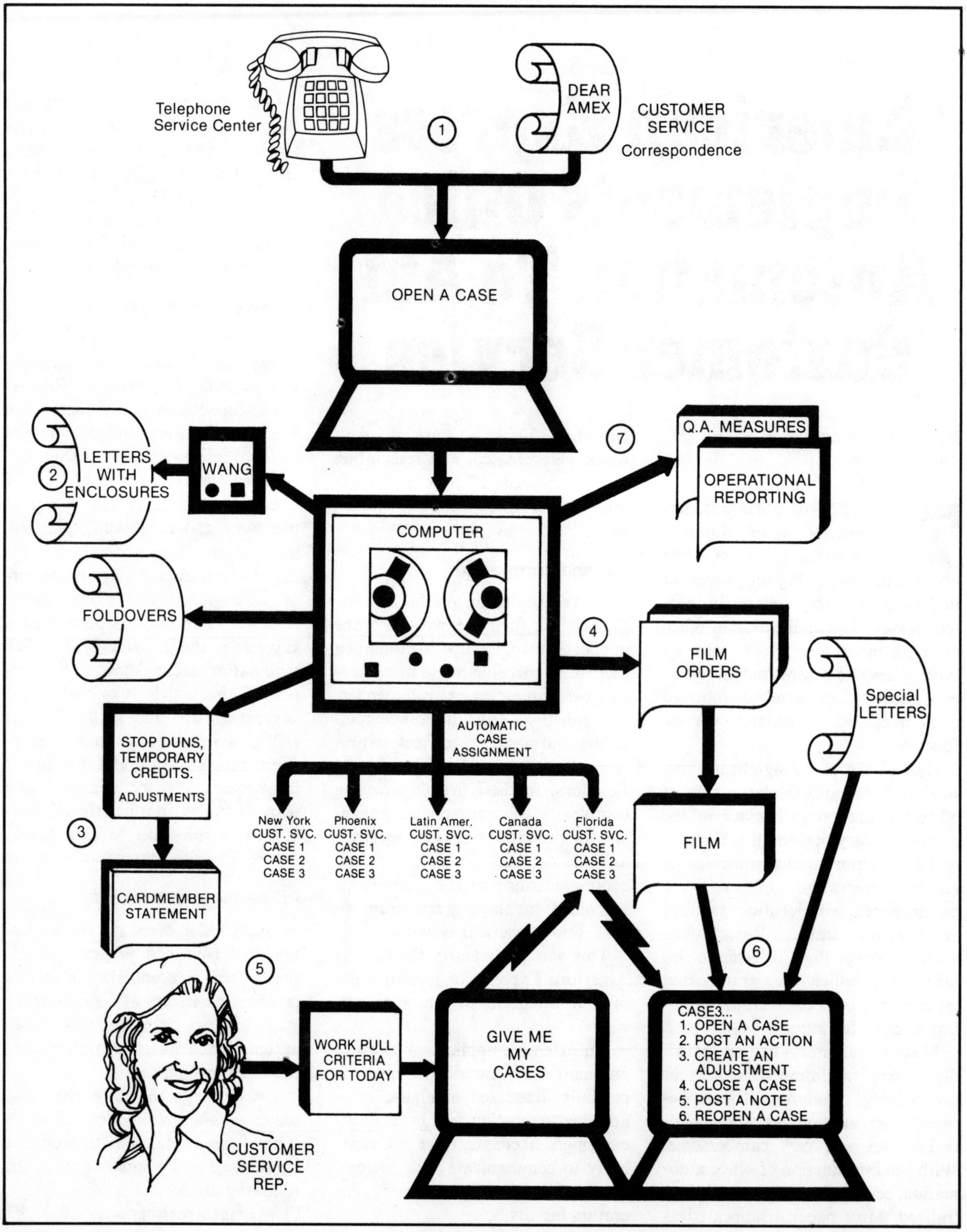

Telephone Service Center

DEAR AMEX — CUSTOMER SERVICE Correspondence

①

OPEN A CASE

② LETTERS WITH ENCLOSURES

WANG

Q.A. MEASURES — OPERATIONAL REPORTING ⑦

COMPUTER

FOLDOVERS

FILM ORDERS ④

Special LETTERS

STOP DUNS, TEMPORARY CREDITS. ADJUSTMENTS

③

AUTOMATIC CASE ASSIGNMENT

New York CUST. SVC.	Phoenix CUST. SVC.	Latin Amer. CUST. SVC.	Canada CUST. SVC.	Florida CUST. SVC.
CASE 1	CASE 1	CASE 1	CASE 1	CASE 1
CASE 2	CASE 2	CASE 2	CASE 2	CASE 2
CASE 3	CASE 3	CASE 3	CASE 3	CASE 3

FILM

CARDMEMBER STATEMENT

⑤ CUSTOMER SERVICE REP.

WORK PULL CRITERIA FOR TODAY

GIVE ME MY CASES

CASE3...
1. OPEN A CASE
2. POST AN ACTION
3. CREATE AN ADJUSTMENT
4. CLOSE A CASE
5. POST A NOTE
6. REOPEN A CASE

⑥

Typical work station in office automation environment at American Express.

□ Dedication of adequate financial and human resources.
□ Validation of major program assumptions.
□ Identification of alternative strategies.
□ Communication to all levels of management on program intent and its potential impact on respective functional areas.

Strategic planning

In an office automation sense, strategic planning comes to grips with policy decisions that will give direction to the tactical planners. One strategic planning decision made at American Express was that it was desirable to utilize the most advanced software and hardware technologies available. Another such element was that office automation was to result in a measurable improvement in customer service.

Other strategic factors encompassed issues such as: the objective to eliminate paper; the application of office automation in servicing the broadest range of American Express customers (cardmembers and service establishments); economic payback; the operational scope (domestic versus international implementation); the rate of new product introduction over the next five years; and the overall rate of growth of the company over the latter time frame.

The IE's role

Provide project management leadership and support: If the objective and scope statement is "the outlining of the territory," the project management phase is the step where we "prepare the detailed map of the territory."

Industrial engineering is the appropriate functional area to coordinate this "mapping" effort since one of its traditional missions has been the development of best operating methods. As the "map" is prepared, it should be done with an eye towards implementing the most effective operating practices that can be anticipated.

Having completed the "map," all of the milestones and events are entered into an automated project management system and a PERT chart is generated. Industrial engineering conducts weekly progress review meetings where open issues are noted and respective responsibilities and due dates are assigned.

Typically, these progress meetings involve conference call participation of a dozen functional departments at a half-dozen locations throughout the U.S. and Canada. The PERT contains more than 2,000 events.

Develop automated management reports: Office automation provides an opportunity to eliminate manual reports required by industrial engineering for productivity analysis. The data can be easily recorded as operator keying is accomplished at a level of detail that would not be practical using manual systems.

The same data can also be used for production scheduling and control purposes—a boon to line management, which can now shift resources, assign workloads and evaluate correspondence inventory status on an hour-by-hour basis.

Prepare new workflows and procedures: Substantial operating efficiencies have been achieved at American Express through the electronic transmission of work that was previously manually transmitted from one work station to another. Further, the availability of a centralized data base has enabled the company to broaden employee training and capabilities so that they can handle a wider range of customer servicing functions than they could previously.

Design ergonomically acceptable workplace: The occasion of introducing office automation provides an opportunity to redesign the workplace. Without this step, there is a risk of incurring substantial employee dissatisfaction along with a loss of potential productivity gains.

Figure 2 illustrates a work place layout developed for a high clerical transaction and analysis function at American Express. In addition to layout considerations, attention was given to lighting, sound levels and glare factors. All required substantial modification compared to manual clerical work stations.

Establish quality measures: At American Express, performance engineering has responsibility for the quality assurance function. Quality is measured in both "timeliness of service" (how quickly we respond to a customer's request) and "accuracy of service" (how accurately we respond either verbally or through correspondence).

Analysts monitor telephone communications and evaluate the appropriateness of correspondence being generated, 24 hours a day, seven days a week. This quality assurance effort has been greatly facilitated through office automation. Timeliness measures are now recorded automatically. Telephone monitors in Q.A. can review the same information on their terminals that telephone service representatives are looking at in communicating with customers.

Additionally, error tracking by individual and by type of error is economically feasible under office automation at American Express.

Conduct acceptance tests of new office automation systems: A most important performance engineering effort in the introduction of all major new computer systems, including office automation, is their acceptance testing prior to their going into production. Here, performance engineering serves as a buffer between the systems, the development group and the user community.

No system is allowed to go into production until it has passed through a simulated production envi-

Figure 2: Work Station Layout For a High Clerical Transaction and Analysis Function

1.	Display Support Table	11.	Tools Caddy
2.	Display Tilt, Swivel	12.	Auxiliary Tools Storage
3.	Keyboard Tilt Tray	13.	Personal Storage
4.	Keyboard Palm Rest	14.	In/Out Storage
5.	Primary Work Process Table	15.	Input Document Holder
6.	Panel Systems	15A.	Secondary Document Holder
7.	Panel Visor	16.	Task Illumination
8.	Case Files Storage	17.	Energy Distribution/Wire Management
9.	Procedures/Systems Manuals Storage	18.	Signage
10.	Forms Caddy	19.	Posture Support

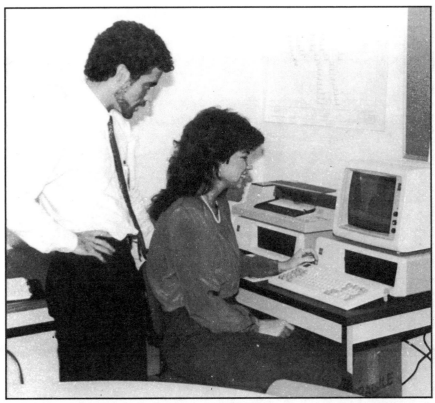

Project manager Bill Leander and project engineer Amy Lyle using project management system in office automation project.

ronment maintained in the performance engineering model office. In this way, we help assure that production departments will be able to deliver a high quality of service in using new systems from the first day that they are put in place.

Provide system implementation support: Performance engineering has played a major role in developing implementation strategies that have helped assure timely and effective introduction of office automation technologies. The following provides some insight into how this was accomplished.

□ *Learning curve analysis*— Throughout the implementation period, performance engineering tracked individual and unit productivity. Special teams composed of industrial engineers, trainers and operating department supervisors closely monitored employee performance. As a result, workflows and training programs were modified and individual needs were given special attention.

□ *Staffing level determination*— Performance engineering determined that while office automation would yield long-term productivity benefits, additional staff would be needed in the short-term. These short-term additions were required to compensate for extensive re-training of the clerical staff (over 50,000 man hours system-wide) and lost production as re-trained personnel went through their learning curve period.

A model was developed to account for the latter influences and to forecast staffing needs for all departments covering the entire implementation period. Accompanying this was an economic analysis that demonstrated the financial payback to the short-term investment in extra staff.

□ *Troubleshooting* — Throughout the implementation period, perfor-

mance engineering coordinated the handling of numerous systems and personnel problems. Each problem was quickly but thoroughly documented with appropriate assignment of engineering and systems personnel to specific corrective tasks. Daily progress meetings were held on these issues during the implementation period to expedite their resolution.

□ *Quality review*—During the entire implementation period, performance engineering's quality assurance staff evaluated 100% of all clerical personnel output. Each individual's work was reviewed with same day feedback on all errors. Based on this information, the training department was able to modify some of its instruction media.

□ *Senior management review*— Each week a senior management review of all productivity and quality performance connected with the introduction of office automation was effected. Performance engineering analysis presented at these meetings included reasons for variances from production plans or acceptable quality and corrective actions being taken.

Preparing for change

Office automation has altered the way in which American Express communicates with its customers. It now does so faster and with greater accuracy. One of the approaches used to accomplish this result was to utilize performance engineering as a key facilitative and coordinative force within a team effort.

The PE perspective helped to blend new computer systems; the most advanced technology available; and a highly qualified work force into a more effective and efficient service instrument.

Almost 10,000 employees have been affected by office automation at American Express, and all have gained. Employee surveys reveal that they find the new systems and equipment easier to work with. They experience fewer frustrations in locating customer information and are able to handle a wider range of inquiries (reducing call transfers).

Senior management has been able to move forward in heavy capital and systems investment in office automation. This has, in part, resulted from the substantial planning efforts conducted by performance engineering, which provided a forecast of break-even points and long-range payback along with an evaluation on the positive, quality impacts.

As it went about its business of preparing the organization for change, performance engineering formed cooperative teams of managers from all functional areas of the organization. No one was left out of the system design or implementation phases. All views were heard with a consensus reached on the best set of compromises.

Both American Express and industrial engineering have taken a giant step forward in introducing significant change within the organization. **IE**

Jay W. Spechler, P.E., is director of performance engineering at American Express, Fort Lauderdale, FL. He is also adjunct professor of industrial engineering at the University of Miami. Spechler was previously director of management services for Florida Power and Light Co., internal consulting manager for Gulf and Western Industries and director of materials and industrial engineering for McCall's Printing Co. Spechler is the author of *Administering the Company Warehouse and Inventory Function,* published by Prentice-Hall. He received a BS in production management from New York University, an MS in industrial engineering from Stevens Institute of Technology and a PhD in business administration and economics from Columbia Pacific University. Spechler is a past member of the Council on Industrial Engineering, an advisory body to IIE, and is currently serving on the Institute's Publication Policy Board.

*Reprinted from **Industrial Engineering**,*
August 1984.

Part 1 Of 2:

Service Organizations Can Use IE Techniques To Improve Productivity Of Professional Workers

**By Doris Magliola-Zoch
and Ronald G. Weiner**
Weiner & Co., CPAs

The service sector is the most rapidly growing area of our economy, and yet there is a dearth of information on how to effectively manage service businesses.

Professional service organizations bring a variety of skills to the resolution of client problems. It is the manager's task to effectively weld these skills together in a manner that provides the client with the greatest return on the dollar, and this, in turn, mandates the effective utilization of a rare resource—the skilled professional.

Planning for change

As in all sectors, productivity improvement will not just crop up by "trying harder." We must first plan on changing. Henry Ford once said that if you don't invest in a piece of equipment that you need, you wind up paying for it even though you don't have it. Similarly, professional services that effectively invest in planning for change will improve productivity. Those that do not will pay for this lack of planning through inadequate results and the cost will inevitably be passed on to, or shared with, the client.

The profitability of the firm is a function of its most valuable resource—the productivity of its people. The plan must curb the nightmare which haunts every professional: the conflict of needs for timeliness, quality and profitability.

If not properly managed, this conflict often becomes a juggling act. The choice is ours. We can let these elements control us and become "reactors," or we can simultaneously achieve these three needs by planning a systematic stream of events to control our services. We should avoid the all too common situation of reacting rather than initiating.

Each assignment, as it moves through the application of professional skills, follows a path largely predetermined by the firm's practices. Each completed phase relieves certain personnel of further responsibility and requires the infusion of others.

This coordination of services to complete the task in a timely, cost efficient fashion is analogous to the objectives of PERT (program evaluation and review techniques) and CPM (critical path methods). These objectives are to coordinate the commencement of the application of skills of a given trade, which can only begin with the completion of another function while other tasks are still taking place.

The problem with the application of network analyses is that they have not been utilized to address the issue of how to manage professional performance. For this we must once again turn to the industrial engineering profession.

Application of scientific principles in a professional environment is practical by using various industrial engineering concepts. The unification of these complementary planning tools has yielded a process which we call PROPLAN for PROfessional PROductivity PLANning.

Time management

Time management in the professional service arena is not approached scientifically as it is in other sectors. The planning process is unrefined, and for the most part, productivity measurements are nonexistent.

This situation exists largely due to the mistaken belief that tasks requiring professional judgment are not subject to the planned controls which industrial engineering has been offering the manufacturing environment during the past 80 years. It can be observed that the more closely the service business parallels the manufacturing business, the better engineered the operation is (*a la* the fast food industry).

However, the further we move into the judgmental sectors of the service industries, the more archaic the planning process becomes. This article will discuss how scientific principles which have been utilized by industrial engineers to increase productivity in manufacturing firms can also be of benefit in the professional services.

While no professional likes to equate his or her function with a production process, there are many similarities. For example, industrial engineers are quite often asked to project "how much time" various new methods of production will take. We, therefore, must be able to measure alternative methods before one is selected and implemented. To do this we must have a systematic approach for providing meaningful time estimates.

The classical industrial engineering approach is utilized to segment the overall operation into components and further decompose each component into its simplest elements. This process reduces the total operation into elemental tasks for which predetermined time standards are readily available.

With standard adjustments for personal time, fatigue and minor delays, the sum of these elemental time values provides the basis for time estimates and facilities and layout planning as well as planning for production, scheduling and control.

Similarly, professional architects, engineers, accountants and others providing services which require the use of skilled judgment are often asked by their clients, "How long? How much?" The professional is able to respond as to when, but to answer how much, a sense of severe discomfort often sets in, for each professional has difficulty estimating the hours of professional time and

the skill levels required to complete a given assignment by a certain date.

This uncertainty arises partly because it is understood that providing professional services requires judgment to be exercised and partly because the tasks one is attempting to estimate are often defined too broadly. When the client poses the question, however, the professional formulates a mental projection of how long the job should reasonably take at a particular skill level.

> " Competitive forces are demanding that the professional service sector provide quality services at a predetermined cost . . . Those committed to providing such service can call upon industrial engineers to scrutinize the efficiency of their practices. "

For example, estimating time for the examination of a financial statement by an independent certified public accountant, the professional subconsciously considers the various components necessary to conduct the audit. Just like the engineer who segments the production process into finite elements, the CPA can also further decompose the various components required in a certified audit into simple sub-tasks.

Although there are no predetermined time standards available, he can confidently provide time estimates and the optimal level of skill for these small sub-tasks. Whereas the overall service may seem difficult to estimate, each independent subtask proves simple to quantify.

Now, more than ever, competitive forces are demanding that the professional service sector provide quality services at a predetermined cost, or if not a finite cost, one which is reasonable in the mind of the client. Professionals committed to providing such service can call upon industrial engi-

Table 1: Sub-Optimal Tasks Performed By Top Level

(1)	(2)	(3)	(4)	(2) × (3)	(2) × (4)	(1) × (3)	(1) × (4)
Billing Rate/Hr.	Scaled Skill Level	Standard Hrs. At Optimum Skill Level	Actual Hours At Skill Levels Performed	Weighted Standard Hours	Weighted Actual Hours	Standard Cost	Actual Cost
$ 20	1	1,000	—	1,000	—	$20,000	—
40	2	—	—	—	—	—	—
60	3	200	—	600	—	12,000	—
80	4	—	—	—	—	—	—
100	5	100	700	500	3,500	10,000	$70,000
Totals:		1,300 hrs.	700 hrs.	2,100 hrs.	3,500 hrs.	$42,000	$70,000
Productivity Percentage = $\frac{\text{Standard}}{\text{Actual}}$ = $\frac{1,300 \text{ hrs.}}{700 \text{ hrs.}}$ = 186%				= $\frac{2,100 \text{ hrs.}}{3,500 \text{ hrs.}}$ = 60%		Actual cost exceeded standard by 67%	

neers to scrutinize the efficiency of their practices.

We are, in essence, experiencing a "professional service revolution." Those who provide professional services should prepare themselves to take advantage of the classical industrial engineering techniques which have evolved from the "industrial revolution." Just as various sciences emerged over the years to meet the changing needs of the times (methods engineering, safety engineering, engineering economy, etc.), the evolutionary process moves on.

PROPLAN was specifically designed to address the dilemma facing the professional service sector. Tailored to scientifically plan and manage non-repetitive, unpredictable and discretionary judgment, the automated system reduces the uncertainty inherent in the budgetary planning process.

It systematically reduces a professional service into a series of quantifiable sub-tasks, enabling the professional to manage his services in a manner which optimizes the cost benefit ratio to the client while providing the professional with fair remuneration.

Prior to investigating the feasibility of PROPLAN as a means of improving productivity, parameters need be defined.

If industrial engineers are to concern themselves with "professional services engineering," they should be equipped with a meaningful definition of the term "professional productivity" as well as a clear understanding of the problems with which the provider of such services is faced.

How can we know that the functions performed by the professional are "worthwhile" in the sense that the client will pay for them? How will the client react to the architect who bills for a considerable amount of time spent on many preliminary drafts that wound up in the wastebasket? Is this considered productive time?

Outside the professional realm, productivity is generally measured by a ratio of output per unit of resource input. Using this ratio, if we were to define input as time (billable hours) and output as the "end product" of the service process, then output would be the financial statement for the accountant, the drawing for the architect, the affidavit or contract for the attorney or the blueprint for the tooling engineer.

Still, to measure productivity solely as a function of time is to discount the bottom line of all professional services—quality judgment at minimal cost. What we really should be measuring, then, is the value that the service ultimately provides the client against the cost to that client. Herein lies the dilemma.

Suppose productivity in a professional environment is loosely inter-preted as the efficiency with which quality judgment is provided by the resources utilized. How do we measure quality judgment? The complexities associated with quality judgment at optimum skill levels force industrial engineers to broaden the scope of their definition.

The unit of input must include not only staff hours, but also the level of skill at which those hours are performed measured against the staff hours at the optimum level of skill for each task. It is further complicated by the harsh reality that the higher the skill level required, the rarer the individual is who possesses such skill.

This concept of measuring productivity with input as a function of two variables (billable hours and skill level) establishes a common thread among all service organizations. Our objective is twofold:

☐ To maximize the billable hours per available skilled hours of input.
☐ To minimize the cost to the client (i.e., task completion at the lowest skill level capable of performing that task).

We will demonstrate one way of "quantifying" skill so as to measure "professional productivity."

In time-critical situations top level people are often heard saying, "I'll do it myself. I can do it in half the time." This already tempting do-it-yourself syndrome becomes more attractive as we face the unfortunate

Table 2: Sub-Optimal Tasks Performed By Next Level

(1)	(2)	(3)	(4)	(2) × (3)	(2) × (4)	(1) × (3)	(1) × (4)
Billing Rate/Hr.	Scaled Skill Level	Standard Hrs. At Optimum Skill Level	Actual Hours At Skill Levels Performed	Weighted Standard Hours	Weighted Actual Hours	Standard Cost	Actual Cost
$ 20	1	1,000	—	1,000	—	$20,000	—
40	2	—	750	—	1,500	—	30,000
60	3	200	—	600	—	12,000	—
80	4	—	150	—	600	—	12,000
100	5	100	100	500	500	10,000	10,000
Totals:		1,300 hrs.	1,000 hrs.	2,100 hrs.	2,600 hrs.	$42,000	$52,000

$$\text{Productivity Percentage} = \frac{\text{Standard}}{\text{Actual}} = \frac{1,300 \text{ hrs.}}{1,000 \text{ hrs.}} = 130\% \qquad = \frac{2,100 \text{ hrs.}}{2,600 \text{ hrs.}} = 80\%$$

Actual cost exceeded standard by 24%

fact that this conception is not an overstatement. In all likelihood, a project manager *can* do his or her staff's tasks in half the time. But, let's examine the cost differential incurred when high level personnel perform tasks which, at optimal performance, can be delegated downward.

For illustration purposes, assume that there exist five levels of skill with billing rates incrementing at $20/hour. We can now create a scale of skill levels proportional to these rates and use this scale to weight the hours.

Suppose the tasks of an engineering project *optimally* require 1,300 hours (i.e., 1,000, 200 and 100 hours of skill levels one, three and five, respectively). Suppose further that all tasks were actually completed by the skill level five person. Assuming he can perform the tasks of levels one and three in half the time (i.e., 500 and 100 hours, respectively), the project would be completed in 700 hours, as illustrated in Table 1.

If we were to simply use hours as our source of input, it would appear that the service was performed at 186% efficiency; i.e., a job estimated at 1,300 hours was completed in 700 hours. On the surface, such performance seems impressive. However, taking into consideration the skill level at which these tasks were performed reveals a grim performance of 60% efficiency. Thus, not utilizing staff at their optimal potential increases the cost to the client by a whopping 67%.

Furthermore, consider the following:

The increase in revenue (even if fully realized at the actual cost to the client of $70,000) would have consumed 700 hours from the highest skilled individual. Suppose each professional were able to generate 1,600 billable hours annually when properly supervised. Then, the level five person could only generate $160,000 of revenues as a result of his own labors (i.e., 1,600 hours @ $100 per hour), whereas if he effectively supervised supporting personnel the optimal revenue he could generate would jump from $160,000 to $672,000 calculated as in Table 3.

The professional manager has a choice—the ease of doing it alone versus leveraging oneself through staff.

A more realistic situation might be that tasks are performed one level higher than the optimum. If the highest skill level can perform the tasks in half the time, we will assume in this example that the next consecutive level can perform the tasks in 75% of the time as illustrated in Table 2.

Here again, measuring input strictly as a function of time is misleading and camouflages opportunities for improvement. Redefining input to be a function of both time and skill level yields a true efficiency of 80% rather than 130%. Since tasks were not fully delegated to their optimal levels of performance, the cost to the client is increased by 24%.

The shortage of qualified personnel at higher levels is the most critical problem shared by most professional firms. In managing our practices, it is crucial that we encourage staff to seek those tasks which can be delegated. For an enlightening and amusing discussion of delegation, see William Oncken, Jr., and Donald L. Wass, "Management Time: Who's Got the Monkey?" (*Harvard Business Review*, November-December 1974, p. 75). As can be seen from Table 1, the proper matching of skill levels to tasks enables a level five

Table 3: Results Of Effectively Supervising Support Personnel

	100 level 5 hours @ $100 per hour =	$ 10,000
	200 level 3 hours @ 60 per hour =	12,000
+	1,000 level 1 hours @ 20 per hour =	20,000
	Revenue generated per 100 level 5 hours =	$ 42,000
	Annualized revenue generated per 1600 level 5 hours =	$672,000

professional to create an additional 600 available hours with which to leverage his skill. Only if a conscious effort is made to analyze each task can we ever hope to utilize our personnel to their fullest potential.

PROPLAN systematically forces this "task evaluation process" from which the derived benefits are realized from three perspectives: The cost to the client is minimized; collectibility and profitability are maximized through leveraging highly skilled personnel; and maximum growth opportunities are available for qualified staff—hence, reduced attrition of talented staff. ⬛

Doris Magliola-Zoch is manager of advisory services for Weiner & Company, Certified Public Accountants where she is responsible for staff productivity, systems, procedures, planning and scheduling. She has also designed and implemented an on-line time management system for professional services and provided management consulting services for a variety of small to international companies. Zoch is a senior member of IIE, holds the office of secretary for the Metropolitan New Jersey chapter and is chairperson of the public relations and facilities committees. She has been appointed to serve on the management advisory services technical and industry consulting practices committee for the American Institute of Certified Public Accountants. Zoch is a member of the management advisory services committees of both the New York and New Jersey State Societies of CPAs. She has a B.A. in mathematics and an M.S. in industrial engineering.

Ronald G. Weiner is managing partner of a certified public accounting firm of 100 people, including seven partners. He is responsible for long range planning, establishment of an independent quality control department, development of internal reporting systems and the creation of a highly qualified staff. Previously, he was client administrator at the firm, aiding small and medium sized businesses to develop the capability for accelerated growth. Weiner has also been a management consultant and accountant in international consulting firms. He is a CPA in New York, New Jersey and Pennsylvania and is a member of the American Institute of Certified Public Accountants.

*Reprinted from **Industrial Engineering**,*
September 1984.

Plan Applies IE Concepts To Improve Productivity And Measure Creative Process Of Professionals

By Doris Magliola-Zoch
and Ronald G. Weiner
Weiner & Co., CPAs

☐ *In part one of their two-part series on managing professional skills in service organizations, Zoch and Weiner looked at how the application of various industrial engineering concepts can improve productivity among this hard-to-measure group of workers. In the following part two, they detail how these concepts were applied specifically to their own certified public accounting firm to measure work and improve productivity there.*

To improve productivity you must first have a benchmark from which to measure. It is a common misconception that the productivity of professional services is immeasurable. The confusion evolves because such measurement entails quantifying an intangible—the creative thought process.

Frank B. and Lillian M. Gilbreth were instrumental in providing formal graphic means of breaking a large problem into smaller problems. Among other things, they originated micromotion study and cyclegraphic records. They stated, "Its skillful application is an art that must be acquired, but its fundamental principles have the exactness of scientific laws which are open to study by everyone." We have applied principles similar to those utilized by the Gilbreths to reduce our problem into smaller ones at our certified public accounting firm.

Problems of growth

Our certified public accounting firm has experienced significant growth since 1976, having grown from a dozen people to approximately 100. During this period of rapid expansion, it became apparent that we were not able to coordinate the scheduling of personnel well enough to meet our client servicing objectives. Inexperienced personnel were always obtainable, but the people who were highly skilled were always in demand.

In our profession, scheduling is generally performed by persons not formally trained in time management. It took approximately one year to identify the profession of industrial engineering as the logical source for an individual of the skills

Figure 1: Breakdown of Function into Tasks

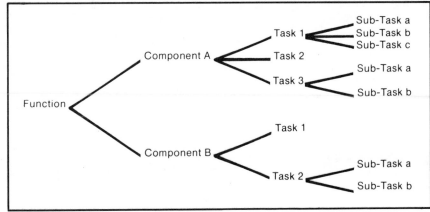

required, six months to convince the firm to try someone from a profession other than public accounting and six months to find the person.

We are now in the fourth year of implementation of the procedures outlined below. We have increased billable hours per professional by 18% and believe we have reduced the amount of time necessary to service clients by about 20%. In addition, our upper level personnel have significantly more discretionary time available than previously.

Budgetary process

It will be demonstrated that by transforming mental projections into a clearly defined detailed time budget, innumerable benefits can be realized. The recommended budgetary process does not guarantee success on every engagement, but it has proven to increase both the efficiency and the effectiveness of the services provided. If the process has not recently been subjected on an objective review, then in all likelihood it can be substantially improved by this exercise.

Before launching into the budgetary process, a standard format should be designed to provide the basis for budgeting any engagement. It should be tailored to precisely accommodate the unique needs, procedures and terminology of your firm.

The approach recommended is a "top-down approach" as opposed to a "bottom-up approach." Rather than sequentially listing those tasks which should be accomplished first, second, etc., work backwards. Starting with the final product, repetitively reduce each function to a simpler and simpler task.

The four steps may seem rather elementary, but don't be deceived by their simplicity. They will provide the foundation for the entire process—the keynote is "be meticulous."

Step One: List all the professional services provided by your firm.

Example: Our CPA firm includes audit, review, compilation, corporate or partnership tax return, individual tax return, consulting, planning, etc.

Step Two: List all the functions which must be performed to provide each of the services in step one.

Example: In order to submit a financial statement to a client, the CPA prepares a report letter, balance sheet, statement of income and retained earnings, statement of changes in financial position, supplementary information and footnotes.

Step Three: Break the functions down into elemental activities. A schematic tree diagram should be structured such that the branches divide each function into their components, components are sub-divided into tasks and tasks into sub-tasks until we reach an element which can be broken down no further. (See Figure 1.) We will call this a "sub-task." It is a necessary condition that sub-tasks be independently cohesive activities.

Step Four: Analyze the hierarchy, carefully verify the constituents of each branch, categorize, indent respectively and assign group numbers. These numbers will be referred to as "workcodes." (See box.)

We have now constructed the foundation for PROPLAN (PROfessional PROductivity PLAnning). The rest is easy. Using the list of workcodes derived in step four, we are ready to prepare a meaningful budget for every engagement in your organization.

For optimal results, the following guidelines should be carefully followed when preparing the detailed time budget. A well established budget is a vital initial condition if your

Example: Workcodes For Branch Analyzation

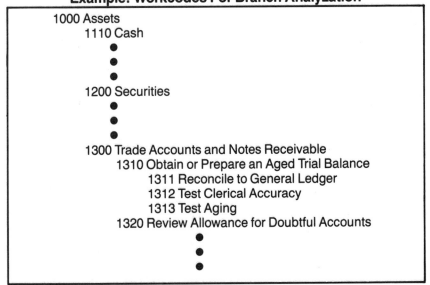

1000 Assets
  1110 Cash
   •
   •
   •
  1200 Securities
   •
   •
   •
  1300 Trade Accounts and Notes Receivable
    1310 Obtain or Prepare an Aged Trial Balance
     1311 Reconcile to General Ledger
     1312 Test Clerical Accuracy
     1313 Test Aging
    1320 Review Allowance for Doubtful Accounts
     •
     •
     •

Table 1: Actual Vs. Budget Exception Report

(0001) Partner Name 10/27/81 PAGE 3			**** INTERACTIVE ACCOUNTING SYSTEM **** THE FOLLOWING JOBS HAVE EXCEEDED THEIR BUDGET BY MORE THAN 10%					% OVER BUDGET		
C/J NO.	C/J NAME	PERIOD CODE	ACTUAL HOURS	BUDGET HOURS	ACTUAL DOLLARS	BUDGET DOLLARS		HOURS	DOLLARS	COMMENT
00050/00000	CLIENT A	00	196	0	7244.59	0.00		****%	****%	
		06	8	0	343.29	0.00		****%	****%	
		07	29	0	2964.39	0.00		****%	****%	
		01	6	0	534.00	0.00		****%	****%	
		02	21	0	588.00	0.00		****%	****%	
		03	3	0	324.00	0.00		****%	****%	
	TOTALS		263	0	11998.30	0.00		****%	****%	
00050/00004	CLIENT B	06	3	0	135.00	0.00		****%	****%	
		07	0	0	18.59	0.00		****%	****%	
	TOTALS		3	0	153.60	0.00		*:**%	****%	

appraisals later are to be intelligible. (See Tables 1-4).

1.) For the service you are attempting to budget, determine the required activities using your list of workcodes.

2.) Ascertain the lowest possible skill level at which each sub-task can be performed.

3.) Assume the lowest level and provide a time estimate for each activity at that level. If any estimate is greater than three hours, budget the time using the next lower indented level of workcodes. Suppose that a component is estimated to take 2.5 hours; then, it is not necessary to budget any time for the tasks and sub-tasks which comprise it.

Little if any scientific data can be found to support the viewpoint that any specific interval of time is best. The optimal interval will vary from situation to situation. Good judgment should be exercised to assure maximum effectiveness. Twenty-minute intervals are not uncommon in a highly repetitive, fast-moving business. Complex operations which are non-repetitive in nature can range as high as four-hour intervals.

From our experience we have found that three-hour intervals are most suitable in our professional service environment. An experienced industrial engineer should be able to guide a firm toward ascertaining an optimum.

4.) Analyze and determine the functional value that each activity contributes to the ultimate goal of the engagement. If a particular task is insignificant, eliminate it. It is suggested that technical support be called upon to add insight when determining unnecessary tasks.

Additional time invested in this stage of planning can easily be recaptured tenfold during the engagement. If, for example, the accountant determines that his client's internal accounting control over cash is strong, he can limit his cash procedures.

5.) Plan the optimal sequence in which the tasks should be performed.

6.) Determine the earliest time frame (week or month) during which each activity can reasonably commence. The accountant can expedite the engagement by performing various tasks prior to the balance sheet date. One example is testing for additions to property and equipment. Although this may be obvious to the supervisor, the junior is inclined to overlook such "tricks of the trade."

In a professional environment it is critical that time budgets be derived participatively. Subordinates who will be responsible for achieving the goal should be asked to provide input. Doing so can avoid the potential situation of setting unrealistic time estimates which would only frustrate the staff and net undesirable effects.

Any individual who has participated in the budgetary planning process feels responsible for seeing that the goals are achieved and is often motivated toward task performance. Rather than channeling his efforts dysfunctionally, he commits his goals to be cohesive with the interests of the firm.

Monitoring the service

An excellent plan poorly monitored is of little value; implementation is critical. Although providing the service is a dynamic process, it can be approached systematically by utilizing the detailed time budget as the vehicle for controlling progress.

A copy of the budget should be distributed to every staff person assigned to the job. It clearly specifies who is assigned to each task, how long each task is expected to take, the anticipated commencement date and the sequence in which these tasks should optimally occur.

The key to success lies in communication. All are aware of their position in the engagement and what is expected from them.

Since all tasks are short intervals, a monstrous project is now reduced to a series of realistically attainable short term goals. Inasmuch as the staff partook in the budgetary process, they feel responsible for its achievement.

Communication is only half the formula for success. The other necessary ingredient is follow-up. A partner in our firm can often be heard saying to his managers and supervisors, "People do not do what is expected... They do what is inspected." Examine the follow-up process and develop a method of reconciling the actual time incurred against the budgeted estimate.

Since many professional services effectively bill based on time incurred, it follows that they must, in some way, identify the time expended on each project. Our firm automated the entire process. The time budgeted for each sub-task is stored in the computer.

During the engagement, the project team records the time spent on each sub-task as it is performed.

Table 2: Efficiency Performance Report

TREND ANALYSIS FOR: _____ SAMPLE EMPLOYEE _____
Name

| Month Ending | CURRENT PERIOD | | | | EFFICIENCY PERCENTAGES | | CUMULATIVE | | | | EFFICIENCY PERCENTAGES | | COMMENT |
	(1) Budget Hrs.	(2) Actual Hrs.	(3) Budget Dollars	(4) Actual Dollars	Time (1) ÷ (2) Budget Hrs. Actual Hrs.	Dollars (3) ÷ (4) Budget $ Actual $	(1) Budget Hrs.	(2) Actual Hrs.	(3) Budget Dollars	(4) Actual Dollars	Time (1) ÷ (2) Budget Hrs. Actual Hrs.	Dollars (3) ÷ (4) Budget $ Actual $	
8 '82													
9 '82	427	489	$21,136	$27,985	87%	76%	427	489	$ 21,136	$ 27,985	87%	76%	
10 '82	171	250	$ 8,788	$15,100	68%	58%	598	739	$ 29,924	$ 43,085	81%	69%	
11 '82	492	576	$26,521	$25,728	85%	103%	1090	1315	$ 56,445	$ 68,813	83%	82%	
12 '82	94	80	$ 4,359	$ 3,479	118%	125%	1184	1395	$ 60,804	$ 72,292	85%	84%	
1 '83	50	48	$ 2,437	$ 2,006	104%	121%	1234	1443	$ 63,241	$ 74,298	86%	85%	
2 '83	1325	1505	$64,455	$79,150	88%	81%	2559	2948	$127,696	$153,448	87%	83%	
3 '83	504	507	$27,191	$26,746	99%	102%	3063	3455	$154,887	$180,194	89%	86%	
4 '83	62	92	$ 3,348	$ 4,226	67%	79%	3125	3547	$158,235	$184,420	88%	86%	
5 '83	123	95	$ 6,328	$ 5,011	129%	126%	3248	3642	$164,563	$189,431	89%	87%	
6 '83	507	413	$26,257	$20,569	123%	128%	3755	4055	$190,820	$210,000	93%	91%	
7 '83	416	422	$20,852	$21,436	99%	97%	4171	4477	$211,672	$231,436	93%	91%	

Time sheets are submitted and keyed semi-monthly. The on-line system produces a "Client Actual versus Budget Report" as needed. This feature reconciles the actual time expended against the budget by task or sub-task, enabling the supervisor to flag deviations.

Periodic surveillance by the supervisor enables one to quickly detect problems and alter the remainder of the plan if necessary. With the use of the Client Actual versus Budget Report, one is equipped to attack the problem and make spontaneous, reliable and effective changes.

The importance of the supervisory role during implementation cannot be over-emphasized. PROPLAN is not intended to be a substitute for weak supervision. It is a tool with which supervisors can effectively monitor time. If during the engagement, the principles of PROPLAN are not adhered to, the budgetary process is pulverized to an academic exercise in futility.

The onus of properly allocating resources lies with management. Negative results can be avoided by planning, re-evaluating and properly utilizing manpower during every stage of the engagement.

Dealing with problems

For example, it is critical that the supervisor continue utilizing the budget in a participatory fashion by jointly identifying problems and solutions rather than finger-pointing. If performance begins to deviate, it is not the time for heads to roll; rather, it is time for the project team to meet, determine why the digression has occurred, and explore alternative solutions.

The PROPLAN approach flags potential problems with sufficient lead time before they mature to the crisis stage. Bear in mind that our objective is to prevent fighting the fires of "undertime" ignited by inadequate controls. Thus, excessive "overtime" at the end of the engagement is avoided, we capitalize on available resources and deadlines are still met.

Although the pressure of rolling heads may prove effective in the short run, it builds negative expecta-

tions. There are sundry dangers ascribable to finger-pointing, all of which stem from a desperate fear to "meet the budget at any cost." Such instinctive reactions of human behavior expose the following four basic risks:

1.) Building cushions in the budget.
2.) Forcing reconciliation by transferring actual time from one work task to another, ensuring that the actual vs. budget reconciles "on paper."
3.) Understating actual time to "look good."
4.) Fudging actual time to preserve the cushion in the budget for the next year.

Once this destructive-type atmosphere sets in, morale drops, confidence lowers, the staff loses interest in its work and future projects are doomed. Within our CPA firm, for example, employees are encouraged to accurately record their time. The partnership is not interested in "camouflaging" our problems. We want to see the results exactly as they are—"warts and all."

Our CPA firm further utilizes the

Table 3: Detailed Time Budget

IN-CHARGE NAME: _____

CLIENT: (01138)—CLIENT XYZ
JOB: (00000)—JOB NAME
PRT: (2010)—PARTNER NAME & MANAGER

BUDGET DESCRIPTION: REVIEW
PERIOD: YEAR END
FYE: 01/31/84

LEVEL	WORKCODE		START DATE	BUDGET HOURS	REVISED HOURS	WORKSPACE ASSIGNED STAFF	COMMENT
AUDIT	0210	FOOTING AND TRACING TO	03/84	2.0			
JUNIOR	0312	PREP. OR TEST BANK REC.	03/84	1.0			
	0331	PREP/OBTAIN/TRACE AR TO	03/84	1.0			
	0336	REVIEW OF SUB CASH RECI	03/84	2.0			
	0390	OTHER ASSETS	03/84	1.5			
	0521	RENTAL EXPENSE	03/84	0.2			
	0522	LEGAL AND PROF EXPENSES	03/84	0.5			
	0524	OTHER TAXES	03/84	0.2			
	0525	MISCELLANEOUS EXPENSES	03/84	1.5			
	0631	PREP REPORT LETTER	03/84	0.5			
	0632	PREP BALANCE SHEET	03/84	0.5			
	0633	PREP INC STATEMENT & R	03/84	0.5			
				11.4	_____		
TOTAL	AUDIT JUNIOR HOURS			410.4			
TOTAL	AUDIT JUNIOR DOLLARS						
AUDIT	0160	PERMANENT FILE UPDATE	03/84	2.0			
SENIOR	0190	SUPERVISION	03/84	2.0			
	0220	POSTING ADJUSTMENTS	03/84	1.0			
	0338	REV ALLOW ACCT & BAD DEBT	03/84	1.0			
	0350	FINAL INVENTORY LISTING	03/84	2.0			
	0360	PREPAID INSURANCE	03/84	1.8			
	0381	PREP OF/OBTAIN SCHEDULE	03/84	2.0			
	0382	RECON OF ADD & DELETION	03/84	0.5			
	0383	DEPR EXP: TIE OUT EXP/AC	03/84	0.8			
	0390	OTHER ASSETS	03/84	1.5			
	0411	PREP & TEST ACCUR: REC	03/84	1.5			
	0415	SEARCHING UNRECORDED LI	03/84	2.0			
	0421	PREP/OBTAIN SCH: RECON G	03/84	2.0			
	0424	INTEREST EXP TIE-OUT	03/84	1.5			
	0430	OTHER LIABILITIES	03/84	1.0			
	0440	PAYROLL	03/84	1.5			
	0461	CURRENT INC. TAX LIAB. &	03/84	1.0			
	0470	STOCKHOLDER'S EQUITY AC	03/84	0.2			
	0494	CLIENT REP. LETTER	03/84	0.5			
	0522	LEGAL AND PROF EXPENSES	03/84	0.3			
	0525	MISCELLANEOUS EXPENSES	03/84	1.0			
	0541	RATIO ANALYSIS	03/84	1.5			
	0613	PREP. OF ENGAGEMENT LET	03/84	0.5			
	0620	PREP INTER-OFFICE MEMO	03/84	1.0			
	0631	PREP REPORT LETTER	03/84	0.2			

data base as a tool for managing the practice and has derived many corollary benefits as well. Since the firm has experienced rapid growth, the on-line system has been an invaluable resource in making intelligent decisions regarding forward planning, scheduling and control.

In doing so, it successfully avoids the common trap of "seat-of-the-pants" type management which inevitably manifests itself in a rapidly growing firm. In an effort to avoid the typical pitfall of generating a voluminous amount of paperwork, we produce various reports on an exception basis. Summarized below are some of the supplementary benefits realized by our firm.

Billing: To assist in billing control, we regularly generate an exception report which highlights those clients having an unbilled amount greater than $3,000 regardless of current period time charges.

Work In Process: Summary reports recap the total charges, billing status, profitability and realization percentage of work in process.

Scheduling: It has been postulated that work expands to the time allotted. Hence, the surest way of over-running a budget is to over-schedule the job initially. To avoid this situation, the scheduling department is furnished with a budget summary report showing total hours budgeted per month by staff level.

Using the budget

Let's assume that the supervisor requests a junior for three days (or 22.5 hours). Using the budget summary report, the scheduler verifies the request against the budget. Not only does this avoid possible over-scheduling "before the fact," but it also confirms that the engagement is staffed at optimal levels. It is a natural innate tendency to request senior staff members even if a junior can reasonably be expected to handle the job.

From our experience, this control has seemingly satisfied two objectives established by the partnership. It fulfills an obligation to the firm's clientele that, wherever practical, functions will be performed at the least expensive dollars. Likewise, it provides a challenging experience to its employees by avoiding the frustrating predicament of having them work beneath their level.

Budget Overruns: An exception report is generated to flag significant deviations from budget. If, for example, a particular engagement has exceeded its budget by more than, say, 10%, it will be noted on the exception report.

Staff Evaluations: The system provides additional insight during the staff evaluation process. For each employee, actual time and percent of total time are accumulated by subtask as well as by client.

At lower levels, these reports have assisted scheduling in assuring that the staff members are being exposed to all functions. At higher levels, the information enables personnel to focus on whether sufficient time is allocated to such areas as supervision, technical issues and client development.

Another interesting by-product of the firm's on-line system is the individual efficiency trend analysis. This report is particularly informative about managers and supervisors. Although a supervisor may have legitimate reasons for going over budget on any given engagement, one should be suspicious if the aggregation of his or her clients is over budget and there is a low realization percentage.

Similarly, the law of large numbers supports the hypothesis that it is not by coincidence that the overall jobs of a "good" supervisor lie within the expected budget and generate a high realization rate.

Forecasting: From the firm's experience, probably the most powerful advantage of the system is the ability to accurately ascertain optimum manpower requirements. Considering the difficulty in obtaining quali-

Table 4: Weiner and Company Client Actual vs. Budget — Current: 1/01/84 TO 4/30/84

	ACTUAL		BUDGET		DIFFERENCE	
	HOURS	DOLLARS	HOURS	DOLLARS	HOURS	DOLLARS
CLIENT: (00672)—CLIENT NAME						
JOB: (00000)—						
330 ACCOUNTS RECEIVABLE 1.17%						
330 364 EMPLOYEE A 2/15/84	1.50	67.50				
CURRENT	1.50	67.50	2.50	100.50	(1.00)	(33.00)
TO DATE	1.50	67.50	2.50	100.50	(1.00)	(33.00)
360 PREPAID INSURANCE 1.17%						
360 359 EMPLOYEE B 2/15/84	1.50	57.00				
CURRENT	1.50	57.00	3.00	129.00	(1.50)	(72.00)
TO DATE	1.50	57.00	3.00	129.00	(1.50)	(72.00)
380 PROPERTY, PLANT AND EQUIPMENT 1.40%						
380 359 EMPLOYEE B 2/15/84	1.80	68.40				
CURRENT	1.80	68.40	7.00	266.00	(5.20)	(197.60)
TO DATE	1.80	68.40	7.00	266.00	(5.20)	(197.60)
390 OTHER ASSETS 4.91%						
390 359 EMPLOYEE B 2/15/84	0.80	30.40				
390 364 EMPLOYEE A 2/15/84	0.50	22.50				
391 359 EMPLOYEE B 2/15/84	1.50	57.00				
391 369 EMPLOYEE C 2/28/84	1.00	71.00				
392 364 EMPLOYEE A 2/15/84	1.50	67.50				
399 364 EMPLOYEE A 2/15/84	1.00	45.00				
CURRENT	6.30	293.40	4.00	154.50	2.30	138.90
TO DATE	6.30	293.40	4.00	154.50	2.30	138.90
410 ACCOUNTS PAYABLE & ACCRUED EXP 2.96%						
410 364 EMPLOYEE A 2/15/84	1.50	67.50				
411 364 EMPLOYEE A 2/15/84	1.00	45.00				
415 364 EMPLOYEE A 2/15/84	1.30	58.50				
CURRENT	3.80	171.00	4.50	162.00	(0.70)	9.00
TO DATE	3.80	171.00	4.50	162.00	(0.70)	9.00

fied experienced personnel, this benefit transforms an attractive feature into a desperate necessity.

Recall that the budgeted time estimates are based on the lowest staff level capable of performing each sub-task and also assume the earliest time frame. Ideally speaking, then, these budgets represent the "best" way to staff the job. Therefore, the sum of all time estimates, for all sub-tasks, over all engagements, yields the billable time demands for each staff level by month.

With minor adjustments for vacation time, sick time, unbudgetable billable time, seminar allowances and growth, the firm is able to determine optimum manpower requirements for one year rolling forward.

Scientific thinking

This movement toward scientific management has continually been poisoned by the pessimism that is normally embedded with "change." While feasible in the long run, scientific management in the short run does not appear attractive with respect to time, cost or acceptability. Scientifically savvy executives, who perceive that these short term factors are investments for long term productivity enhancement, often get little support from their associates.

A glimmer of hope is that although pessimism spreads rapidly, it loses potency with time. A determined executive realizes that neither difficulty, pessimism nor the socio-political considerations constitute sufficient reasons for neglecting a productive opportunity.

A progressive executive from a growing firm has no choice. He must address a wide spectrum of demands daily. He must constantly consider the legitimate needs of his clients as well as the practical concerns of his staff. He must adhere to the standard of providing quality services and maintaining sound controls.

He must resolve the inevitable conflicts that arise in a manner that will be in the overall best interests of client and firm alike. He must carve out the time to tend to fires as they ignite, anticipate problems before they arise and maintain control across the board.

Yet, through it all, he must plan for future growth and prepare the firm to "comfortably" take the next step. Scientific thinking enables the professional to curb this nightmare. It is a movement to "manage" rather than "juggle" this conflict of needs between quality, timeliness and profitability. Chargeable time is the life-blood of a professional practice. And scientific thinking cultivates it.

Tomorrow?

Sources of revenue in our post-industrial society have dramatically shifted from labor intensive to knowledge intensive. Yet, are we positioned to effectively manage knowledge and information skills that characterize this intellectual society in which we live?

Is our concern for improving professional productivity any different from that which triggered national concern for improving the efficiency of American industry in the late 1800's and the early 1900's? We think not.

We believe that we are in a professional service revolution—one that will allow us to witness the unveiling of another branch of science: professional services engineering. The result? Professional vehicles such as PROPLAN. ▣

Doris Magliola-Zoch is manager of advisory services for Weiner & Company, Certified Public Accountants where she is responsible for staff productivity, systems, procedures, planning and scheduling. She has also designed and implemented an on-line time management system for professional services and provided management consulting services for a variety of small to international companies. Zoch is a senior member of IIE, holds the office of secretary for the Metropolitan New Jersey chapter and is chairperson of the public relations and facilities committees. She has been appointed to serve on the management advisory services technical and industry consulting practices committee for the American Institute of Certified Public Accountants. Zoch is a member of the management advisory services committees of both the New York and New Jersey State Societies of CPAs. She has a B.A. in mathematics and an M.S. in industrial engineering.

Ronald G. Weiner is managing partner of a certified public accounting firm of 100 people, including seven partners. He is responsible for long range planning, establishment of an independent quality control department, development of internal reporting systems and the creation of a highly qualified staff. Previously, he was client administrator at the firm, aiding small and medium sized businesses to develop the capability for accelerated growth. Weiner has also been a management consultant and accountant in international consulting firms. He is a CPA in New York, New Jersey and Pennsylvania and is a member of the American Institute of Certified Public Accountants.

*Reprinted from **Industrial Engineering**,*
January 1984.

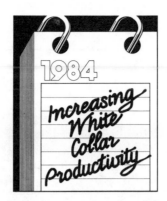

IEs Measure Work, Write Standards For White Collar Workers At Financial Institution

By G. Michael Anthony
Seattle-First National Bank

Productivity for white collar employees is becoming an increasingly important concern. The United States has already achieved high levels of productivity in the farm and manufacturing sectors and must now look towards achieving comparable results in the white collar area.

According to the Bureau of Labor Statistics, the white collar sector is about 64% of the total labor force; the blue collar is about 33%; and farm workers are about 3%. Yet only about 20% of industrial engineers are engaged in analyzing white collar productivity, which indicates a vast opportunity for the profession.

The standard classifications for white collar workers are clerical (36%), sales (12%), professional/technical (31%) and managerial/administrative (21%). In banks, the clerical group includes tellers, proof operators, file clerks and other paper-processing employees. Banking customer service representatives are included in the sales group. The professional/technical group includes bank lending officers, recruiters, bond and money market traders, loan examiners, trust officers, etc.

At Seafirst, we are now also writing standards for members of the managerial/administrative group the frontline supervisors and managers. We are achieving good results, although the process is still in the prototype stage. The professional/technical and managerial/administrative areas we have studied include the commercial banking, corporate trust, mortgage marketing, energy, employment, training and development, staff relations, staff activities, public affairs, and loan review and examination departments.

Using a straightforward, traditional process, we enter an area, examine the tasks being performed and see if the work can be done more easily. Then we measure the work.

After we have measured it, we generally find that the user can increase production, or improve quality, or both—thereby achieving higher productivity. Although many people think that professional activities are non-routine and non-repetitive, we have found that if the scale of reference is expanded, they reoccur on a predictable basis.

Basic information

The three basic pieces of information we need to accumulate are descriptions, volumes and times. Descriptions are gathered according to the appropriate level of detail.

For example, it would not be appropriate to describe hand motions if the task is "study credits suitable for syndication," which takes 81 minutes per credit, or "prepare speeches," which takes 104 minutes per speech.

The process of description starts with the purpose, a one-sentence description of the area's mission; this is followed by function, activity, task, step and motion, as appropriate (see Figures 1 through 5). First we analyze from purpose down to motion; later we synthesize from motion up to purpose.

To gather volumes and times for professional/technical staff, we use time diaries, estimates, work sampling and direct observation.

The loan review and examination area provides a good example of how we have used these familiar tools to achieve excellent results. This area performs the quality control function for our loans.

When we interviewed area staff members, we obtained a list of their jobs, coded them and identified count points. We next installed time diaries for each task, on which staff members entered start time, stop time, job code and number of units completed. These time diaries were later computer processed using a simple program which we developed internally.

Most staff members are requested to fill out time diaries for four to six weeks; however, for areas with longer natural cycles, we model a smaller sample for a longer period. This will be discussed later.

Figure 1: Analysis Of Purpose

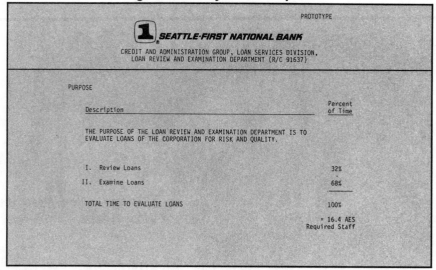

After compiling the data, we analyze total time, total units, time per unit, and the total time percentages for each task. One reason for obtaining a percentage profile is that we want to pick the most significant tasks and return to perform direct observations.

Leveling

Information gathered during time diaries can be validated through direct observation. Because these are long-cycle activities, we do not observe them in their entirety necessarily; rather, we observe selected tasks. Usually, direct observation reveals that such tasks require less time than is indicated by the time diaries.

If, after directly observing other tasks within the process, we experience similar results, we reduce all tasks within the process by a constant percent. We call this technique *leveling*.

By applying this technique, we can obtain reasonable standards for long-cycle tasks without actually observing the entire lengthy process. Watching the complete process of hiring an executive, which would take about 30 calendar days, is not feasible. Therefore, we use selected observational samples and leveling to obtain our standards.

Modeling

The activities of some white collar employees vary according to the time of year. Once we have obtained and leveled a one- or two-month sample, we "model" it for a 12-month period or for whatever the appropriate cycle is. Modeling means spreading the small sample over a year, adding in tasks that didn't occur during the sample period, then weighting them appropriately.

Modeling can be difficult, which is why we frequently return to the area to perform maintenance ahead of our normal schedule. We know that the imponderables are innumerable in such areas.

At each level of description we are looking for action and object; e.g., "The purpose of the loan review and examination (LR & E) department is to *evaluate loans* of the corporation for risk and quality (purpose)," "*sort* and *edit* LR & E reference journal (IC6)" and "*Obtain binder* from shelf (C6b)." Also, we need clear-cut input and output steps; e.g., "*receive* LR & E reference journal ... (C6a)," "... *replace* binder on shelf (C6h)." In addition, we need

Figure 2: Analysis Of Functions

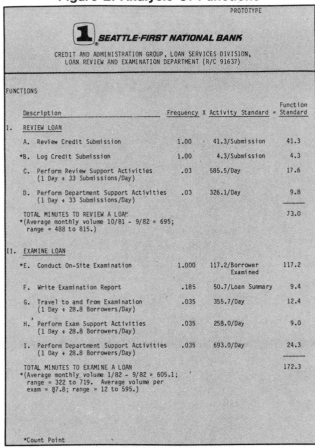

PROTOTYPE

SEATTLE-FIRST NATIONAL BANK

CREDIT AND ADMINISTRATION GROUP, LOAN SERVICES DIVISION,
LOAN REVIEW AND EXAMINATION DEPARTMENT (R/C 91637)

FUNCTIONS

Description	Frequency X	Activity Standard =	Function Standard
I. REVIEW LOAN			
A. Review Credit Submission	1.00	41.3/Submission	41.3
*B. Log Credit Submission	1.00	4.3/Submission	4.3
C. Perform Review Support Activities (1 Day ÷ 33 Submissions/Day)	.03	585.5/Day	17.6
D. Perform Department Support Activities (1 Day ÷ 33 Submissions/Day)	.03	326.1/Day	9.8
TOTAL MINUTES TO REVIEW A LOAN			73.0
*(Average monthly volume 10/81 - 9/82 = 695; range = 488 to 815.)			
II. EXAMINE LOAN			
*E. Conduct On-Site Examination	1.000	117.2/Borrower Examined	117.2
F. Write Examination Report	.185	50.7/Loan Summary	9.4
G. Travel to and from Examination (1 Day ÷ 28.8 Borrowers/Day)	.035	355.7/Day	12.4
H. Perform Exam Support Activities (1 Day ÷ 28.8 Borrowers/Day)	.035	258.0/Day	9.0
I. Perform Department Support Activities (1 Day ÷ 28.8 Borrowers/Day)	.035	693.0/Day	24.3
TOTAL MINUTES TO EXAMINE A LOAN			172.3
*(Average monthly volume 1/82 - 9/82 = 605.1; range = 322 to 719. Average volume per exam = 87.8; range = 12 to 595.)			

*Count Point

Figure 3: Analysis Of Activities

PROTOTYPE

SEATTLE-FIRST NATIONAL BANK

CREDIT AND ADMINISTRATION GROUP, LOAN SERVICES DIVISION,
LOAN REVIEW AND EXAMINATION DEPARTMENT (R/C 91637)

ACTIVITIES

Description	Frequency X	Task Standard =	Activity Standard
B. LOG CREDIT SUBMISSION (CONT.)			
*10. Log out Credit Submission (Manual)	.10	2.3/Submission Out	0.2
11. Batch 3142's	.53	0.8/Batch	0.4
TOTAL MINUTES TO LOG CREDIT SUBMISSION			4.2 USE 4.3
*(Average monthly volume 10/81-9/82 = 695; range = 488 to 815.)			
C. PERFORM REVIEW SUPPORT ACTIVITIES			
6. Sort and Edit LR&E Reference Journal	.200	45/Week	9.0
12. Compile Monthly Statistics (1 Month ÷ 21 Days)	.048	240/Month	11.5
13. Type Review-Related Material	1.000	90/Day	90.0
14. File Credit Submissions and Workpapers	1.000	45/Day	45.0
16. Perform Clerical Activities	1.000	120/Day	120.0
17. Perform Loan Review Supervisor Activities	1.000	60/Day	60.0
43. Maintain Reference Material (25 AES X 1.000)	25.000	10/Staff-Day	250.0
*TOTAL MINUTES TO PERFORM REVIEW SUPPORT ACTIVITIES/DAY			585.5

*Count Point

a good count point; e.g., task: "log credit submission (IB);" count point: "*submission.*" Further, there will be two types of steps: fixed and variable. Examples: fixed: "Aside error 3142's for resubmission with end-of-week input;" variable: "*If* systematic errors are noted, notify ACL."

The number of variable steps will make a substantial difference in the average amount of time it takes to do a task. We include them in an overall time if they are few; or we time these steps separately; or we break them out as a separate task if they are many.

Finally, the levels of precision and accuracy will vary according to each task.

Precision and accuracy

By precision, I mean how many decimal places we use. By accuracy, I mean how close we are, plus or minus, to the average.

Precision will generally depend on the cycle time: short, medium or long. The "Complete Loan Review Worksheet," which is long-cycle, has a cycle time of 51 minutes per submission. Most often, the level of precision for long-cycle will be to the nearest minute.

The accuracy sought will generally depend on the volume of work. If we have only a tenth of a person doing the job, we do not need the same level of accuracy as we would need for 120 people doing that job. We determine our number of observations accordingly.

Product standards

Earlier, I explained that after analyzing the work from macro to micro, simplifying each layer as much as possible, we synthesize the work from micro to macro, simplifying each layer and the relationships between layers. Steps are constructed from motions, tasks from steps, activities from tasks, functions from activities and purpose from functions.

The result is a clear work flow with standards or time profiles. For example, a vertical slice, macro to micro, would be as follows:

"The purpose of the loan review and examination department is to evaluate loans of the corporation for risk and quality."

☐ Function I is "review loans."

☐ Activity I.C. is "perform review support activities."

☐ Task I.C.6. is "sort and edit LR&E reference journal."

☐ Step I.C.6.b. is "obtain binder from shelf."

Figure 4: Analysis Of Tasks

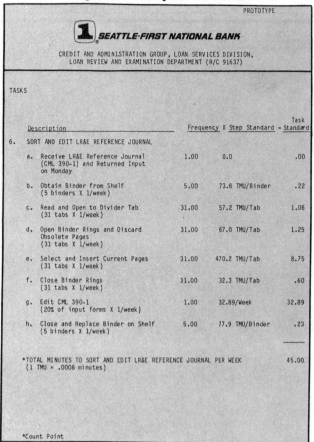

PROTOTYPE

1 SEATTLE·FIRST NATIONAL BANK

CREDIT AND ADMINISTRATION GROUP, LOAN SERVICES DIVISION,
LOAN REVIEW AND EXAMINATION DEPARTMENT (R/C 91637)

TASKS

	Description	Frequency X	Step Standard =	Task Standard
6.	SORT AND EDIT LR&E REFERENCE JOURNAL			
a.	Receive LR&E Reference Journal (CML 390-1) and Returned Input on Monday	1.00	0.0	.00
b.	Obtain Binder from Shelf (5 binders X 1/week)	5.00	73.8 TMU/Binder	.22
c.	Read and Open to Divider Tab (31 tabs X 1/week)	31.00	57.2 TMU/Tab	1.06
d.	Open Binder Rings and Discard Obsolete Pages (31 tabs X 1/week)	31.00	67.0 TMU/Tab	1.25
e.	Select and Insert Current Pages (31 tabs X 1/week)	31.00	470.2 TMU/Tab	8.75
f.	Close Binder Rings (31 tabs X 1/week)	31.00	32.3 TMU/Tab	.60
g.	Edit CML 390-1 (20% of input forms X 1/week)	1.00	32.89/Week	32.89
h.	Close and Replace Binder on Shelf (5 binders X 1/week)	5.00	77.9 TMU/Binder	.23

*TOTAL MINUTES TO SORT AND EDIT LR&E REFERENCE JOURNAL PER WEEK 45.00
(1 TMU = .0006 minutes)

*Count Point

Figure 5: Analysis Of Steps

PROTOTYPE

1 SEATTLE·FIRST NATIONAL BANK

CREDIT AND ADMINISTRATION GROUP, LOAN SERVICES DIVISION,
LOAN REVIEW AND EXAMINATION DEPARTMENT (R/C 91637)

STEPS

	Description	Step Standard
6a.	RECEIVE LR&E REFERENCE JOURNAL (CML 390-1) AND RETURNED INPUT ON MONDAY	0.0
6b.	OBTAIN BINDER FROM SHELF	73.8 TMU/Binder
i.	Reach to binder R24B	
ii.	Grasp G1A	
iii.	Move to table M24B	
iv.	Release RL1	
v.	Reach to binder cover R10B	
vi.	Grasp G1A	
vii.	Open cover M10B	
viii.	Release RL1	
6c.	READ AND OPEN TO DIVIDER TAB	57.2 TMU/Tab
i.	Look to tabs ET 8/16	
ii.	Read tabs EF x 3	
iii.	Reach to tab R10B	
iv.	Grasp G1A	
v.	Move to open M10B	
vi.	Release tab RL1	
6d.	OPEN BINDER RINGS AND DISCARD OBSOLETE PAGES	67.0 TMU/Tab
i.	Reach to ring release R8B	
ii.	Grasp G1A	
iii.	Press down AP1	
iv.	Open MfA	
v.	Release RL1	
vi.	Reach to pages R8B	
vii.	Grasp G1A	
viii.	Move to trash M24B	
ix.	Release RL1	

☐ Motion I.C.6.b.i. is "reach to binder."

A horizontal slice at the activity level is shown by activities A through I in Figure 2's function breakout.

Therefore, all work can be defined and interconnected with as much detail as is necessary. Each item appears as part (except purpose) and whole (except motion).

The final product is a manual, of which Figures 1 through 5 are an extract.

The primary use of standards is to establish appropriate staffing indicators that can go from a simple, one-time application to a continuous, volume-sensitive reporting system, identifying problems and opportunities.

In addition, there are many secondary uses throughout the corporation in such areas as compensation, systems, organization and costing analysis; writing of job descriptions, procedures and operating manuals; and structuring jobs that don't yet exist.

Conclusion

As we turn to a service-based economy, particularly in the rapidly changing financial industry, we have the same need for work rationalization as other industries. We have adapted the traditional techniques not only to service tasks in general, but also to the more complex technical/professional and managerial/administrative functions.

Simply put, standards for white collar employees differ in degree, but not in kind, from standards for manufacturing employees. **IE**

G. Michael Anthony is an assistant vice president at Seattle-First National Bank. Prior to joining Seafirst, he worked for two different consulting firms. Anthony has a BA from St. John's College, Annapolis, MD, and an MA from the University of Oregon. He was editor of *IIE News: Banking and Financial Services* last year. This article was adapted from a lecture given to the Puget Sound chapter in September 1982.

*Reprinted from 1984 Fall Industrial
Engineering Conference Proceedings*

THE FUNDAMENTALS OF AN OPERATIONS IMPROVEMENT TASK FORCE

Robert M. Cowdrick Jr.
Xerox Computer Services

Dennis Wormington
Texas Tech University

ABSTRACT

An operational improvement task force is an
internal corporate group of individuals with
varied talents who act as catalysts for improving
productivity and profitability. These
individuals, lead by a team leader, attack
multidiciplinary problems with a goal of overall
operational cost savings. This paper presents how
to implement an operational improvement task
force, who should be on the team, what areas to
investigate, tools to use, and the benefits that
can be received from a task force of this type.

INTRODUCTION

A task force steps beyond the concept of the line
or staff organization and brings together
fundamental principles with manufacturing
technologies. A typical stratified organization
concentrates on problems within it's area of
control. The operational improvement team has the
goal of overall improved productivity while trying
to combine these organizational areas.

Objectives of the task force include attitude,
automation, and cooperation with the over all goal
of operational cost savings.

Individuals with varied backgrounds and technical
expertise in specialty areas are lead by a team
leader who acts as the project coordinator. The
project coordinator, an individual who is capable
of organizing and managing numerous projects,
reports directly to the plant manager.

An evaluation of each potential operations
improvement project is made with respect to the
return on investment that the project will yield.
The project is also rated with respect to the
areas that it impacts and any conflicts that it
may impose upon currently active projects or
task forces

Various tools are available for the task force's
usage. Besides drawing upon a wide variety of
skills from the team members other tools can be
used to answer "what if" questions, Kepner-Tregoe
Analysis could be used to quantify decisions based
on many external parameters. Likewise Value

Analysis can aid in cost reduction of
manufactured product lines. Materials
Requirements Planning Systems can be utilized for
evaluating long term requirements and Capacity
Requirements Planning enable an analyst to
evaluate plant capacity.

The benefits that the team can expect include
overhead reductions, lower direct costs, quality
at the source, improved productivity, and in turn
profitability. Many corporations have
implemented an operational improvement task force
with significant benefits realized.

THE TASK FORCE

Concern for quick profit requires nearsightedness.
Concern for people and corporate survival requires
long term vision. Economics often forces
executives to squeeze financial results into the
short term horizon recardless of how they feel
for the long term considerations and goals.

Productivity improvement requires long term vision
and when coupled with factory automation and
quality circles, becomes fundamental to not only
the growth of American manufacturing but its
survival. For productivity to take place, the
long term manufacturing strategies have to be
tied into the overall business or corporate
strategic plan.

For a productivity improvement program to be
successful, a rational approach to productivity
planning, implementation, measurement, and control
are essential. Each organization must first be
able to assess the aspects of their organization
that makes them unique and gives them a
competative edge in their own marketplace. The
success of an organization must be based on the
products, markets, existing technology,
costs/pricing relationships, and the competitive
enviroment under which it operates. Today in
most companies only a small number of factors
(10-12) dramatically effect an organization's
success.

Once the business relationships are defined, a
strategic or business plan is developed with goals
based on the assessment evaluation. From this
plan evolve clearly defined tasks, Figure 1.

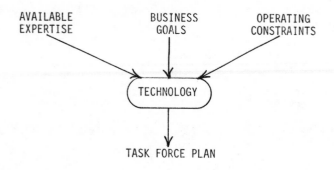

AVAILABLE   BUSINESS   OPERATING
EXPERTISE   GOALS   CONSTRAINTS

TECHNOLOGY

TASK FORCE PLAN

FIGURE 1: Task Force Planning Criteria

The term productivity means doing tasks differently to achieve the highest level of performance with the lowest possible use of resources. This concept includes aspects such as management style to utilizing state of the art technology.

Developing managerial attitude and style that will maximize people and productivity, benefits both the individual and the organization. By giving each employee the right and respect to express their ideas, a rich resource of ideas, inspiration and wisdom can be utilized. When management openly commits itself to a Theory Y style of developemnt and management the burden of success or failure falls on management's shoulders. However, managerial attitudes alone are not enough to bring about productive change within an organization; government and labor must participate.

Legislative action must continue to make sense out of the maze of conflicting industrial policy through the vehicle of the court system. Regulations must be precise but yet flexible for adaptability and effectiveness. Likewise organizations must meet the productivity challange by being able to meet with government in consensus agreements on manufacturing strategies and primarily have their own operations in a poised state. The American labor system must be revamped to eliminate the current labor-management conflicts and the archaic work standards and practices of the 1900's. Worthwhile productivity improvement is near impossible without a revamping of the labor system.

With a positive organizational attitude being developed and the benefits of labor laws being revised, a company can proceed to investigate opportuunities for improvement within its own organization.

This type of effort is most commonly and effectively carried out in a task force approach. A task force is a group of individuals with varied backgrounds and technical expertise who are led

by a team leader. The concept of participation is used to mean involvememt in a team with joint responsibility for success. The team attacks cost-beneficial projects which the organization has never addressed or the stratified structure of the company has prevented from addressing.

Participation in a task force is however not always the most prefered method for undertaking a task or assignment. The use of a team does however use:

* new sources of expertise

* benefits from collaboration that multiplies an individual's effort

* tackles problems that no one person owns, and

* develops and educates people through participation

Other advantages of a task force approach include focusing on specific objectives, using talent from all staff areas, tackles problems that effect many areas, and can be dissolved when necessary.

THE TEAM

The secret of task force success lies in the ad hoc nature of problem awareness and resolution. No individual is suspicious of changes they help initiate. Instead they are convinced such measures can be healthy and positive in nature. Employee involvement not only eliminates potential resistance but brings about assistance. This is the basis behind productivity improvement through a task force approach.

The committee makeup can be enhanced by involving individuals whose personal needs for advancement and esteem are of lesser desire then the attention paid to accomplishing the tasks and working effectively withing a group setting. Group members who are motivated to cooperate are more favorable in their impressions and more involved in the task seeking greater satisfaction.

Research findings concerning committee size indicate that the ideal size of a group should contain five members who possess the necessary skill to solve the problems. This number of five may have a marginal influence on planning task force size but the emphasis is certainly on skills and a small united group. Although team efforts are vital, individual effort is perhaps paramount.

But what should the team consist of? Most active productivity task forces agree that speciality areas of computer science, machine tool specialists, material handling engineers, and individuals with strong financial backgrounds. These individuals should be led by a task force leader. All individuals should be trained in problem-solving techniques and project managment.

Innovation and change require a mixture of leadership, consensus, autonomy and participation. This task force leader should have both task and social leadership abilities. An ideal balance is often sought but never achieved. The team leader must seek a balance between task force goals, resources available and the implementation strategy. The leader's emphasis on implementation begins at the earliest phase of task force planning. The team leader must be a well respected individual with a wide variety of experiences but primarily an excellent long-term project manager.

The early stages of a task force development may include time for education and training of the project leaders and the members. In participation task forces, someone must initiate the process since grass roots activism in highly unlikely. The leader must initially promote the concept and provide the initial thrust. Later the change process will evolve into a true participatory nature as individuals sieze ideas and feel a sense of ownership (1).

Outside individuals must also become team players. Vendors, consultants, and governmental officials need to play a role. These individuals usually are joined with the task force for solving certain problems where specialized knowledge is required. However these individuals do not play an active role in the task force.

IMPLEMENTATION

A good organization is one which has a working environment that supports, motivates, and encourgages the productivity of work places. But often no driving force is there to provide the catalyst of the effort. Management support can fullfill this role but a well planned implementation is still the key toward a successful productivity task force effort.

With the business plan formulated and managerial support providing the catalyst for the group, implementation can begin. There is no one correct way to carry out a productivity task force but some key elements need to be considered.

The stages of implementation can be defined as:

* plan - establish a sequential plan to bring about change, publicize, solicit support,

* measure - develop criteria to reference at a later date,

* evaluate - review and critique the project's progress, and

* control - modify and re-align the project path if necessary.

The stages of implementation should always include a feedback mechanism to close the loop on the implementation process, Figure 2. Through a feedback mechanism the course of the project direction can be continually adjusted the stay in line with the business goals and constraints. Project planning, the technique used to set measures of success and completion dates for all tasks, is one method of keeping a project on course.

For example, if one of the goals of the task force is to help employees feel good about their jobs, their organization, and themselves, this the following steps should be considered:

* job enrichment,
* delineated career paths,
* technology advances, and
* competative compensation.

Resistance to change takes the form of hostility or lack of assistance and cooperation. Resistance may be concealed, expressed as counterproductivity, apathy, or decreasing job performance and occurs when individuals envision their freedom to carry on desired behaviors as being threatened or eliminated (2).

Causes of resistance may include threat of self interest, distortion of ideas, objective disagreement, and psychological reactance. Dealing with resistance can be handled through participation, education and incentives where necessary.

The implementation strategies most commonly used include demonstration, power, persuasion, and participation. Participation strategy is based on the assumption that involving participants in the planning will have far-reaching benefits in the implementation stage of the tasks developed. Thus the concept of a task force is used.

Effective participators are made and not born. It takes knowledge and experience to be able to contribute effectively. Individuals with more information than others will tend to dominate the group until all team members are informed to the same level. Also early in the implementation, stage participation will be slow in development. Individuals will base their levels of decision making on prior constraints. Until it is known that some of these constraints can be changed, prior decisions may limit the teams effectivity.

Successful implementation not always means changing the organizational structure. In reality, established managerial habits usually exist even with organization resturcturing. Where change has been successful is in the areas of involvement in implementation,timing of various moves, and securing the participatioon neccessary for success.

The degree to which an implementation occurs is the degree to which the changes intended to take place do happen. This occurs when behavior of individuals and the organization become normal or accepted routine happenings.

Take a step, walk, then run! Understand the impacts, resolve conflicts, then proceed to the ultimate goal of full implementation; while always providing a feedback mechanism.

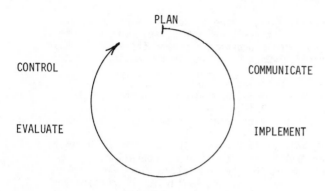

FIGURE 2: The Closed Loop Implementation

TOOLS

The success of the task force effort is based not only upon a well developed implementation and a speciality team, but on using tools appropiately to get the job done. Some tools that may be used include the following.

Economic analysis can be used to evaluate project benefits and rank alternatives for a investment return basis. Return on investment, internal rate of return, and capital budget planning can all be used for financial analysis. Currently available spreadsheet software packages can be used as a tool to quickly evaluate numerous alternatives.

Systems will be the backbone of productivity. Systems will provide the information necessary for problem determination and evaluation. They will provide the link between task evaluation and implementation. Materials Requirements Planning Systems will assist in the planning of manufactured and purchased requirements based on a prior developed business schedule. Capacity Requirements will determine restrictions on manpower or machinery that may limit the plan.

Computer simulation and modeling can be used to evaluate the "what if" questions and to evaluate and summarize various alternatives quickly. Whether it be on a personal computer or mainframe, simulation and modeling can be used effectively. Take a concept, apply relationships in the form of mathematical equations, vary the constraining parameters, and summarize the results. Simulation

can examine many different alternatives quickly and accurately.

Operations Research (OR) is the scientific approach to decision making. OR focuses on problems of conducting and coordinating activities within an organization. Elements of OR that are of special help in problem solving include:

* linear programming,
* queueing theory,
* optimization, and
* sensitivity analysis.

A survey by the American Society for Quality Control conducted by Bozella and Jacobs revealed that 76% of American consumers feel that American industry cares about producing quality products but only 35% feel that American workers care about quality. Quality Circles can renew product quality and consumer confidence.

ATTACKING THE PROBLEMS

Many of the problems that a potential task force will investigate are unique to each company and based on the business plan developed. But throughout American industry, many of the same problems surface again and again.

One major goal of factory automation all experts agree upon is minimizing raw and reducing work-in-process inventory levels. Parts sitting in stores or by an assembly line that are not currently needed ties up cash and costs the company in interest expenses. Seeking the Just-in-Time inventory levels are new goals for the entire organization. More control ultimately will reside with the production worker.

Attacking the area of quality control suggests the question of an approach to the quality of the products or the quality of the designs and processes. Emphasis of product quality itself leads to statistical inspection of incoming products and inturn a improved overall manufactured product quality. The strategic plan requires quality assurance and this drives the paramenters of quality control. If a cost effective approach is taken toward quality assurance in the design engineering, design reviews, and manufacturing processes stages of a product development cycle then in the long run, the manufacturing operation will be a productive entity of the company.

Automated storage systems are made up of three elements: storage, transportation and control. The transportation and control elements are the keys to where productivity can be achieved. Automated handling speed materials to processing workcenters thereby reducing handling between processing centers with a goal of minimizing associated labor costs.

Standardization of packaging and storage is worthy of an exploratory investigation. With standardization of storage facilities, comes reduction of diversity among items in order to receive economies of scale in automated storage and handling systems. Also with standardization comes greater utilization of space.

The most overlooked area in factory automation is the proper use of the appropiate people within an organization. The organization that runs "lean and mean", with skills and talents spread across a united group will function most effectively.

CASE STUDY

Deere and Company, a major manufacturer of agricultural machinery and equipment in the Midwest, developed an Operations Improvement Department to attack productivity problems during 1983. This concept was first started at their engine facility and is composed or approximately thirty individuals with varied skills and talents. The department reports directly to the plant manager.

The reasons behind the development of the department are numerous but stem primarily from an erosion of market base and the need to attack productivity agressively. Net sales of farm and industrial equipment fell to $4.6 millions in 1983 and employment was also reduced to 45,700 from a high level of 61,000 in 1980. With decreasing production and staffing requirements came the need to provide alternatives for utilization of manpower. Traditional layoffs of blue colar workers accounted for the majority of the reduction in employment during the three year period. And a small percentage of white collar layoffs also impacted the corporate employment base.

With top management supports the task force was developed. The group started with a team of approximately thirty individuals and focused on a wide range of potential cost saving areas. Renewed emphasis in quality of design and manufacturing processes were stressed.

Productivity improvement suggestions are made to the department manager who then creates internal departmental teams who attack the problem through first an investigation into the problem followed by a task list for attacking the problems. Recent areas that have been investigated include:

* vendor scheduling   using a weekly generated MRP, shipments were scheduled on a weekly or daily basis from vendors.

* manufacturing processes - some heat treating process were eliminated due to material improvements.

* load consolidation - with daily deliveries being tested to meet a "just-in-time" goal, inbound load consolidation was implemented for transportation cost savings.

* production scheduling - using computer simulation, runsizes and inventory levels were planned based on output from MRP.

Deere relies on a strong integrated network of advanced systems to provide the necessary information and make some state of the art improvement ideas payoff on the required large scale basis. Manufacturing and materials systems feed an accounting network of cost and financial tracking for project evaluation and control.

Benefits from the program are significant. Besides utilizing individuals to seek out potential cost savings never envisioned before, inventory levels have fallen, direct manufacturing costs have been reduced, and the program continues to be a success.

SUMMARY

A productivity task forcewill work best when it is well managed. This includes a clearly designated project goals, assigned tasks and responsibilities, education, and a means of providing recognition and rewards for the team's effort. Managing participation is a matter of balancing resources and talent within operating constraints. Thus a productivity task force can provide a means of improving productivity and in turn profitability within an organization dedicated to long term success.

REFERENCES

(1) Kanter, Rosabeth Moss, "The Changemasters: Innovation For Productivity in the American Corporation", Simon and Schuster, 1983.

(2) New and Singer, "Understanding Why People Reject New Ideas Help IEs Convert Resistance Into Acceptance", IE, p.51-57, May 1983.

BIBLIOGRAPHY

Robert M. Cowdrick Jr., CPIM, is a Business
Consultant for Xerox Computer Services in
Atlanta, GA. He is responsible for
installation of integrated manufacturing systems
with Xerox clients. Bob holds a BSAE from the
University of Maine, a MSIE from Purdue
University, and has prior experience with John
Deere and Reed Industries.

Dennis Wormington is a Teaching Assistant for the
Small Business Development Center at Texas Tech
University in Lubbock, Texas. He has a BSIE
degree from the University of Missouri and is
currently obtaining a Master of Science degree in
Production and Operations Management at Texas
Tech. Denny has five years of engineering
experience with Deere and Company in Reliability
and Industrial Engineering.

Reprinted from 1984 Fall Industrial
Engineering Conference Proceedings

PRODUCTIVITY PROGRAMS IN AN ELECTRIC UTILITY

Richard L. Fancher

Salt River Project

Successful productivity improvement efforts depend on the involvement of employees. Effective involvement requires (1) awareness and trust, (2) processes and (3) tracking systems.

A few years ago, Time magazine had as it's cover story a major article entitled "Johnny Can't Read." It was followed by "Johnny Can't Write" and "Johnny Can't Compute". Today we pick up a current national news magazine and see the cover story "Johnny Can't Produce". It would seem Johnny has moved into the work place. Now everyone wants to know how to get Johnny to be more productive. As the education system discovered in attempting to improve their system, so industry is learning in its attempts of improvement: There are no easy answers, there is no single answer, and there seems to be a multitude of fads and consultants marketing them. Frankly, more seems to be written about productivity than is known.

Developing successful productivity programs in an electric utility is no easy task. Most people in the business perceive that the controllable expenses are such a small part of the total costs that the time and trauma involved in cost control program development and administration are not worth it.

Typical Utility Cost Breakdown

Uncontrollable Costs

- Fuel and purchased power	27%
- Payment of interest on loans	25%
- Taxes	16%
- Reployment of principle bonds	5%
	73%

Controllable Costs

- O&M	20%
- Reinvestment in plant	7%
	27%

Another barrier to convincing people that they can have an impact on the utility's bottom line is the business of measuring service type output. How do you count satisfied customers?

Thus, the banes of all productivity improvement efforts are present:

1. Awareness that an issue exists;

2. Proof that controllable expenses have an impact; and

3. Demonstration that efforts can be measured.

New emphasis on productivity at the Salt River Project (SRP) was undertaken in April of 1981 when the General Manager asked that a presentation be given to his staff. The presentation was to address the issue of how SRP employees can be convinced that top management is serious about productivity improvement and cost reduction.

To put the issue of productivity at the SRP in perspective, it is important to know something about the Salt River Project.

The Salt River Project is an organization of two entities - a private corporation and a municipality. The private corporation, known as the Salt River Project Water Users Association, was formed in 1902 to provide a reliable source of irrigation water to the Salt River Valley, through a canal system supplied by what was to become a series of four dams on the Salt River. Hydroelectric power was later added to the dams.

The Salt River Agricultural Improvement and Power District was formed in 1937 as a political subdivision of the State of Arizona. Its purpose is to manage the electric system and facilities. With annual revenues of $800 million and 2200 megawatts of generating capacity, consisting predominately of coal fired generation on the Navajo Indian reservation, SRP is now a pretty good sized utility. Our primary concerns are encompassed in our statement of corporate purpose: "Provide a reliable and adequate supply of Water and Power at the lowest reasonable cost".

We realized at an early date that no one program would suffice for all departments. Further, lasting improvements would not be made if programs were mandated. "This too shall pass" was a statement we hoped to avoid. One of the most effective tools, of course, is a well entrenched MBO (management by objectives) program and a salary adjustment program which depends on performance against objectives, which in turn contain productivity improvement objectives. SRP already had such a salary and MBO program in place.

As the first step in making employees aware of the need to improve productivity, we developed a videotape and a manual. The tape is primarily designed to convince people that there is an issue we need to deal with. The productivity manual was prepared to provide basic information relative to the design of productivity improvement programs. It contains a list of "thou shalts" and "thou shalt nots", and a list of the tools (primarily evaluation) that are available to managers and supervisors wishing to deal with specific issues.

Implementing productivity improvement programs usually depends in a large part on one's ability to see something that can be counted. There are many operations where it is not possible to count things. This often turns managers and supervisors against the entire effort. To overcome this problem, we needed more than a manual and tape.

Our results have been positive. The most visible elements of the results have been the quality circle program, the total firm productivity model and our "barriers consulting".

Quality Circles

We started a quality circle program and now have 25 circles in operation. This program has provided an opportunity for the people doing the work to identify problems and improve the quality of their work by solving these problems. The program has been well received in the company and has enjoyed great support from upper management. We feel a key to the success of a quality circle program is to have facilitators who are industrial engineering oriented. Since these are the people most skilled in resolving work place problems, they are also best suited to help guide others in doing the same. Of course, not every I.E. is cut out for this type of work. It takes additional skills with people and group dynamics. But we feel that we must start with an I.E. background and build from there.

Total Firm Productivity

We developed a total firm productivity model to help management see the entire picture of how productivity relates to profitability. The model is based on one developed by the American Productivity Center in Houston, Texas. It compares the current year to a base year. As a utility, we had to adapt the model designed for manufacturing to fit our needs. We also modified the model to allow us to test our five-year financial forecast for its productivity implications. There are now actually two models: one

reporting historical values against targets set by the budget process and the other which looks at long-term forecasts.

Barriers Consulting

Our "barriers consulting" approach to productivity improvement originated from a training module on management control systems which we developed and presented originally to the executives of SRP. Initially, we took a very pedestrian approach to the issue. We had borrowed liberally from a book prepared by James Hershauer and "other authors" and developed a five-step approach to developing productivity indicators. The five steps are:

1. Define the output. (What is done and why?)

2. Define the deviations in output. (What can go wrong?)

3. Define the system (process) that produces the output.

4. Identify causes of these deviations and tie them into the system.

5. Develop indicators to minimize occurrence, of the deviations, and to help keep problems solved that have been solved.

A second productivity videotape was developed to help teach this "how to" approach, and training was given to all middle managers and supervisors. Logical as the process seemed, it was still only getting lukewarm response. More people than not would conclude that of the entire operation only a very small portion of it could be counted and productivity improvement efforts would be wasted. We re-thought the entire approach and concluded that the five-step process, though correct, was too academic! It needed to address something more near and dear to all their hearts: namely, the problems and issues they manage. Our job should not be to create more problems for them; by creating demands that served no useful purpose; rather we should be developing a process that would help.

What we did was turn steps 1 and 2 around. We started by asking the managers and employees what their problems were - what barriers keep them from reaching their targets of performance. This, when properly done, usually opens the door to a long list of problems and catches their interest. (Note that this is step 2, "defining the deviations".)

After zeroing in on a specific problem to address, the next step is to tie the problem to the output it affects and to validate the need for that output. (This is step 1 of the original process.) The last three steps are followed in the same order as the original process.

Using this modified approach, we have been able to achieve our goal of providing productivity improvement tools that can be tailored to each area. The typical sequence goes like this:

1. A manager approaches our department and requests help in improving productivity.

2. The first meeting is used to agree on objectives and to schedule a set of follow-up meetings with the entire group. (A key point is that this is a multilevel activity designed to open lines of communication.)

3. In the first group meeting, we attempt to develop a list of problems. Normally forty or fifty problems or barriers are generated, copied after the meeting and distributed to all participants.

4. In the second meeting, the list of problems is reviewed and clarified. Then it is narrowed down to a list of five to seven major problems. Finally, through rank ordering and discussion, a concensus is reached on the first problem to address.

5. From the third meeting on, group problem solving processes are employed to develop solutions. Indicators are then identified to assure that the problem stays solved.

6. The other major problems are approached using the same process, until all issues have been resolved.

Key points that have been found to be helpful in ensuring the success of this process include:

1. The session leader needs to be skilled in group dynamics, group problem solving, and the details of this process. We have found that an experienced I.E. with training in quality circles or group problem solving makes a good meeting leader.

2. The process needs to be participative, meaning it needs to involve all individuals who have a part in the problem/solution. Sometimes the initial group, usually consisting of the manager and subordinates must be expanded either vertically, horizontally, or both.

3. The process takes time. Each meeting mentioned above takes an hour or two, and many meetings may be held before a problem is finally resolved. Remember, these are not easy problems or they would have already been resolved.

4. The manager needs to be open and non-defensive during the process. Often a manager is surprised at the length and nature of the problems identified. If the manager resorts to defending the present situation, rationalizing, or interrogating the individual who suggested the problem, the participants will stop contributing ideas and the process will cease to be effective. The manager must make an honest effort to seek improvement by working with everyone to objectively resolve the issue.

The results have been very encouraging. Productivity improvement no longer has a negative connotation. It has positive overtones. It makes life easier. We're trying to solve problems and keep them solved.

SUMMARY

We made people aware that a problem existed and that we can successfully address it. In fact, the distribution of controllable and uncontrollable expenses is:

Uncontrollable

- Capital	28%
- Fuel & Purchased Power	22%
	50%

Controllable

- Labor	29%
- Participation Generation	12%
- Materials	8%
- Energy	1%
	50%

Efforts can be tracked, but only if the organization develops their indicators in the context of barrier removal and measures their results in the context of the customer's opinion of the product or service that is provided.

costs.

7. The profitability ratio of sales to costs can thus be converted to a ratio of output quantities times prices in the numerator to input quantities times unit costs in the denominator.

8. This ratio can be converted into the product of two ratios: the first is a ratio of output quantities to input quantities, the second a ratio of output prices to input unit costs.

These relationships are illustrated in the following equation:

$$\text{Profitability} = \frac{\text{Sales}}{\text{Costs}} \quad \begin{array}{l}\text{Where Sales = Output Qtys.}\\ \text{x Unit Prices and Costs =}\\ \text{Input Qtys. x Unit Costs}\end{array}$$

$$= \frac{\text{Output Quantities x Prices}}{\text{Input Quantities x Unit Costs}}$$

$$= \frac{\text{Output Quantities}}{\text{Input Quantities}} \quad \text{x} \quad \frac{\text{Prices}}{\text{Unit Costs}}$$

The first of these two ratios, output quantity over input quantity, is productivity. It is an output to input ratio in terms of units rather than prices, and it tells the amount of resources consumed to produce the firm's output.

The second ratio, output prices over input unit costs, is price recovery. Changes in this ratio over time indicate whether changes in input costs are absorbed, passed on, or overcompensated for in the prices charged for the firm's output. A ratio of less than one means the firm is helping to reduce inflation while greater than one would indicate it is contributing to inflation increases.

The equation can be restated (in English) as:

Profitability = Productivity x Price Recovery

In these terms, a change in profitability from one period to the next must be explained by a change in productivity, a change in price recovery, or a combination of the two. In other words, for a firm to increase its profitability during the next period: it can produce more output per unit of input consumed or it can raise its prices faster than its suppliers are raising theirs, or some combination of these two.

An example of how the model is being used at SRP is shown below:

Table 1 shows the corporate outputs for the base year and for the upcoming budget year. These outputs are labeled as Electricity, measured in terms of kilowatt hours sold, Water, measured in terms of acre-feet of water delivered (sold) to users, and Capital, measured in terms of dollars invested. The third figure is included as a corporate output because it is a source of significant revenues to the company even though it is not the primary business of the company.

APPENDIX A

TOTAL FIRM PRODUCTIVITY MODEL

SRP began development efforts on a total firm productivity measurement system during 1981. At that time, management had no overall productivity measurement system for tracking corporate productivity performance. This appendix will briefly explain the measurement system developed at SRP, and provide an example of the results put out by the system.

Several criteria were important in the selection of a system to track corporate performance. First, the system should include all aspects of the company, both outputs and inputs. Second, it should be a relative, rather than an absolute measure, to allow direct comparisons from one year to the next, or from any of several years against a base year. Finally, the system should be capable of measuring productivity independent of inflation, so that actual performance would not be clouded by it.

Search of various possible total firm productivity measurement systems eventually led to one developed by the American Productivity Center in Houston, Texas. An article by Dr. William Ruch, Professor of Management at Arizona State University, brought this model to the attention of the SRP employees assigned to develop the measurement system. With some modifications of the original model, it has been applied successfully to SRP.

The following assumptions serve as the basis for the measurement system:

1. Output can be measured by sales.

2. Inputs can be measured by costs.

3. Profit is the difference between sales and costs.

4. Profitability can be defined and measured as the ratio of sales to costs.

5. Sales can be broken down into quantities sold (of perhaps many different items) times their respective prices.

6. Costs can be broken down into quantities (such as hours of labor, units of materials, kilowatt hours of energy) times their respective unit

TABLE 1

CORPORATE OUTPUTS

	Base Year (000)	Budget Year (000)	% Increase
Electricity Output			
KWH	12,500,000	13,300,000	6.4
$	677,900	747,100	10.2
Water Output			
A-F	1,200	1,200	0
$	5,300	7,400	32.8
Capital Output			
$ Invested	189,200	255,700	35.1
Interest Earned	35,100	47,400	35.0

Table 2 shows the corporate inputs, measured in
various physical quantities, and the costs
associated with each input are also provided.

TABLE 2

CORPORATE INPUTS

	Base Year (000)	Budget Year (000)	% Increase
Materials			
Unit	1,000	1,100	10.0
$	49,300	55,400	12.4
Labor			
Man-Hr.	11,400	11,900	4.4
$	188,000	207,000	10.7
Gen. Fuel			
BTU	64,436,000	78,405,000	
$	114,500	145,000	
Energy			
KWH	311,000	342,000	10.0
$	6,100	7,600	24.6
Purchased Power			
KWH	2,300,000	2,200,000	4.3
$	26,500	28,500	7.5
Pwr.Plnt.Cptl.			
KW	2,400	2,400	0.0
$	169,900	180,900	6.4
Water Capital			
A-F	2,800	2,800	0.0
$	6,900	9,600	39.1
Part.Gen.			
KWH	5,000,000	5,300,000	6.0
$	76,600	83,100	8.5

Table 3 shows the final results of the model for
SRP. This example shows that the budget for the
next year is set up in such a manner that the
company is expecting to produce a productivity
increase of 1.1% for the year, and a price recovery
drop of approximately 1.8%. The overall result of
this will be a drop in corporate profitability of
.7%. Interpreting this information leads to the
conclusion that, while the company will be more
efficient in converting its inputs into outputs,
profitability will suffer because the output prices
for the company's products will not be raised as
fast as the costs of inputs will go up due to
inflation. SRP would be absorbing this inflation
for its customers by not passing on all cost
increases.

TABLE 3

TOTAL FIRM PRODUCTIVITY RESULTS - BUDGET YEAR

	Profit-Ability	Produc-Tivity	Price Recovery
Materials	99.3	98.0	101.4
Labor	101.4	103.2	98.2
Generation Fuel	88.2	88.6	99.5
Energy	89.6	98.0	91.4
Purchased Power	103.8	112.7	92.1
Power Capital	104.9	107.8	97.3
Water Capital	80.2	107.8	74.5
Partic. Gen.	102.9	101.7	101.2
TOTAL	99.3	101.1	98.2

Each of the separate input categories has a distinct
effect on the overall total firm productivity.
Looking at the individual inputs can help us to
understand the final result for total firm productivity. For example, in the labor category, a
productivity improvement of 3.2% is expected, but
the price recovery is expected to drop by 1.8%. The
combined contribution of labor to profitability is
therefore a 1.4% improvement. This means that some
of the other input categories must be causing a
decrease in profitability that is greater than the
increase caused by labor.

Since in the total firm calculation, each of the
inputs is weighted by its total value, several of
the inputs could exhibit improvement, and still be
overshadowed by the decline in profitability of one
or two large inputs. This is the case here, where
four of the eight inputs show improved profitability,
but the total firm result is a decline.

By reviewing the details of the results for the
various inputs, management can determine what areas
are the problem areas, and help to plan for future
actions in those areas. The example presented here
is for the budgeted year, so changes could be made
before the year has passed. During the year,
actual results will be tracked to show how closely
SRP is keeping to its plan. This will also help
management to react quicker to apparent problems
in productivity or price recovery.

Reprinted from 1984 Annual International Industrial Engineering Conference Proceedings.

PRODUCTIVITY ON A SHOESTRING
THE UTAH EXPERIENCE

Roger W. Black, Deputy Director
State of Utah
Department of Administrative Services

ABSTRACT

The State of Utah has had encouraging success with initiatives to improve productivity. The same forces that have led other governments to focus on productivity improvement stimulated action in Utah and the response shows that much can be done with very few resources. Examination of this experience leads to observations and recommendations that may be useful to public entities having productivity problems but little budget flexibility to absorb new expenses.

BACKGROUND

The rhetoric of professional public administration, to say nothing of political campaigns, has long included attention to the need for improved effectiveness and efficiency in government. In recent years, events (in the last four fiscal years, the state's budget has been reduced seven times) have intensified this preoccupation, and also given us a new vocabulary. We now speak of "productivity improvement" and "enhancing the quality of working life." Our attention is attracted to successful private sector organizations and our spirits are lifted by advice to develop a "bias for action", to get "close to the customer", to promote "autonomy and entrepreneurship" and otherwise nurture the characteristics of excellent organizations. The State of Utah has certainly been touched by these currents.

Utah has indeed faced very sobering realities. With one of the nation's fastest growing populations, during a period of recession, our bank accounts have barely been adequate to maintain minimal service levels. So there was nothing extraordinary about Governor Matheson's determination to initiate a productivity improvement program. What may be unusual is the way in which the program has unfolded.

Typical responses

The productivity improvement literature has consistently extolled the virtues of attacking the productivity problem as a high priority business strategy--suggesting approaches that are comprehensive, integrated, and energetic. For example, groups like the American Productivity Center identify some ten elements of a comprehensive productivity program, and argue that the best chance of success exists when an organization pays specific attention to each of these elements. Some writers have even gone so far as to suggest that failure to attack the problem on all fronts with careful planning and consistent implementation is worse than doing nothing at all.

Several of our sister public organizations across the country have demonstrated that there is much wisdom in this position. The State of North Carolina, has successfully implemented a number of innovations which, taken together, constitute a productivity improvement strategy of very impressive dimensions. The cities of Phoenix, Arizona, and San Diego, California, are other public entities whose leaders have seen in a comprehensive productivity initiative, an investment opportunity paying rich dividends to residents and taxpayers. Even in difficult times, the management of these public entities have evaluated the resource requirements of productivity improvement interventions in terms of anticipated return on investment, and have been willing to allocate scarce resources, dollars and staff, to pursue productivity improvement opportunities.

Mitigating circumstances

There are many public sector organizations however, where the climate for such dispassionate business-like, evaluation of investment opportunities has not been auspicious. Political and philosophical differences between executive and legislative branches and declining real tax revenues are some the conditions which get in the way. Certainly this has been our lot in Utah. Our experience, however, may give cause for optimism for what can be accomplished even when virtually no resources are available for investment in productivity improvement. Our story illustrates some of what can be accomplished through the simple raising of management awareness, the high payback of what can be called "cultivated spontaneity" and the benefits of doing something--anything--even within the limits of a shoestring budget.

The Utah response

Similarities - Our story is not unlike that told by many other organizations. It starts with the awareness of top management. Shortly after his election to a second term as Governor, Scott Matheson recognized that the difficult economic times we were experiencing would continue for some years into the future. Therefore, meeting increasing demand for services through traditional budget increases was simply not possible. Improved productivity offered a solution.

The Governor appointed a citizens committee representing universities, organized labor, professional societies including AIIE, the business community and state management. He gave the committee the responsibility to define opportunities for improvement. The committee undertook it's assignment with vigor and in a few short months, presented its findings. The group found three major problems: organizational support and resource commitments to productivity were insufficient to the task; a disturbing number of elements in the institution of state government functioned as strong disincentives for productivity improvement; and accountability for change could be strengthened dramatically. Their recommendations addressed each of these problems, and they received the Governor's active endorsement.

Constraints - But as words became actions, significant differences emerged.

For one thing, we determined it would be counter-productive to push productivity improvement as a new thrust or program. Executive branch agencies had already absorbed a new budgeting system, a new personnel system, a new performance appraisal system, a new accounting system, a new planning process and a new mechanism for managing data processing resources. We feared that launching a productivity program with lots of hoopla, grandiose promises, and the trappings of a new institution would not only seriously compromise its chances for success, but could well lead to management indigestion.

A second constraint was the necessity of launching the program without additional resources. No new staff were to be hired, and no new funds would be available for training and consultant services.

Finally, we had to launch the experiment with no expectations that we could change any of the institutional rules and practices that were then operational.

Advantages - What we did have going in our favor, however, was a set of circumstances that ultimately proved to be very helpful.

The first of these was a very clear gap between the level of services our people expected and the financial resources available to provide them. This gap had the effect of galvanizing management attention on the absolute need to exploit every possible opportunity for getting maximum results from limited resources.

Secondly, there was a considerable amount of management commitment and energy to support the search for successful interventions. The Governor himself was an electric cheerleader for the cause in his several meetings with management. He even went so far as to accept the advise of the citizens committee to modify his own supervisory behavior providing an example for his managers to emulate.

Thirdly, there was a wealth of information about productivity improvement ideas which had been tried and proven in both public and private sector organizations. We had gathered several boxes of journals, reports, case studies and other accounts which our agencies were encouraged to draw upon.

Fourthly, we were surprised to discover a number of voluntary resources who were willing to assist us with productivity innovations. One of the more significant of these being the Salt Lake Chapter of the American Institute of Industrial Engineers. The president of that group had served on the citizens study committee, and the chapter expressed a willingness to provide public service time to state agencies. In addition, there were people at the Utah State University Center for Productivity Improvement and Quality of Working Life, at Hill Air Force Base, and, perhaps not surprisingly, in our work force itself, who were eager and willing to assist in meeting the challenge to improve the operation of our programs.

Finally, and perhaps most importantly, there was an established tradition of experimentation and innovation in state government management. These were the elements which came together to comprise the State of Utah's productivity improvement program.

Steps - That program unfolded in the following sequence of events:

As mentioned, the citizens committee completed its work and made a report to the Governor. This report specifically recommended five actions. First, the adoption of goals for productivity improvement. These were posed in terms of improved worker performance, exploitation of technology, streamlining of work methods and procedures, strengthening management skills, and improving the quality of working life. Secondly, articulation by the Governor of a commitment to productivity improvement. Thirdly, assignment of responsibility for productivity improvement efforts to a state official, and creation of a permanent organizational home for this assignment. Fourthly, development of a strategy to eliminate disincentives to productivity improvement. And, finally, strengthening the accountability processes of state government. With some modifications, each of these recommendations was adopted

The adoption of the program took form in a directive from the Governor to his program managers, couched in terms of re-emphasis and new articulation of established policies and commitments. This directive also announced the appointment of a state productivity coordinator who was to have the responsibility of providing technical assistance, coordinating awareness and training programs, identifying disincentives (including working through the legislature for their elimination), and assisting the Governor's management team to improve performance measurement systems. The directive also spoke about the need for creating a productivity ethic, and appealed to managers to cultivate a value system that would have a permanent impact on state government.

Following this directive, the productivity coordinator embarked upon a period of missionary service to state agencies. Each program entity received a presentation explaining the Governor's commitment and expectations, stimulating interest in the prospect of improved productivity and identifying the support resources available to assist. Our objectives here were to clarify the Governor's policy, to create an awareness in management of the richness and attractiveness of proven productivity interventions, and to underscore the Governor's determination that productivity improvement become part of the warp and the woof of agency management responsibility. This was not an announcement of new obligations, but rather the marketing of new resources and ideas which program managers could use to satisfy standing commitments.

These itinerant camp meetings helped to identify individuals inclined to take up the cause with their colleagues and/or who received an assignment to do so by the agency head. These individuals became a part of a loose, informal coalition of productivity promoters-- the focus of the next stage of the program. We held discussion groups at which these promoters gathered to get better acquainted with one another and to participate in work shops focusing on productivity ideas. We had presentations, for example, on quality circles, on socio-technical systems, and on the application of statistical quality control ideas to public sector applications. These discussion groups further led to the holding of a productivity fair, attracting representatives of all levels of government. This was an opportunity to hear nationally recognized leaders address topics such as worker involvement, measurement, and leadership.

In response to this studied, soft-sell, voluntary pursuit of productivity improvement, we saw a veritable kaleidoscope of experimentation take place. Agencies started job sharing programs, alternate work schedules, flextime, bonus and incentive systems, organization development interventions, quality circles and a proliferation of word processing and data processing systems affecting the entire work force, up to and including the Governor. Many of these efforts had started before the citizens committee made its recommendations. But all of them came into focus as reasonable and logical responses to the challenges before us under the banner of productivity improvement. It was as though the Governor's directive galvanized management attention and gave legitimacy to inclinations that otherwise would have seemed out of synch with the economic realities of state government. It is also clear that managers tried things they would have not pursued if the program had not gotten off the ground.

ANALYSIS OF OUR EXPERIENCE

Weaknesses

We acknowledge weaknesses in the our productivity efforts.

Measurement - For one thing, measurement is very, very soft. The investment in office automation, the supervisory training, quality circles, team building and other organization development interventions have all been justified intuitively. Managers have supported the cost of these ideas by squeezing funds out of their regular program appropriations. Managers believe that their programs have improved, that they have been able to absorb new workload without having to hire additional staff, but there is virtually no hard evidence to support this observation.

Creative chaos - Another weakness is the seeming chaos of these management initiatives. There is no guiding strategy, no integrating rationale, no explicit sense of priority or return-on-investment type analysis of the ideas and interventions being employed. By the same token, this very chaos and spontaneity is also very much a grass-- roots undertaking, and very little central direction or control has been required or exerted to cause these changes to occur and to perpetuate them after they have been started. We have reason to believe that the successful practices will continue with or without continued emphasis from the top.

Training - A third weakness has to be the inadequacies of our training efforts in productivity improvement ideas. Change has occured in those organizations where serendipity created conditions that make them most prepared for change. But a substantial portion of our work force is out in the cold. Additionally, where management by experimentation can be very creative and innovative, it can also be very disconserting to rank and file workers who may become disillusioned and disaffected. Training, obviously, could help minimize these untoward consequences of our innovative spirit, and it is unfortunate that we have not been able to support training needs as well as we would like.

Results

Gross indicators - On the other side of the ledger, the evidence we have examined leads us to conclude that, on balance, the program has been more of a success than a failure. Global statistics clearly show that state government is more productive now than it was four years ago. State employment has remained virtually constant over that period of time, whereas our population has increased 12 percent causing the ratio of state workers to population served to decline continuously.

Anecdotal evidence - Additionally, we have collected anecdotal evidence to suggest that productivity has indeed improved. Agencies report being able to re-direct budget, for example, from clerical activities to professional activities, as they have implemented office automation capabilities. Quality circles have created practical ideas that have taken days off of time required to repair heavy equipment, reduced the turn-- around time for the issuance of driver licenses, and cut staff required to handle increasing case-loads of juvenile probationers. And organizations have experienced a new vibrancy of commitment as a consequence of paying attention to quality of working life issues.

Management outlook - We have also noticed a change of management attitude toward the characteristics of productive organizations. There is a greater recognition of the contribution that employees can make to the decision making process, and a greater willingness of managers and employees alike to cooperate and become less protective of turf in grappling with the service demands that our population places on us.

Momentum - A final result, in which we take some pride, is the momentum that has built up behind the productivity initiatives I have described. Because of the approach that we took, virtually all of the initiatives have started at the bottom and have been encouraged by upper management, rather that the other way around. We think this momentum will sustain the initiatives through the pending change of governors. We also see an increased appreciation for the ongoing value of attention to productivity concerns. We have recently hired an industrial engineer onto our staff, who has begun to work with the agencies, teaching them to apply rigorous analytic techniques to the management of their program responsibilities. As with the computer, we

believe when management understands how valuable good measurement tools can be in husbanding scarce resources, that it will not require continuing push and impetus from the Governor to keep these initiatives moving. Productivity improvement is really becoming a part of the way we do business.

CONCLUSIONS AND RECOMMENDATIONS

In surveying our experience, particularly comparing it with the strategies that others have tried, I think it points to some conclusions that perhaps could be offered in the form of recommendations.

Do something

The first is that public sector organizations owe it to themselves to do something--anything--that has the promise of improved productivity. In many ways our failures have been just as valuable as our successes. They have provided us information that has been helpful in shaping new responses to the circumstances we face. The opportunity for public sector organizations to become productive is almost without limit, but none of this can be realized if there is not a willingness to at least try something different and see what happens.

Aim low

Second, it is helpful to aim low. If expectations are realistically cautious, and we avoid grandiose promises, there is very little face to save when an experiment goes awry. And the probability of success is much greater. Moving from one small success to another, cumulatively can have a very profound impact on organizations. This has certainly been our experience.

Nurture readiness

Third, recognize the need for conscious nurturing of institutional readiness. Not all organizations or sub-organizations are equally ready to try new ideas. Force-feeding new ideas to an entity that is unprepared for change is worse than counter-productive. Much better to paint a picture and trust the intrinsic drawing power of the new idea than to insist that it be tried before adequate preparation has been made.

Voluntarism

Fourthly, voluntarism is essential for permanent change to occur. Groups that are first to volunteer, have the best chance of succeeding, and thereby reduce the barriers for other groups to try the same idea. Moreover, this approach requires considerably less time, attention and concern at the top than does an approach that requires all parties to get in line and stay there.

Top management protection

Finally, and perhaps most importantly, it is essential for top management to consistently support innovations and experiments. If Governor Matheson were not so consistent in his public and private support of all that state managers were doing, it is certain that much less would be done.

BIOGRAPHICAL SKETCH

Roger W. Black is Deputy Director of the Utah Department of Administrative Services and also serves as State Productivity Coordinator. He has had previous experience in human relations training, consulting and city and county government. He is a member of the editorial board of the Public Productivity Review; served as a panelist in the 1983 White House Conference on Productivity; and teaches courses in public administration and public sector productivity as adjunct faculty at the University of Utah.

*Reprinted from 1984 Annual International
Industrial Engineering Conference Proceedings.*

FLEET MANAGEMENT OPERATIONS BENEFIT FROM A PRODUCTIVITY IMPROVEMENT PROGRAM

LINCOLN H. FORBES, P.E.
CHIEF, MANAGEMENT ANALYSIS
METROPOLITAN DADE COUNTY
PUBLIC WORKS
MIAMI, FLORIDA

This paper addresses the application of a Productivity Improvement Program in the Maintenance and Repair facilities of a major bus transit operation. The program has been ongoing from 1980 to the present and has resulted in improved fleet reliability, lower operating costs and improved availability of spare parts.

The following are provided:

a) A brief overview of the analysis of the routine maintenance operations in the "Central Garage" main facility.

b) An in-depth report on the analysis of

. the major "subsystem" overhaul activities in the "Unit Room"; and the impact on spare parts availability.

. The Bus Rehabilitation program; and its impact on fleet reliability and service levels.

INTRODUCTION

The Metropolitan Dade County Transportation Administration (MDTA) is head-quartered in Miami, Florida and operates a fleet of approximately 600 buses with close to 2,000 employees and an annual operating budget of over $80 million. The maintenance operation is staffed by over 400 employees and represents an annual operating cost of over $30 million. Logistically, MDTA is comprised of a central garage and four satellite garages of which two are operating garages and the other two are repair facilities dedicated to the major rehabilitation of aged buses.

MDTA is embarking on a major rapid transit operation and the bus fleet is to be increased in size and interfaced with the rail operation as a feeder system, starting in 1984. Partly because of short falls in government funding, the current bus fleet is smaller than the planned size, underscoring the need for deriving the highest level of efficiency that is practically attainable.

WHY A PRODUCTIVITY IMPROVEMENT PROGRAM?

When the Productivity Improvement Program (PIP) was started in 1980, several factors were evident to justify its inception.

. Vehicle reliability was below the desired levels and many buses broke down in service. (Over 100 "road calls" per day).

. Service needs (i.e. the number of buses required by the Transportation Division to meet service peaks) were not always met.

. Although a basic Preventative Maintenance Program was in existence, buses scheduled for PM work often had to be diverted back to active duty to meet service needs.

. The cosmetic appearance of the fleet was poorer than was desired.

. The duration of body repairs was unplanned and in order to maintain adequate availability such repairs were often unduly deferred.

. Management was uncomfortable with the amount of employee overtime required to maintain the fleet adequately.

. Many sub-system failures appeared to be premature, contributing to reduced fleet reliability. These failures pointed to inadequacies in the "Unit Room".

. The bus rehabilitation program was not meeting the planned targets. Repair completion dates were unanticipated and the associated costs were creeping above the target levels.

. Record-keeping of vehicle history and status information was mostly inadequate for diagnostic purposes. Repairs tended to be partial as a result.

CONDUCT OF THE PRODUCTIVITY IMPROVEMENT PROGRAM (PIP) AND ASSOCIATED ANALYSIS

SCOPE:

The first phase of the analysis was conducted in early 1980 by an analyst team from Metro-Dade's then Productivity Analysis Unit (P.A.U.). Subsequent phases were conducted by MDTA's Management Services Unit from late 1980 onwards and the writer was a member of both teams.

The P.A.U. study was limited to the central garage facility as it represented the greatest payoff. The Management Services Unit, in future phases, conducted analyses on Unit Room Operations and the Bus Rehabilitation facility, and coordinated the implementation of the primary recommendations.

PROJECT PLAN

The approach generally used in carrying a Productivity Improvement Program are as follows:

a)   Orientation.

b)   Operations review.

c)   Methods Analysis Improvement.

d)   Manpower Standards Development.

e)   Manpower Budgeting.

f)   Work Planning and Scheduling.

g)   Implementation.

Improved procedures/methods/systems are put in place. (This step may be integrated with any of the foregoing steps).

h)   Performance Reporting and Fine Tuning.

Performance reporting systems are designed to provide feed back to management on the extent of improvement. If necessary adjustments are made through the "fine tuning".

OUTLINE OF PRELIMINARY ANALYSIS - CENTRAL REPAIR FACILITY

As part of the Operation's Review, a preliminary study was carried out in the Central Garage to get a "feel" for the greatest opportunity areas within this shop. This preliminary work consisted of a ratio-delay study and an analysis of defect cards (used to log mechanical problems).

RATIO-DELAY STUDY

Three categories of employee activities were selected for the purpose of the ratio-delay study - working, delayed stationary and delayed walking. Various points were selected in the garage for making observations of employee activity. Approximately seven thousand (7,000) observations were made in a one-week period, and the results were as follows:

Total Observations   - 7,162
Percentage Working   - 56.6%
Percentage Delayed Stationary - 31.4%
Percentage Delayed Walking   - 12.0%

It was inferred from the ratio delay study that the percentage working (56.6%) represented room for significant improvement in a maintenance operation. The percentage delayed was attributed to a need for better supervisory control, work planning and scheduling.

DEFECT CARD ANALYSIS

A study was conducted to examine defect cards. This study identified the most frequently occurring defects and subsequent method improvements were focused on these items in order to obtain the maximum benefit.

SUMMARY OF SELECTED METHODS IMPROVEMENTS IMPLEMENTED IN THE INITIAL STUDY

1.   Paint and Body Shop Work Order System.

This system was developed to improve the control of coach work repairs in the body shop. Essentially "damage information" was recorded and tracked, needed repair parts were listed and staged prior to repair commencement, thereby reducing costly delays. This system was combined with improved estimating methods and performance reporting systems to indicate labor productivity etc.

2.   A new Preventative Maintenance Program was developed to better conform to manufacturers' specifications.

3.   The Production Coordinator function was established. The methods improvements required the initiation and handling of maintenance related information for planning and decision-making purposes. This staff position now serves to provide the analytical and planning input needed to support these relatively sophisticated systems. The more operations-oriented Superintendents are thereby freed to conduct day-to-day shop management activities.

Each Production Coordinator is supported in turn by a Maintenance Clerk who is primarily responsible for record keeping.

ANALYSIS OF THE UNIT ROOM OPERATIONS

Background

The Unit Room is a most important facet of maintenance operations at MDTA. Whereas other sections of the garage represent remedial repair work to operating buses, Unit Room activity is devoted to the complete rebuilding of subsystems such as engines, transmissions, drive train components, air conditioning compressors, starters and alternators. These subsytems are indispensable for proper bus operation, hence the quality of repair work performed on them obviously influences the reliability of the fleet. Transit facility shops have traditionally done extensive rebuilds because of the limited availability of replacement parts that customarily exists in the industry.

The Management Services Unit (MSU) carried out an in-depth study of Unit Room operations between May and June of 1981 as a prelude to developing a Unit Room work Order System. It was considered that the Work Order System would provide a vehicle for making several necessary improvements in the Unit Room in conjunction with the following provisions:

. Management policy decisions on the mix and volume of repair work.

. Manpower budgeting to meet these requirements.

. The application of equitable work standards.

. Better supervision, quality control and work methods.

Operations Review

An Operations Review was conducted in the Unit Room. Repair activities were observed and discussions were held with the Foreman and some mechanics to learn of problem areas as well as recommendations for improvement. Unit Room related problems were also discussed with the Division Superintendent of Maintenance, Maintenance Administration staff, Engineering staff and Stores staff.

General Description of Unit Room Activities

The Unit Room is staffed by a Superintendent, 20 Bus Mechanic II's and a General Helper. Approximately 180 different items are repaired there.

Only about 20% of this total are repaired frequently, major examples being engines, transmissions, air conditioning systems, air systems and electrical machines. Certain mechanics repair engines and other subsystems almost exclusively but the remainder repair a wide variety of items.

In general, the mechanics are assigned to repair work under a number of main categories as follows:

SUBSYSTEM	NO. OF MECHANICS
Engines	6
Transmissions	3
Brake Systems	3
Electrical Machines	4
Air Systems	2
Air Conditioning	2

Work Output

Records were examined to determine the production output between October 1980 and early June 1981.

ITEM	OUTPUT (8 mths)	AVERAGE MONTHLY OUTPUT
Engines	63 (9 mths)	7
Transmissions	108	13.5
Differentials	34	4.25
A/C Compressors		
AMG/GMC	191	24
RTS	137	17

Maintenance Administation

Coach records were briefly reviewed to get a "feel" for the frequency of engine changes. These records appear to be either incorrect or incomplete in many instances. Such deficiencies could be due to a lack of information from the General Repair Sections.

Synopsis of Findings from the Operations Review

1) Fleet operations were severely affected from time to time by shortages of rebuilt replacement units.

2) No Production Plan existed. Actual production of repaired unit was not being monitored. Efforts to do this monitoring were frustrated by users practice of collecting freshly repaired items outside of any inventory control procedures, because of urgency.

3. There was a frequent lack of components for doing rebuild work.

4. Time standards were not being used.

5. Many repairs took significantly longer than accepted standards e.g. engine repairs took 143 hours vs. a standard of 80 hours. Dismantling took a disproportionately long 2 days of a Mechanic II's time.

6. Many major subsystems were not tagged with the date of repair, making "age" determination more difficult.

7. Many minor tasks e.g. dismantling engines, were being done by highly skilled workmen.

8. There was no control of the inventory of the defective components awaiting repairs. Shrinkage was very difficult to detect as a result.

9. Staff size was not based on manpower budgeting principles.

10. Rebuilt items were often subject to early failure, indicating a lack of proper quality control.

11. Users (minor repair mechanics) displayed a low level of confidence in repaired items.

12. The cost of rebuilds was not monitored, neither for parts nor labor.

13. Due to lack of cost information, make-buy decisions could not be quantified.

14. There was only limited "tracking" of rebuilt items.

15. A work sampling study indicated a low level of work productivity - 56% working, 44% delayed walking or stationery.

DEVELOPMENT AND IMPLEMENTATION OF METHODS IMPROVEMENTS

1. An ABC analysis was carried out to determine which rebuilt items were most crucial to operations in terms of quantity and value. The "A" items identified were given high priority in a Production Plan.

2. Tags were selected for the labeling of defective components. These were solvent proof and were provided with spaces to record source and defect information.

3. A Unit Room work order system was developed, with "Component Rebuild" cards serving as input documents. These cards were to be used by mechanics in the rebuild shop, i.e. one card for each job lasting in excess of four hours. Shorter jobs were to be based on "batches" of completed items. Parts used and labor hours expended were to be written on each such job card.

4. Production Report. This was based on the cards utilized for each period - daily, weekly, etc.

5. An inventory control system was introduced for the management of rebuilt item transactions.

a) All components were issued on an exchange basis only, i.e. to obtain a repaired unit, a mechanic submits the defective item at the parts counter.

b) All defective items were tagged with all available defect information.

c) Defective items are washed and placed in a holding area which provides a buffer.

6. Materials requirements planning. An ongoing analysis was carried out on the "Component Rebuild cards" to determine the frequency of usage for each replacement part in each typical repair job. From this analysis it was predetermined what mix and quality of parts would be needed for the planned production.

7. Usage forecast. All garage managers were asked to forecast their need for rebuilt items during a specified time period (fiscal year) based on the projected fleet miles. They were required to enter this information on a worksheet, indicating the expected monthly need (reflecting seasonal trends) for each rebuilt item. all the forecasts were consolidated to determine the rebuilt item needs for the entire organization.

8. Production Plan Development. The composite forecast was presented to a committee comprising the Materials Management Department and the manager of the Unit Rebuild shop. The latter manager submitted a bid on the portion of yearly requirements that could be met by the Unit Rebuild shop. The Materials Management Department was asked to obtain the remaining yearly needs from outside vendors.

9. Production schedule development. This schedule was developed based on the skill of the repair staff. Allowances were made for annual vacations and expected absences to tailor production to meet the predetermined monthly quotas. In the absence of standards, production volume was based on historical production data.

10. Capacity Improvement requirements were determined. These improvements were implemented (a) work stations were tailored to meet expected production levels, e.g. additional engine stands (b) lighting levels were increased over the benches with the most detailed repair tasks (c) Manpower budgeting was used to tailor work force

size to the expected production.

11. **Asset control was implemented.** Rigid procedures were set up to improve the security of repaired items. An audit trail was set up to document all production and to pinpoint any inaccuracies in the reporting of transactions for the receipt of defective items or the production of repaired items.

12. **A Production Coordinator function was established.** This position now serves to assist the manager of the Unit Repair shop to do short interval scheduling, monitor production and to perform continuing analytical work.

13. **Computerised Equipment Management System (EMS).** Subsequent to the initial Unit Room study in 1980, a computerized equipment management system was acquired to improve the accessibility of fleet status information. Currently, this system is being interfaced with the Unit Room Work Order system and will provide the following features:

- Production reports.
- Historically-based repair time standards.
- Production forecasting.
- Quality control monitoring.
- Repair Cost monitoring and analysis.
- Labor productivity reporting.

BRIEF DISCUSSION OF BENEFITS

1. Additional production was obtained with minor reassignments within the same size work force.

2. Stockouts of rebuilt items were reduced by 75% with the same size of work force. These stockout occurrences were monitored at the parts counter via reporting forms before and after the implementation of the methods improvements.

3. Overall rebuilt item needs were determined for the first time. Variance reporting was set up to advise garage managers of their actual usage versus their forecasts.

FIGURE I ORGANIZATIONAL SECTION ASSIGNMENTS-MANAGEMENT OF REBUILDABLE/REBUILT ITEMS

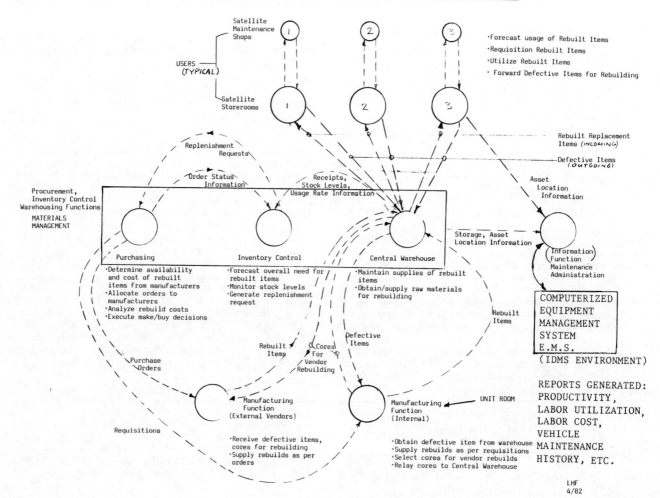

4. Make or buy decisions became possible with the production cost reporting system. Several items were dropped from internal production and purchased externally at a lower cost.

5. Improved quality control resulted in fewer failures of rebuilt items, increasing the confidence and morale of the line mechanics who use rebuilt items.

6. Repair times have reduced in some cases, e.g. transmissions because defects are communicated to the rebuild shop staff via the tags.

7. Failure reports on components defects can now indicate whether a specific vehicle has systematically recurring problems.

8. Standards development has begun for rebuild tasks based partly on the increasingly accurate historical information. These standards will be negotiated based on current performance versus "nationally accepted" repair times.

9. Manpower budgeting will be introduced based on the new standards. External acquisitions of rebuilt items is being reduced based on the increase in internal productivity.

10. Inventory carrying cost has been lowered, based on the analysis for the rebuilt items as well as for their component parts.

11. Costly emergency purchases have been almost eliminated due to the improved availability of components.

UPDATE OF THE UNIT ROOM OPERATIONS

The analysis and the ongoing monitoring of the level of productivity and the cost effectiveness of Unit Room operations have led to a major change in Top Management strategy. Production costs for the rebuilding of sub-systems have been compared with the prices of established remanufacturers, on an item-by-item basis, and the cheaper alternatives selected.

At the present time, Unit Room Operations have been terminated and a new facility is being constructed. Future in-house production will be based on the foregoing criteria.

FIGURE 2 PARETO ANALYSIS UNIT ROOM 14:10 TUESDAY, OCTOBER 4, 1983 1
FREQUENCY BAR CHART

ITEM		FREQ	PERCENT	CUM. PERCENT
65.1	Ixxxxxxxxxxxxxxxxxxxxxxxxxxxxxxxxxxxxxxxxxxxxxxxxxxxxxxxxxxxxxxxxxxxxxxxxxxxxxxxxxxxxxxxxxxxxxxxxxxxx	96	12.44	12.44
59.0	Ixxxxxxxxxxxxxxxxxxxxxxxxxxxxxxxxxxxxxxxxxxxxxxxxxxxxxxxxxxxxxxxxxxxx	66	8.55	20.98
58.0	Ixxxxxxxxxxxxxxxxxxxxxxxxxxxxxxxxxxxxxxxxxxxxxxxxxxxxxxxxxxxxxxxxxx	64	8.29	29.27
27.0	Ixxxxxxxxxxxxxxxxxxxxxxxxxxxxxxxxxxxxxxxxxxxxxxxxxxxxxxxxxxxxxx	61	7.90	37.18
56.0	Ixxxxxxxxxxxxxxxxxxxxxxxxxxxxxxxxxxxxxxxxxxxxxxxxxxxxxx	55	7.12	44.30
72.0	Ixxxxxxxxxxxxxxxxxxxxxxxxxxxxxxxxxx	34	4.40	48.70
24.0	Ixxxxxxxxxxxxxxxxxxxxxxxxxxxxxx	30	3.89	52.59
149.0	Ixxxxxxxxxxxxxxxxxxxxxxxxxxx	27	3.50	56.09
26.0	Ixxxxxxxxxxxxxxxxxxxxxxxxxx	26	3.37	59.46
7.0	Ixxxxxxxxxxxxxxxxxxxxxxx	23	2.98	62.44
97.0	Ixxxxxxxxxxxxxxxxxxxxxx	22	2.85	65.28
71.0	Ixxxxxxxxxxxxxxxxxxxxx	20	2.59	67.88
89.0	Ixxxxxxxxxxxxxxxxx	17	2.20	70.08
146.0	Ixxxxxxxxxxxxxxx	15	1.94	72.02
78.0	Ixxxxxxxxxxxxxxx	15	1.94	73.96
57.0	Ixxxxxxxxxxxxxxx	15	1.94	75.91
114.0	Ixxxxxxxxxxxx	12	1.55	77.46
3.0	Ixxxxxxxxxxxx	12	1.55	79.02
52.0	Ixxxxxxxxxxx	11	1.42	80.44
85.0	Ixxxxxxxxx	9	1.17	81.61
84.0	Ixxxxxxxx	8	1.04	82.64
113.0	Ixxxxxxx	7	0.91	83.55
98.0	Ixxxxxxx	7	0.91	84.46
48.0	Ixxxxxx	6	0.78	85.23
150.0	Ixxxxx	5	0.65	85.88
105.0	Ixxxxx	5	0.65	86.53
51.1	Ixxxxx	5	0.65	87.18
9.0	Ixxxxx	5	0.65	87.82
151.0	Ixxxx	4	0.52	88.34
110.0	Ixxxx	4	0.52	88.86
80.0	Ixxxx	4	0.52	89.38
79.0	Ixxxx	4	0.52	89.90
53.0	Ixxxx	4	0.52	90.41
51.0	Ixxxx	4	0.52	90.93
37.0	Ixxxx	4	0.52	91.45
147.0	Ixxx	3	0.39	91.84
109.0	Ixxx	3	0.39	92.23
107.0	Ixxx	3	0.39	92.62
90.0	Ixxx	3	0.39	93.01
50.0	Ixxx	3	0.39	93.39

FIGURE 3 UNIT ROOM REPORT #3 14:33 MONDAY, OCTOBER 3, 1983

ITEM	SUB	AVERAGE UNITS PER WEEK	ANNUALIZED FORECAST UNITS	ARITHMETIC AVERAGE HRS/UNIT	MEAN HRS PER UNIT	STANDARD DEVIATION MEAN HRS	AVERAGE HRS/WEEK	ANNUAL HOURS
1		0.8667	45.07	0.615	0.615		0.5333	27.73
3		3.2000	166.40	4.052	4.096	2.1263	12.9667	674.27
4		2.0000	104.00	0.917	0.944	0.1102	1.8333	95.33
5		18.0000	936.00	0.141	0.134	0.0571	2.5333	131.73
6		12.2000	634.40	0.126	0.134	0.0518	1.5333	79.73
7		4.4000	228.80	5.356	5.358	1.0965	23.5667	1225.47
8		0.4667	24.27	5.500	5.500	1.1180	2.5667	133.47
9		1.0667	55.47	5.375	5.854	2.1151	5.7333	298.13

ANALYSIS OF THE BUS REHABILITATION PROGRAM

Background

The Bus Rehabilitation Program was started at MDTA in 1979 with the objective of increasing the size of the operating fleet and thereby improving the level of safety, reliability and rider attractiveness by refurbishing older coaches to meet near-new specifications. In addition, the program would assist in meeting the growing increases in ridership demand over the years, and especially the future demand required by the interfacing with the rapid transit system.

The program was partially funded by the Federal Government and was administered under strict guidelines, e.g.

1) On a life cycle costing basis, a rehabilitated bus should cost a maximum of 70% of the cost of a new bus amortized over a 12-year life.

2) Each bus should be at least 12 years old or have in excess of 500,000 accumulated miles.

3) Buses should be selected with similar rehabilitation needs - mileage, extent of corrosion, etc.

4) Rehabilitation should extend service life to twenty (20) years from the date of original entry into service, with a minimum "Rehabilitated Life" of 5 years.

5) The repair work should be substantial enough to restore major sub-systems, e.g. engine, transmission, drive train, brakes, A/C, electrics, as well as body-related components to near original specifications.

It must be understood that transit buses are both very expensive (over $130,000.00 each) and are very long lead time purchases, with delivery schedules frequently in excess of one year, hence the program is a highly beneficial one, in principle.

Description of the Rehab Operations

The Rehab facility was set up as a separate building from the routine maintenance operations. The building comprised four internal repair bays of which three were devoted to body work and one to mechanical repairs. A fifth external bay was used for preliminary inspections as well as for final adjustments. Part of the building was used for the storage of spare parts and basic materials such as plywood for flooring and plexyglass sheeting for windows, etc.

Staffing

The Rehab staff comprised a Division Superintendent, a Maintenance Clerk and approximately 25 repairmen. This latter group was subdivided into approximately five (5) mechanics and 20 body repairmen. In turn, five crews were created, i.e. one body work crew at each of the three body bays and a mechanical crew at each of the two mechanical bays. Each crew had a designated lead worker or "crew chief".

Extent of Repairs

Typically, repairs included the following:

. Removal and replacement of seats, floormats and all interior paneling.

. Repair and/or replacement of floor boards.

. Complete repairs to all coachwork, correcting corrosion and structural damage.

. Complete overhaul of the brake, suspension and steering systems.

. Replacement of the engine and transmission systems with refurbished units.

Problems with the Rehab Program

. The production of repaired buses fell far short of expectations, i.e. one and one-half buses per month versus four buses (scheduled).

. The repaired units were proving to be unreliable.

. The actual repair costs were beginning to exceed the maximum planned levels. (Part of these additional costs would ultimately impact the operating budget).

. Because of the long "pipeline time" required to repair a bus, the regular fleet was being slightly depleted by the units "in process" in the Rehab Program.

CONDUCT OF THE STUDY

Orientation

Orientation of the shop staff was conducted, with the assistance of the Rehab Division Superintendent and Foreman, to apprise them of the purpose of the study and to seek their co-operation.

OPERATIONS REVIEW

Ratio-Delay Study

A ratio delay study was conducted to get a perception of the level of employee utilization in the facility. Three categories of activities were established for this study: working; working delayed; and not working delayed.

Individual work units were selected as the point of observation, i.e. each of the three coachwork repair bays and the two mechanical repair bays.

Discussion of results of the Ratio-delay Study

For the body bays, employees were observed working 67.3% of the time. However, for the mechanical bays, the percentages were only 45.9%. It was evident that mechanical labor utilization was much lower than desired. In addition, overall labor utilization for both trades dropped during overtime hours - 38.3% not working vs. 28% not working during normal hours.

Analysis of service records indicated that the rehabilitated buses, instead of adding to the overall quality of the fleet, were experiencing a higher rate of road-calls than the fleet average. In a follow-up study, using a S.A.S. program, the Management Services Unit was able to develop a Roadcall Summary exclusively for Rehab buses, indicating "occurrences" and "mean miles between roadcalls". (See fig. 5). This summary, together with other reports is currently serving as a diagnostic tool to detect generic problems.

Diary Reporting

Self-reporting forms were designed to ascertain the distribution of labor hours over the various body repair tasks, to identify those with the best potential for methods improvement benefits. The forms were given to each of the three bodywork crew chiefs. They were instructed to list the man hours expended on the broad categories of work on a daily basis.

Discussion of Findings

The work distribution chart (WD),(see fig. 4) indicated that 58.7% of the labor hours were being expended on the following:
floor repairs - 18.7%
bodywork exterior - 16.3%
window work openings - 14.1%
window work frames - 9.6%

It was decided to concentrate on developing methods improvements for the foregoing activities.

METHODS ANALYSIS

Data was collected on work methods by several means: employee interviews, work sampling and reviews of repair records.

A summary of pertinent findings is as follows:

Duration of repairs	Percentage
1-2 months	25%
2-3 months	9%
3-4 months	33%
4-5 months	33%

. Documentation on completed jobs indicated much greater complexity of repairs than the preliminary estimates would indicate. Labor standards were not being used and man hours needed were not predetermined, ostensibly because of the great detail involved in Rehab repairs.

. Because of the lack of forward planning manpower needs could not be pre-planned and job completion dates were not estimated.

. Many repairs were delayed because of the lack of certain crucial parts.

. Time was lost in spray painting interiors because of the need for the other crews to evacuate the building. In addition, as many as 5 colors were applied on separate occasions.

. Job completion times varied widely on buses with similar labor and material costs, indicating a lack of proper scheduling.

. Analysis indicated a shortage of specialized tools and equipment such as automatic wheelhoists, rivet guns, routers and air conditioning hose wrenches.

. There was a lack of standard repair procedures and therefore, repairs were not conducted in a reliable sequence.

. Certain tasks were exceeding the expected repair standards; e.g. repairs of removable bus windows often consumed over 6 man hours, versus the expected standard of two and one-half hours each in a "glass shop" located elsewhere.

Recommended Methods Improvements

Based on the methods analysis, a number of recommendations were made, of which some were implemented with significant results.

. Some detailed tasks were terminated and transferred instead of other better equipped facilities, e.g. the repainting of small parts and the repairing of the removable window frames.

. A more detailed estimating process was developed and conducted earlier in the repair cycle to allow more time for parts acquisition.

. Parts for bus repairs were "staged" to a greater extent than before to reduce process delays.
. Approximate quality standards were developed to establish "cut off" points for repair work.

. A scheduling system was established to maximize resource allocation to complete jobs on a timely basis.

. The repair cycle was redesigned to include buses waiting in the yard, prior to being brought into the shop. Preliminary cleaning and stripping were initiated on these buses to create more "in process" inventory.

. Eventually, repair costs for engine and transmission replacements were reduced. It was determined that these units could be repaired by private companies instead of in-house at lower cost.

. A line-balancing process was introduced to better "dovetail" mechanical work with coachwork to avoid the underutilization of the mechanics observed in the ratio-delay study.

SUMMARY OF BENEFITS FROM THE REHAB PROGRAM STUDY

In the course of the 2-year period since studies were initiated, the output of the Rehab program doubled from one and one-half to 3 buses per month. In addition, the quality of output from the repair facility has improved and refurbished buses are proving to be more reliable than before.

FIGURE 4

REHAB DIVISION

WORK DISTRIBUTION CHART

(For Body Work Bays 1,2 & 3 Only)

Task/ Activity Number	Task/Activity Description	Total Man Hours	% of Total
1	Stripping Interior	166	4.6
2	Floor Repairs	672	18.7
3	Bodywork-Interior	217	6.0
4	Seat Repairs	308	8.6
5	Step wells	88	2.4
6	Heater, A/C, (Bodywork)	52.7	1.5
7	Painting	130	3.6
8	Refitting Interior	64	1.8
9	Window work (Frames)	344	9.6
10	Window work (openings)	506.5	14.1
11	Bodywork-Exterior	585	16.3
12	Front Cap/Bulkhead Repairs	307.8	8.6
13	Door Repairs	93	2.6
14	General Shop Work	58	1.6
		3592	100%

FIGURE 5

ROADCALL SUMMARY FOR REHAB BUSES FROM APRIL

	MONTH					
5/83		6/83		7/83		
JEN- COUNT	I- IN- ID- MERC IEX	FREQUEN- CY COUNT	I- IN- ID- MERC IEX	FREQUEN- CY COUNT	I- IN- ID- MERC IEX	
M	IS- CALC IUM	SUM	IS- CALC IUM	SUM	IS- CALC IUM	
BUSNO	COMPL- IETE					
195	111782	.	1844 0	5	387 0	
196	42683	10	414 0	11	299 0	
197	122282	2	1421 0	6	629 0	
205	120181	3	1229 0	7	458 0	
208	20281	.	1657 0	7	338 0	
212	10181	1	1700 0	7	356 0	
220	60182	.	. .	.	. .	

The improvements in the program are evident by the incremental addition to the fleet of attractive, roadworthy, reconditioned units, which have increased fleet availability, whereas many of these units have previously been inoperable.

From a cost standpoint, the applied overhead per unit has been effectively halved to $5,400 by a doubling of the production rate, for a potential saving of $193,000 per year. (See fig. 6). Savings have also been realized because of the gradual reductions in costly "road calls" sustained by the refurbished units.

Improved performance monitoring systems have been implemented. The "Roadcall Summary" together with other problem-oriented reports, is currently serving as a diagnostic tool to detect generic equipment malfunctions, so that repair procedures can be gradually upgraded.

CONCLUSION

The Productivity Improvement Program at MDTA has proven to be highly successful during the four years of its existence. Major dollar savings have been derived as well as intangible benefits such as improved bus operator and passenger morale.

While a program of this nature does not entail extremely sophisticated analytical techniques, it must be recognized that it represents a significant advance in the transit industry which has been lagging far behind other industries in terms of the adoption of modern management methods. It is gratifying that the program has stimulated the evolution of the MDTA transit operation towards computerized information handling and decision-making, gradually replacing the "seat of the pants" management model.

ACKNOWLEDGMENTS:

MDTA MANAGEMENT SERVICES UNIT:

    Mr. Terry McKinley
    Mr. Art Tillberg

METRO DADE COUNTY - OFFICE OF PRODUCTIVITY MANAGEMENT - formerly the P.A.U.:

    Mr. R. Slocum

FIGURE 6 - ANALYSIS OF REHAB PROGRAM OVERHEAD COSTS

Salaries of Supervisory and support staff plus fringes	- $168,000
Misc. overheads: Utilities etc.	- 25,000
TOTAL:	- $193,000

Production rate before the study = 18 buses per yr.

Applied overhead per unit - $\dfrac{\$193,000}{18} = \$ 10,722$

Production rate after the study = 36 buses per yr.

Applied overhead per unit - $\dfrac{\$193,000}{36} = \$ 5,361$

ESTIMATED SAVINGS IN OVERHEADS
36 x $5,361 = $193,000

ABOUT THE AUTHOR

MR. LINCOLN FORBES, P.E. is Chief of the Management Analysis Division of the Department of Public Works, Metropolitan Dade County, in which capacity he is responsible for productivity improvement, performance planning and other diverse management support functions department-wide. Previously, he was a Senior Management Analyst with the Dade County Transportation Administration and a Productivity Analyst with the Metropolitan Dade Productivity Analysis Unit.

In earlier positions, Mr. Forbes worked in Jamaica, West Indies from 1968 to 1977 as a Senior Engineer with Attas, De Four, Benghiat Consultants, specializing in Electrical/Mechanical services, and as a Plant Engineer with Alcoa Minerals from 1968 to 1969. He holds a Masters Degree in Industrial Engineering and an M.B.A. from the University of Miami, having majored in Management Engineering and Operations Management respectively. He also holds a B.S. in Electrical Engineering and is a member of ALPHA PI MU, a Foundation Member of the Jamaica Institution of Engineers and is a Registered Professional Engineer in the State of Florida. Mr. Forbes is currently President of the Miami Chapter and has also served the Institute as Chairperson (Chapter Development) for Region 4.

Reprinted from 1984 Annual International Industrial Engineering Conference Proceedings.

EMPLOYEE PARTICIPATION PROGRAMS
THAT IMPROVE PRODUCTIVITY AND QUALITY

Ross McManus/Frank Sentore
Eaton Corporation-Controls Division

ABSTRACT

This paper focuses on a 5 year employee participation program at our divisional plants. During this
period we have expanded our program from one Productivity Team to 7 Quality Circle Teams. This was
accomplished in a small plant environment, within a
division where plants are situated throughout a wide
geographic range. Both types of participative programs are covered, focusing on the format, and objective of each, along with their results.

INTRODUCTION

It has been apparent that lagging productivity and
stiff foreign competition poses a serious problem to
U.S. industries. Folklore has it that U.S. manufacturing plants cannot excel in both quality and
productivity. But the 1980's has brought a new era
of enlightenment, a period in which productivity and
quality cannot be seen in seperate terms, but as a
synonymous goal. In todays competitive environment, all resources must be utilized to their ut-
most. Eaton's most valuable resource is people.
The initiation of Productivity and Quality Circle
Teams at our plants was the catalyst needed to blend
together, productivity and quality. Our participative programs were an added dimension in opening
lines of communication and solving problems, which
resulted in increased productivity and improved
quality.

In an effort to increase productivity, the Controls
Division established a Productivity Team at one of
their manufacturing plants in the Spring of 1979.
This team included plant engineers, supervisor, and
one hourly assembler. The team studied all aspects
of one particular manufacturing process. The results of this team's efforts was a work station,
which was easier for the operator to assemble piece
parts, while also, resulting in a cost savings.

The success of this Productivity Team prompted management to add one more team. It was after the ini-
ation of the second team that the Productivity Council began investigating other types of participation
groups. They were interested in a program which
would offer greater participation, as well as, increase their scope of activity. In 1981, Control's
Division, in lieu of Productivity Teams, opted to
establish Quality Circles in three of the nine
plants. Unlike the Productivity Teams, Quality
Circles would comprise of the supervisor (leader)

and 5 to 7 hourly employees. Today, we have 7 circles, with a continuing commitment to add additional
teams.

PRODUCTIVITY TEAMS

Productivity teams were instituted for the primary
reason of gathering together employees of varied
skills to improve productivity at one particular
manufacturing operation. This operation was assign-
by a joint consensus of the Productivity Council,
and plant managment. This first project received
"fish bowl" attention. The Productivity Council
was just as concerned with how the team functioned
as a group, as with the actual productivity improv-
ment. The first team was formed at one of our divisional plants for the primary purpose of improving the workplace layout at an assembly operation.
This project, as well as, all future projects
would be assigned by the Productivity Council.

Training

The Productivity Team consisted of a Manufacturing
Engineer, Industrial Engineer, Supervisor and an
Assembler. Members of the team. along with division support people, trained the team members in
method, and workplace layout, as it pertained to
the specific process they were investigating. They
also, investigated into different types of parts
containers which were on the market, as well as,
designing and building in-house custom containers.
All training was localized to the particular project they were analyzing.

Scope

This first Productivity Team concentrated on making
the job easier for an operator at an assembly operation. This was accomplished by reducing reaches,
improving parts layout, and where necessary, building new containers. This first team adopted the
philosophy "work smarter, not harder". The results
of the teams efforts was a workplace which was more
efficient, with an increase in productivity.

Later a second Productivity Team was formed, only
with a different format. This team consisted of a
supervisor and 4 assemblers. The reason for this
change was the positive response exhibited by the
employees. The success of the first team prompted

interest among other employees. For this reason, the Productivity Council, in conjunction with plant personnel, thought a second team was warranted, but with more hourly employee involvement. The second team received the same training as the first team, while investigating a similar manufacturing operation.

TRANSITION PERIOD

The success of Productivity Teams cannot be measured only in the productivity improvements realized, but in the enthusiasm generated among the employees. Their desire to contribute their knowledge and work experience with management to increase productivity was evident. Although, Productivity Teams were a good vehicle to involve production personnel, as well as, supervisors and engineers in making workplace improvements, it was limited in its scope and involvement. The Productivity Council did not want to limit the participation of these teams to only productivity improvements, nor did it want to limit their responsibility to pre-assigned projects. With this in mind, an investigation into other types of participative programs was initiated.

It was learned, that quality circles were intended to function for a long period of time, whereas, productivity teams are formed to solve specific problems. The training for both programs also varied. Quality circle training covered many parameters, preparing team members to analyze a variety of different problems. Unlike productivity teams, quality circles would involve more employees, with a broader base of responsibility. Quality circles would concentrate on improving quality related problems, in conjunction with productivity improvements. This approach emphasizes a "people building philosophy", which allows team members to meet on a regular basis, to analyze and solve a wide variety of problems. With this in mind, a transition was made from productivity teams to quality circle teams.

QUALITY CIRCLES

A quality circle is a group of people from a common work area, who voluntarily meet on a regular basis, to identify and solve problems. This Quality Circle approach would serve two basic functions, which were not being addressed under the Productivity Team concept, these are:

1. Greater hourly employee participation.

2. Increase scope of involvement.

Through structured training program, each team would be capable of analyzing a variety of different problems, ranging from method improvements, tooling, revisions, scrap reductions, to quality improvements.

Format

Our Quality Circle Teams were established as a strictly voluntary group, comprised of hourly employees from a designated work area. The leader of the team is also their supervisor. The number of circles in each plant is dependent on the plant size as well as, employee response. Each plant has a facilitator who acts as the overseer for the plant circles. The size of each circle varies from 5 to 7 employees. This size helps to assure balanced participation by all members. The most important concept of a quality circle team, is that they are free to work on any problem they choose.

Facilitator-The facilitator is the overseer of Quality Circle Teams, he or she assures the team is receiving support from plant management, staff personnel and the Productivity Council. They should be ready to assist the leader during those lull periods all circles will at one time experience. Because, no plant in our division has more than two quality circles, it would be difficult to justify a full time facilitator. For this reason, the plant Industrial Engineer acts as part time facilitator. This dual role has worked out very well for our division. This success can be attributed to both "hats" reporting to the same individual, the Manager of Industrial Engineering, who is also a member of the Productivity Council. Because, of this arrangement, we have not experienced any conflict of interest.

Leader-The key to any successful circle is the team leader and it becomes more apparent where part time facilitators are utilized. Because, of the dual responsibilities of part time facilitator, there are times when engineering projects may take presidence over the facilitating responsibilities. For this reason, our Quality Circle leaders take on more responsibility than circles with full time facilitators. No one knows the circle members better than their own supervisor, who is also, their team leader. These leaders are able to guide and mold the team in a direction, which is conducive to the members varied personalities and goals. Each leader will reach their own level of leadership and it is up to the facilitator to nourish and fill this leadership gap, in the form of:

1. Continued educational material.

2. Assure support and assistance of plant middle management.

3. Intermediary, between the Quality Circles, and Productivity Council.

4. Obtain circle information from other circles, and International Association of Quality Circles.

Training-There are many different avenues open for establishing a quality circle training program. The type of training program selected, depends on the circle goals, as well as, the plant environment. Our Divisional Productivity Council opted to do their own training with the aid of a consultant's training manual and video tapes. This training covered the folowing areas:

1. Brainstorming.

2. Problem Solving Techniques.

3. Sampling Techniques.

4. Pareto Charts.

5. Histograms.

6. Cause and Effect Analysis

7. Management Presentation.

The plant facilitators were trained at Divisional Headquarters by the Manager of Industrial Engineering. The facilitators then went to their respective plants to train both the leaders and team members. The training of the leader was concurrent with member training and lasts approximately 10 weeks. After this intitial training period, the facilitator will gradually reduce the number of meetings attended. The number of meetings attended are dependent on the leader's, leadership ability. The first 10 weeks of training is only the beginning of a continued educational process. During the course of the circle's life, there will be periodic training in such areas as:

1. Statistical Process Control.

2. Value Engineering.

3. Guest Speakers.

4. Computer Training.

Scope

The transition from Productivity Teams to Quality Circles, brought with it, not only greater numbers of hourly people participating, but also, an increase in the variety of projects available to them. Quality Circles are more than just a means to solve problems and increase productivity. It is a program to further open communications between management and labor. This will ultimately enhance the trust between both parties and improve the quality of work life for all concerned.

The scope of projects available to quality circles is limited only by their own creativity. However, these projects should be limited to their own work area, and within their own area of expertise. Some projects, quality circles can become involved in are as follows:

1. Workplace and Method Improvements.

2. Tooling and Equipment Revisions

3. Quality Improvements.

4. Scrap Reductions.

Regardless, of the type of problems circle members are analyzing, they should be encouraged to exhaust all available resources in solving this problem. Solicit help from engineers, supervisors and last but not least, fellow employees. The more a circle involves other employees and management in their projects, the easier it will be for them to gain future assistance and support.

SUMMARY

In the five years Controls Division has been involved in employee participation groups, we have gone from one goal oriented group of Productivity Teams, to the total involvement of Quality Circles. Productivity Teams worked well for what they were intended, improving productivity at a specific work area. But, we soon discovered, that with the enthusiasm generated, combined with the variety of ideas flowing from the employees, this participative approach was too narrowly defined. For this reason, we made the transition to Quality Circle Teams, an approach with a broader concept, which would allow for more employee involvement.

Although, the division did realize productivity improvements, along with some significant cost savings, the real success cannot be measured in dollars. The flow of communication between labor and management has not been better. The most rewarding changes are the ones which are most visible. Individuals, who at one time would not think of contributing their knowledge to management, are now concerned with productivity and quality. Others, who were too intimidated to apply for an inspectors or auditors position, are now applying for these positions due to their involvement in Quality Circles. Employee participation groups are more than just a means to increase productivity and solve quality problems, they are a people building organization. They are a vital necessity in today's competitive environment, and an organization which does not utilize all their human resources will be left behind.

BIOGRAPHICAL SKETCH

Ross McManus is the Industrial Engineer at the Eaton Controls Division, Warren Plant. Part time Facilitator for plant Quality Circle Teams. Member of the Blackhawk Chapter of Illinois.

Frank Sentore is the Industrial Engineer at the Eaton Controls Division, Savanna Plant. Part time Facilitator for plant Quality Circle Teams. Member and Director of the Blackhawk Chapter of Illinois.

Reprinted from 1984 Annual International Industrial Engineering Conference Proceedings.

POWER PLANT MAINTENANCE
PRODUCTIVITY IMPROVEMENT PROGRAMS

R. H. Seemer
Florida Power and Light Company
Management Services Department
Miami, Florida 33102

ABSTRACT

This paper presents four programs currently underway in Florida Power and Light's generating stations. All have had significant impact upon productivity. It is also emphasized that although these programs utilize many of the principles of "Scientific Management", they rarely provide the solutions. Solutions are achieved when people know what needs to be done, are organized effectively and provided the appropriate resources.

INTRODUCTION

In 1979, Florida Power and Light embarked upon a comprehensive program to improve operations in the power plant maintenance management area. The initial Operations Review conducted jointly by the Management Services and Power Resources departments identified several major opportunities and established timetables for development and implementation.

Major power plant projects completed, currently under development or part of an on-going effort include:

. Centralized Equipment Repair Shop Feasibility Study,

. Engineered Performance Standards for Maintenance,

. Equipment Coding System Project,

. Maintenance Planning Study,

. Material Coding System Project,

. Nuclear Control Room Equipment Placement and Traffic Flow Analysis,

. Nuclear Plant Startup and Maintenance Staffing Study,

. Nuclear Staff Reviews for the Administrative, Health Physics, Maintenance and Security Departments,

. Operations Reviews for Various Fossil Plant Organizations,

. Reactivation, Startup and Maintenance Staffing Study of a Retired Fossil Plant,

. Reliability Centered Maintenance Pilot Program,

. Staff Reviews for Plant Organizations,

. Work Sampling of Routine Maintenance and Scheduled Outages.

PROGRAMS

Four programs identified by our management as having significant impact upon current and long term productivity include;

. Engineered Performance Standards for Plant Maintenance,

. Work Sampling of Routine Maintenance and Scheduled Outages,

. Operations Reviews and Maintenance Planning Studies, and

. Staff Reviews of Plant Organizations.

Following are brief overviews and some results of these programs:

Engineered Performance Standards (EPS)

This program consists of the development, application and on-going support of job time standards for power plant maintenance. The system utilizes the universally accepted principles of benchmarks, ranges of time and job slotting based upon work content. The data base of elemental job times was acquired from the Central Electricity Generating Board (CEGB). It has also been successfully implemented at other utilities in the U.S. such as Commonwealth Edison, Tennessee Valley Authority and Wisconsin Electric Power.

Florida Power and Light's program is nearing completion of the development phase and is well into implementation. To date, over 2960 benchmarks have been developed in ten major task areas with over 85,000 slotted job standards currently in use throughout our twelve plants. Presently, the program has been fully implemented at three plants with the others scheduled during 1984.

Other phases of the EPS program currently under development include revision of our performance reporting systems, standards validation and auditing procedures and the automation of our total Job Planning System.

Areas in which specific productivity gains are expected or have been achieved with EPS include:

. Reduction of planned overhaul turbine main-tenance manhours by 30% on specific 800 MW units, 20% on specific 400 MW units,

. Increases in worker "efficiency on standard" of 12 to 30% in routine maintenance at the pilot plants,

. Identification of Specific Plant Work Order deficiencies which inhibit main-tenance standards coverage,

. Identification of deficiencies in the equipment maintenance history coding system which inhibit data accuracy and require additional planning time.

Work Sampling

Work Sampling continues to prove itself as a simple but effective work measurement system with a variety of applications. Typical uses of work sampling in the power plant environment include the study of outage and routine maintenance activities. It is also applied in conjunction with Operations Reviews and is used to validate values and results from other time management and work measurement systems. When applied systematically, work sampling has proven in-valuable in the determination of staffing levels for many classifications (discussed in Staff Reviews, below).

Routine Maintenance and Scheduled Outages

Uses when applied in this area include:

. Determination of worker utilization by craft and job activity,

. Identification of productivity inhibitors such as delays attributed to planning such as materials, tools, craft coordination,

. Delays attributed to workers such as idle time, unplanned breaks, late starts, early quits, absence from area and excessive discussion,

. Delays attributed to supervision such as too many discussions with the crews, unavailabi-lity for consultation or absence from the area,

. Delays attributed to other departments such as Health Physics, Security, Quality Control, Stores and the other maintenance departments.

The greatest improvements in worker utili-zation result from immediate interpretation and feedback of the findings and posting the summaries in an area frequented by all personnel. Data collected from Monday through Friday is summarized via remote terminal and printer and presented to plant management that evening or prior to the next Monday morning meeting. Each supervisor is briefed regarding findings in his specific area. Contractors are also evaluated, there-by providing additional criteria used in the contractor selection process for large projects.

Validating Data

Work sampling also played an important role in the development of Personal, Fatigue and Delay (PF&D) allowances during the initial development of the Engineered Performance Standards (EPS) program for plant maintenan-ce. By identifying the extent of delays in the health physics (radiation work permits, decon functions, etc), security (check points), operations (clearances) and quali-ty control checks, better assessments could be made in determining reasonable time values for these functions. In addition, methods improvements were noted for further study in each area.

When used in conjunction with our Outage Management Scheduling System (Project/2) it has verified the existence of slack human resources. In each case, these per-sonnel were shifted to other activities in our goal to decrease plant outage duration and total job costs.

The joint application of these two systems has been successful at each of the three outages in which they have been employed together.

Although work sampling is invaluable in pointing out problem areas, it does not offer many solu-tions. We must usually develop them through other means.

Operations Reviews and Maintenance Planning Study

Operations reviews are a means of assessing the status of existing conditions and comparing them to predetermined expectations.

A recent review consisted of the maintenance planning departments at eleven fossil plants. It focused upon several areas that would indicate the current "level of planning" at each plant. It also identified specific deficiencies that inhibited the planning effort and planner and crew productivity.

Audits were performed upon the work order files at eleven fossil generating plants in our system. The primary emphasis of the audits included the assessment of each plant's performance in the following areas:

- Work order coverage (specific vs. blankets),

- Job definition,

- Equipment/Work order coding,

- Priority assignment,

- Work type designation,

- Use of job time standards,

- Maintenance history system coding error experience,

- History of plant's outage report,

- Completeness of attachments (material and tool requirements, drawings, procedures, etc.),

- Productivity of the planners, and the

- Productivity of the work force for which they plan.

With weights applied to each of the above areas an overall grade was applied. Also, since our planning procedures are standardized throughout the system, plant by plant comparisons of each area to other plants and system means were easily made.

Using random Work Sampling to determine crew utilization, it was determined that the crews of those plants with the lowest planning scores also had the lowest percent "Direct Work".

In addition, the crews' percentages of time in "Transport", "Travel" and "Work Wait" were always higher at those plants, with the results of each category correlating highly with the planning efforts expended by each plant in the individual areas noted above.

Specific short and long term objectives have been established for each of the above areas at each plant. Resources required to attain those objectives were determined during the Staff Review process discussed below.

Applying factor analysis to data from this study, the on-going Work Sampling Programs, EPS and the Staff Review projects (discussed in Staff Reviews below), efforts are underway to answer the questions most often asked by our management of the maintenance area; "How much planning is enough and which areas of emphasis provide the greatest payback?"

Staff Reviews

Staff reviews have recently been completed within all of our fossil generating plants to determine the appropriate personnel staffing levels for;

- Clerical,

- Planning, and

- Supervisory functions.

These studies required the application of several techniques including;

- Work Flow Analysis,

- Random Work Sampling,

- Systematic (Fixed Interval) Sampling,

- Organization Analysis (including interviews and surveys),

- Regression Analysis.

The above techniques were useful in determining if people were organized effectively, knew their jobs and what was expected of them. In addition, value judgements were made of the level of utilization of each employee classification to determine if workers were interested in doing their jobs.

The systematic sampling process revealed many interesting results, among them;

- Supervisors allocated significantly less time to direct supervision functions than management expected,

- One "hybrid" job classification between planner and supervisor allocated a small percentage of time to the functions of either classification,

- Planners consistently worked outside their prescribed job classifications resulting in a lower level of planning,

- Planners spent a low percentage of time identifying job material requirements (leaving it up to the foreman or crew personnel) although this is a primary job function,

- General meetings and miscellaneous paperwork consumed an inordinate amount of time on everybody's part, and

- Maintenance and use of a sophisticated equipment history system required less time than plant personnel originally expressed (actual was 14%, implied was 48%).

Through the quantification of work outputs and the factoring of data from the results of the random work sampling efforts, work determinants were developed for the major job functions. In the case of planners, 83.6% of their total job time was factored. This information was used in the regression models to develop the basis for the final staffing criteria.

For supervision, objectives were established for how much time should be spent in the field. In addition, many time consuming extraneous tasks were identified and either reduced, delegated to other employee classifications or eliminated.

A clerical staffing model was developed based upon work determinants and has proved to be easy to apply and understand, yet sensitive to cyclical trends encountered during the course of the year.

SUMMARY

Each of the above programs represents approaches by which opportunities for productivity improvements can be identified. In many cases, "Scientific Management" tools are instrumental in the identification of productivity inhibitors. In all of the above cases improvements were achieved when people knew what was to be done, were organized effectively and were provided the appropriate resources.

BIOGRAPHICAL SKETCH

Robert H. Seemer is a Senior Management Analyst in the Management Services Department of the Florida Power and Light Company in Miami, Florida. Bob's primary responsibilities include the development and implementation of productivity improvement programs in the fossil and nuclear generation system, specializing in the maintenance area. He has also held Industrial Engineering, Supervisory Industrial Engineering and Line Management positions with the Boeing Commercial Airplane Company and United Parcel Service. Mr. Seemer holds a Bachelor of Science degree from the Georgia Institute of Technology, 1971, and is nearing completion of the MBA program at the Florida Institute of Technology.

*Reprinted from Industrial Engineering,
February 1983.*

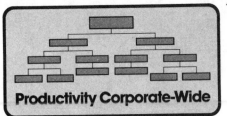

Productivity Corporate-Wide

Putting A Productivity Improvement Program Into Action: A Six-Step Plan

By Shafique Jamali
Public Service Co. of New Mexico

☐ *For a productivity improvement program to be successful, a rational approach to productivity measurement, evaluation, planning and improvement is essential. A considerable amount of attention has been given to productivity improvement. However, this is only one of several considerations related to productivity management—as is shown in the basic model, Figure 1. For a productivity program to be successful, equal attention must be paid to all aspects of the model.*

Productivity encompasses virtually everything that concerns managers in the running of their departments. It is important to emphasize that productivity improvement is not just a part of or an addition to the manager's job, but rather it is the job itself. In other words, the productivity program should not be considered a replacement for sound and disciplined management.

The management services department of the Public Service Co. of New Mexico has developed simple, step-by-step guidelines for implementing a productivity program. This article presents an overview of those guidelines.

The implementation guidelines are designed to capture all phases of the basic model of productivity management as shown in Figure 1. Normal business planning for input, processing and output is integrated into the productivity planning process.

The guidelines are structured as follows:

☐ *Step 1—Creating awareness.*
☐ *Step 2—Productivity measurement.*
☐ *Step 3—Productivity evaluation.*
☐ *Step 4—Productivity planning.*
☐ *Step 5—Productivity improvement.*
☐ *Step 6—Control reporting.*

These guidelines are a planned series of steps from input through processing to output with a feedback mechanism for determining how effectively company resources are being used.

Step 1: Creating Awareness

Awareness of productivity in an organization can be created by first explaining and defining what is meant by the term "productivity" and then emphasizing the importance of productivity so that both management and employees can understand and appreciate it.

Productivity definition

The term productivity means different things to different people; for example: "Doing right things and working right;" "more bang for the buck;" "working smarter and harder."

A widely accepted formal definition is the following: "Productivity is reaching the highest level of performance with the lowest possible expenditure of resources."

It is also important to understand what is *not* meant by

Figure 1: Basic Model of Productivity Management

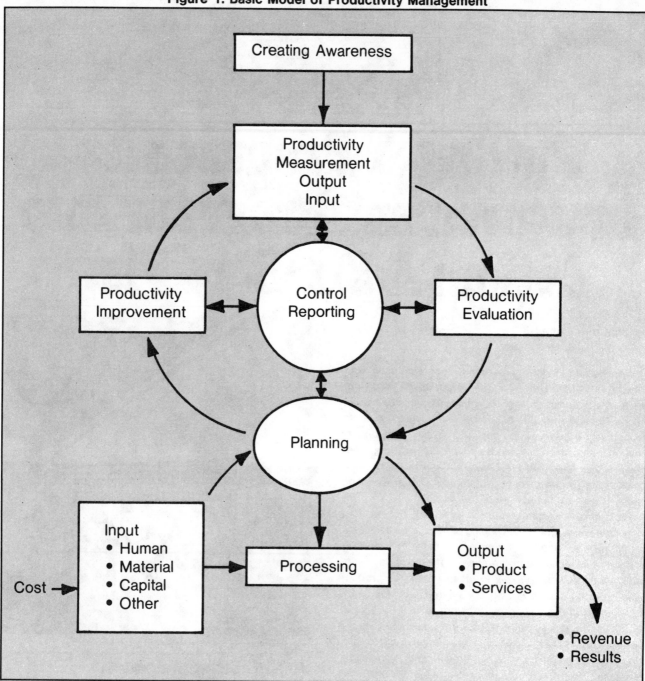

the term productivity. Table 1 provides a comparison that helps illuminate this point.

Importance of productivity

In a simplistic sense, productivity is important because if employees' salary increases are not matched with corresponding increases in productivity, a higher unit labor cost will result.

More essentially, the importance of productivity can be emphasized by identifying its benefits to management and employees.

Benefits to management

From management's standpoint, increasing productivity:

☐ Provides better organization and control of the department.

☐ Allows more effective planning and allocation of the company's resources.

☐ Creates a basis for effective supervision.

☐ Helps eliminate backlog and overtime and create a consistency of work flow.

☐ Improves work scheduling.

Benefits to employees

Employees can benefit from productivity enhancement programs because they:

☐ Provide opportunities for participation in the decision-making process (for example, in quality circles).

☐ Give recognition for achievements.

Table 1: What Is Productivity?

Productivity is:	Productivity is not:
☐ Techniques to improve employee performance and work environment. Through quantitative and qualitative techniques employees are encouraged to work smarter and harder. ☐ The ratio of the quality and quantity of production to the resources utilized. Both output and input are important factors. ☐ A factor useful in profit planning. Keeping the input resources constant and increasing productivity results in increased revenue. ☐ One means of improving quality.	☐ A dehumanizing and cunningly devised technique to make workers speed up and sweat. ☐ A yardstick to measure production quantity. Increasing output may or may not result in improving the productivity. ☐ An indicator of profitability. Under certain circumstances projects having low productivity can be profitable. ☐ The same as quality. An increase in productivity does not ensure better quality.

☐ Enable job enlargement, enrichment and satisfaction.

☐ Reward motivated efforts applied towards achieving the goals and objectives of the organization.

☐ Improve working conditions.

The above lists are in no way exhaustive, but contain some of the more recognizable benefits that can result from an effective productivity program.

Finally, on a national level, productivity improvement is reflected in improvements in the standard of living, reduced inflation and reduction of unemployment.

Step 2: Productivity Measurement

Productivity measurement involves the systematic selection of productivity measurement approaches to be used (productivity measures) and inputs and outputs for the company or department. These are explained below.

In measuring productivity, several factors must be considered, including the nature of the inputs and outputs, the complexity of the operations and the level on which productivity is being measured (company, department, individual, etc.). The following productivity measures incorporate these factors.

1.) *Overall ratio* (total factor productivity) is the ratio of total output to total input, or:

$$\frac{\text{Total Output}}{\text{Labor + Materials + Energy + Capital + Miscellaneous Input}}$$

This measures takes into account the use of all resources under management control and is therefore more suitable for assessing the productivity of a company or department.

An example of the total factor productivity of an electric utility company is shown in Table 2.

2.) *Indicator/outcomes* (objectivity measure) indicate a measure of achievement of an individual manager or a department at the end of a given time period. For example:

Productivity outcome desired = Satisfied customers, better quality, etc.

The degree of satisfaction of the customers may be monitored by the ratio of:

$$\frac{\text{Sales lost in a given period}}{\text{Number of customer complaints in same period}}$$

A gauge of the quality of a product or service may be customers' opinions about it.

Examples of *productivity indicators* are the number of employees per 1,000 customers served and the number of kWh sold per 100 employees.

3.) *Partial factor productivity ratio:* This measure can be used when the productivity of a single input factor is desired, such as the ratio of total output to total input or of total output to materials input. For example (figures used are from Table 2):

$$\text{Labor productivity ratio (Period 1)} = \frac{196,120}{21,250} = 9.22.$$

Material productivity ratio (Period 1) =

$$\frac{196,120}{74,240} = 2.64.$$

4.) *Individual measures/work standards:* This is the ratio of individual output to expected output. For example:

$$\frac{\text{Actual number of power meters read in a given period}}{\text{Number of meters scheduled to be read in same period}}$$

5.) *Time ratio/timeliness:* This is a measure of output relative to a deadline for completion of a project or effort. It can be used where it is impractical to quantify the inputs and/or outputs. For example:

$$\frac{\text{Number of projects actually completed within schedule and budget}}{\text{Number of projects scheduled to be completed}}$$

6.) *Management by objectives (MBO) measure:* This measure is used to assess the productivity of a department or individual. The limitation of this method is that it requires a well-structured approach to setting objectives, and usually it is not very effective where quantification of the input/output is not possible.
Example: The objective of the transmission planning department is "to redesign the power transmission line route between two southwestern cities to reduce mileage from 240 to 210 miles by April, 1983."

Inputs and outputs

We are concerned with the following:

Table 2: Total Factor Productivity of an Electric Company

	Base Period (1978)	Period 1 (1979)	Period 2 (1980)
Output (sale of electric power):	$166,820	$196,120	$215,030
Inputs:			
— Labor	21,250	23,650	27,540
— Material (fuel and supplies)	59,650	74,240	80,900
— Energy	150	190	180
— Capital (depreciation of equipment)	14,450	16,530	20,810
— Other (interest and taxes)	40,110	46,200	51,500
Total input:	$135,630	$160,910	$180,930
Total factor productivity ratio:	1.23	1.21	1.18

Note: All figures in thousands; 1979 and 1980 figures in constant 1978 dollars.

□ Types of outputs and inputs to be used and where to get them.
Examples:
—Output to be measured—power generated in kWh; data source—power operations department.
—Input—Amount of coal/gas used; data source—fuels management department.
□ Methods of measuring outputs and inputs.
Examples: Use of historical data, development of time standards, use of work sampling.
□ Converting outputs and inputs into a common denominator if they are measured in different units.
Example: The input factors of the total factor productivity ratio are labor (man-hours), energy (kWh), etc. These should be converted to a common denominator (dollars) before being added together to get the total amount of input required for calculating the ratio.

Selecting productivity measures

After the basic concept is understood, the next question is how do we select a measure? The three most common methods are:
□ Consultative approach.
□ Nominal group techniques (employee participation).
□ Multifactor productivity measurement.
These approaches are explained below.

Consultative approach

With the consultative approach, the type of productivity measure to be used is determined by management or a consultant.
Example: The manager of the power meter maintenance department may decide to measure the productivity of the department using two measures: the ratio of the number of meters repaired to total man-hours used and the ratio of actual expenditures of the department to amount of authorized budget.

Nominal group techniques

A group of employees discuss various productivity measures and select one they think is appropriate for their organization. Since the employees are involved in the decision-making process, there is a high probability that the productivity measure will succeed.
Example: The employees in the power meter maintenance department may decide that the productivity measure of their department should be the ratio of actual man-hours used for repairing a meter to the standard number of man-hours allowed for this repair.

Multifactor productivity measure

This type of method for determining productivity measure/ratio is based on the premise that profitability is

238

Figure 2: Productivity, Profitability and Price Recovery Ratio Trends

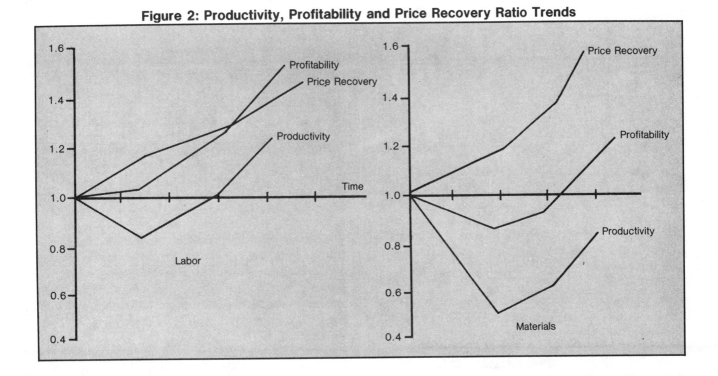

a function of productivity and price recovery (manipulation of the prices of products or services); that is, profitability ratio = productivity ratio × price recovery ratio.

By rearranging the terms we get:

$$\text{Productivity ratio} = \frac{\text{Profitability ratio}}{\text{Price recovery ratio}}$$

Such a ratio may be determined for several periods. This method requires simple arithmetic calculation. Among other facilities, the productivity center at Oklahoma State University has developed a computer model for this purpose. Interested readers may wish to contact this source for further information.

After selecting any combination of productivity measures from among the six described above, depending on what is needed in your individual situation, develop them for each key function and/or department.

Tabulating indicators

The next step is tabulating productivity ratio indicators from the base period to date.

Choose a base year and tabulate the productivity measures from the base year to the present time. If the multifactor productivity measurement is used, informa-

tion on sales, costs and profits by company or department will be needed.

Next, analyze the trends of the ratios and retain the documentation for use as input to the productivity evaluation to be carried out in Step 3.

As an example, some of the ratios calculated using the multifactor productivity measure are plotted and shown in Figure 2.

Point of View

The time has come to focus on the human element in the productivity equation. Innovation holds the answer here as well. What was effective 20 or even 10 years ago will not necessarily work today. Circumstances change. Attitudes change. Motivational plans such as incentive pay systems may be outmoded, at least in their present forms. In a society of increasing affluence for a greater number of people, survival of the individual is no longer the basic driving force.

James Skinner, President, SKF Industries Inc.

Point of View

Engineering and management imagination provide the only keys we have to compete more effectively in world markets and to restore economic strength and world leadership.

Robert L. Sproull, President, University of Rochester.

	Period 1	Period 2
Productivity Index	$\dfrac{1.21}{1.23} = 0.98$	$\dfrac{1.18}{1.23} = 0.95$

Step 3: Productivity Evaluation

The evaluation process is a transitory stage between measurement and planning. It produces the data and provides the tools which enable managers and supervisors to gain more direct control over their operations.

This process helps identify the need for prompt and positive action to correct poor performance or alter unfavorable trends. Without an effective evaluation process the program has no meaning, and instead of bringing rewarding profits it will merely become a useless and costly mass of paper.

The evaluation focuses on analyzing productivity ratios and identifying unproductive methods and performance by departments, units and individuals for the purpose of determining problems and opportunities for improvement.

The guidelines that follow provide the basis for the evaluation process.

Guidelines for productivity evaluation

1.) Calculate and compare the productivity indexes of two or more periods to detect trends. A productivity index =

$$\frac{\text{Productivity ratio of period 1}}{\text{Productivity ratio of base period}}, \text{etc.}$$

An example of the use of this index is as follows (figures used are from Table 2, Step 2):

If other measures, such as productivity indicator/outcomes, are used, compare the values of the selected periods with the values of the base period.

2.) If the trend is declining, conduct a management audit to identify problems and opportunities for improving basic management processes such as planning, organizing, controlling, etc.

There are two primary reasons for conducting a management audit. One is that identifying a decline in the numerical productivity ratio compared to previous periods does not automatically identify its causes. A manager still needs to know what is causing productivity to fall and what actions are needed to correct it. A management audit provides this information and helps identify opportunities for improvement.

Second, top management has traditionally treated productivity as a production problem, something to be addressed at the shop level by the foremen or supervisors. Increased productivity means effective utilization of *all* the resources of a company, and a management audit can reveal how top management is managing its fundamentals (planning, organizing, controlling, etc.).

The management audit process starts with getting a broad, strategic, integrated perspective of top management by assessing its overall effectiveness in setting and meeting goals and objectives. For example, are objectives clear, measurable and result-oriented? Do they help motivate employees? Are they subject to periodic evaluation?

The organizational structure must be evaluated to determine whether the present structure effectively maintains coordination between departments, promotes open communication among all levels of the organization, ensures balanced delegation of responsibility and authority, etc.

Organizational reporting systems should also be assessed to determine whether they provide the feedback necessary for managers to analyze the extent and causes of deviations from the plan. This analysis is crucial if managers are to know what measures are necessary for successful achievement of the desired objectives.

Another essential evaluation is that of supervisory skills, specifically in the following areas:

☐ Technical skills—use of functional techniques and productivity methods applicable to particular jobs.

☐ Work improvement skills—work methods, problem solving, decision making, operation planning, etc.

☐ Interaction skills—motivation, communication, leadership style, etc.

Other types of problems could exist, depending on the type of organization, and may need to be included in the supervisory evaluation. They include:

☐ Unequal distribution of work.

☐ Ineffective scheduling or dispatching of resources.

☐ Use of unproductive individuals.

☐ Failure to use proven, effective methods.

☐ Cumbersome work flow.

☐ Unnecessary use of overtime.

Finally, if applicable, individual employees' performance of identified tasks should be evaluated to determine the causes of poor performance and ways of improving that performance. Some common causes of poor performance are:

☐ Poor communication between employees and supervisor.

☐ Lack of understanding of how the employee's performance is measured.

☐ Improper use of tools and techniques.

A survey might be conducted to determine employees' attitudes regarding productivity and organization/management practices.

3.) Summarize the results of the evaluation and list the actions to be taken to correct problems. This written summary will be the input for the planning process, Step 4, and will become the basis upon which the productivity improvement plan will be developed.

Preparation of this summary requires that a manager assess the results of an evaluation and develop specific objectives which should result in productivity improvement.

For example, the results of a management audit of the plant engineering and operations department may indicate that due to the absence of a work management program, the manager and supervisors are not able to plan and schedule activities and tasks. This results in a backlog of work, unproductive delays and poor coordination of activities.

The objective, therefore, should be to develop a work management program. In the planning process, this objective is detailed in an action plan that identifies activities to be performed, scheduled completion dates and responsibilities of the personnel assigned.

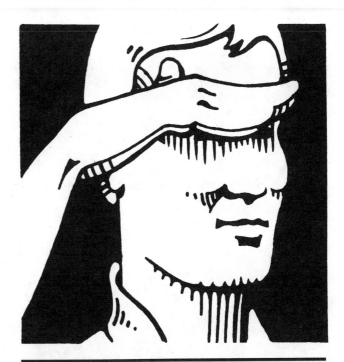

Step 4: Productivity Planning

The productivity measurement and evaluation steps must be followed by scheduled actions which are understood and supported by both management and employees. Many fine productivity improvement programs have failed to produce the results they were designed to achieve because of a lack of effective planning and follow-up.

For this reason, the manager and supervisor must document their plan and schedule a sequence of actions.

Guidelines for productivity planning

Review the results of the productivity evaluation—Step 3—to develop an action plan for the objective selected. For example, if the objective as identified in the evaluation step is to be accomplished in-house, an action plan similar to the one shown in Table 3 might be developed.

The following assumptions and constraint were used during development of the action plan.

Assumptions:

☐ The management services department of the company is willing to provide personnel to act as internal consultants.

☐ A sufficient number of personnel are available in the Plant operations and engineering department to be trained as analysts.

Constraint:

☐ A desired cost-to-benefit ratio has been set for

**Table 3: Action Plan for Developing and Implementing
a Work Management Program by December, 1983**

Action Steps/Tasks	Responsibility	Deliverables	Comp. Date
1. Select from existing staff personnel to be trained as analysts	Plant manager	Names of employees	Jan. 1983
2. Train analysts	Internal consultants	Analysts trained in the use of industrial engineering tools/techniques, such as work simplification, work standard development and process charting	March 1983
3. Supervisor/managerial seminar	Internal consultants	Manager/supervisor gets overview of the work management program	Feb. 1983
4. Work force orientation	Plant manager, internal consultants	Written goals and objectives of the program circulated to the employees involved	Feb. 1983
5. Analysts start documenting present plant's work activities	Analysts	Documentation of: a. What is done? b. What processes exist? c. How is workload distributed?	May 1983
6. Analysis of current work	Analysts	Identification of: a. How is the work done? b. Causes of work backlog and frequent work force delays Work improvement methods	July 1983
7. Development of work standards (use of selected and acceptable measurement techniques such as MTM, predetermined time standard, etc.)	Analysts	Work standards for identified work activities/tasks	Aug. 1983
8. Analysis of present work planning and scheduling procedures	Analysts	Recommended improved planning and scheduling procedures	Aug. 1983
9. Work management program implementation and performance control reporting	Analysts	A reporting system capable of: a. Collecting actual work data b. Comparing actual with standards c. Disseminating the performance results among concerned personnel d. Identifying activities/tasks for improving unfavorable performance and recommending positive correction actions	Sept. 1983

achievement under the program.

Success criteria for the work management program might include the following:

☐ Improvement in overall personnel utilization—Increase overall utilization of the power plant engineering and operations department's personnel from 75% to 90% in the six-month period following implementation of the work management program.

☐ Effective planning and scheduling of work—Use of 90-day planning horizon through the application of short-interval scheduling techniques.

☐ Elimination of backlog and unproductive delays—Reduce backlog from 15% of work orders to 5%.

☐ Improvement in overall operating results—Period 5 (1983) must show a 15% increase in the productivity index over period 2 (1980).

The most important point here is that a step-by-step plan of activities is needed to identify the tasks that are to be performed and deliverables from them and establish completion dates.

Step 5: Productivity Improvement

Productivity improvement can be broken down into the following objectives:

☐ Improving effectiveness ("doing right things").

☐ Improving efficiency ("working right").

☐ Improving quality of work life.

Improving effectiveness

Basically, effectiveness means achieving the desired objectives. To make this possible, the objectives must be:

☐ Clearly stated.

☐ Measurable.

☐ Consistent with the overall organization's goals.

☐ Timely.

☐ Specific.

☐ Reasonably obtainable.

☐ Result oriented.

Improving efficiency

A broad-based, traditional industrial engineering approach is the best approach to improving efficiency. This involves the auditing and analysis of work, work simplification, work standard development, use of technology and determination of resource requirements.

Audit/analysis of the work:

☐ Employees' work distribution and workload volume.

☐ Flow of work.

☐ Physical design of work place.

☐ Backlog if any.

☐ Methods of forecasting work volume.

☐ Methods used for scheduling the work.

☐ Procedure for and utilization of forms.

☐ Information retrieval system.

☐ Impact of internal and external environment on work assignment.

☐ Material and other resources usage.

Work simplification: For each step of the activity or process, consider the possibility of:

☐ Eliminating steps that are redundant or unessential.

☐ Combining steps.

☐ Simplifying steps.

Work standards development: For activities whose outputs can be quantified, work standards can be set by any of the following methods, depending upon the nature of the activities.:

☐ Time studies.

☐ Work sampling.

☐ Predetermined time standard.

☐ Activity logging.

Use of technology: Technology can contribute to improved efficiency. However, before any technology is introduced it is essential that work processes be simplified and measured. Otherwise, its use may result in unwitting commitment to inefficiency or automating of non-essential activities or functions.

Determining resource requirements: The human resources requirement is determined by:

☐ The work volume.

☐ Work standards.

☐ The number of people on the current payroll.

☐ The work deadline requirement (activities schedule).

Requirements for other resources (material, energy, etc.) are determined by the same factors.

Improving quality of work life

Improving the quality of work life, in general terms, means maintaining management commitment to and support of employees' involvement in the decision-making process. This involves such factors as:

☐ Participation in decision making through quality circles.

☐ Career opportunities.

☐ Training and development.

☐ Improvements in work climate conditions.

☐ Recognition and sense of accomplishment.

☐ Communication and trust between employees and management.

☐ Rewards and incentives.

☐ Safe physical environment.

☐ Team building.

It should be emphasized here that a quality-of-work-life program is not simply a "happiness" program, because a "happy" employee is not necessarily a productive employee.

Improving the quality of work life can lead to increased productivity. Individual guidelines and specific quality-of-work-life factors can be defined only after an on-sight review of the organization by a trained professional.

Step 6: Control Reporting

One of the principal purposes of a productivity improvement program is to establish and maintain a balance between volume of work and utilization of resources. The degree to which this balance is achieved reflects how well resources are being utilized and is, therefore, an indication of the overall effectiveness of the department or company.

The following concepts and techniques, when properly applied, will enable managers and supervisors to maintain a direct influence over the factors which determine this balance and, in turn, to optimize to a greater degree the utilization of the resources under their control.

Guidelines for control reporting

The control reporting guidelines are in two parts. The first deals with reporting progress on the implementation of the productivity program. This occurs *once* when the program is being implemented for the first time.

The second part provides ongoing information for use

Table 4: Progress of Program Implementation

Steps/Milestones	Deliverables
Creating productivity awareness	Summary of actions taken
Productivity measurement	Productivity measures selected
Productivity evaluation	Results of evaluation
Productivity planning	Copy of the planning abstract
Productivity improvement	Method selected for productivity improvement

in controlling the performance of the department or company after the implementation of the productivity program.

1.) Guidelines for monitoring the progress of implementation of the productivity program: Each productivity area manager should report progress in taking the steps or reaching the milestones indicated in Table 4.

2.) Guidelines for ongoing performance reporting:

☐ Establish means for reporting:

—Achievement of the desired performance criteria by the department/employees.

—How the department's employees spend their time.

—Volume of work processed by the employees.

—Amounts of other resources used in processing the work (such as material).

☐ Prepare productivity report after determining a productivity ratio (output/input) or productivity indicators as applicable.

☐ Prepare productivity/performance trend chart (on selected time period basis).

☐ Manager or supervisor prepares a report discussing significant problems and actions taken to serve as feedback on the productivity measurement, evaluation, planning and improvement steps.

☐ Manager prepares a report summarizing significant resources data and performance for sector executives.

For further reading:

Buehler, Vernon M., and Shetty, Ykrishna, *Productivity Improvements,* New York, AMACOM, 1981.

Gaither, Normal, *Production and Operations Management,* Hinsdale, IL, the Dryden Press, 1980.

Mali, Paul, *Improving Total Productivity,* New York, John Wiley and Sons, 1978.

Ranftl, Robert M., *Study Report—R&D Productivity,* Los Angeles, Robert Matthea Ronftl, 1978.

Riggs, James L., *Production Systems: Planning, Analysis, and Control,* Wiley Series in Management, 1980.

Ross, Joel E., *Productivity, People, and Profits,* Reston, VA, Reston Publishing Co., 1981.

245

Ross, Joel E., *Managing Productivity,* Reston, VA, 1977.
Vough, Clar F., and Asbel, Bernard, *Productivity: A Practical Program for Improving Efficiency,* New York, AMACOM, 1979. **IE**

 **Shafique Jamali** is an operations analyst in the management services department of the Public Service Co. of New Mexico. He is responsible for conducting analyses of the company's operations and providing technical consultations to staff members. Most recently, he has been involved in the development and implementation of productivity measurement and improvement programs. Jamali received a BS in mathematics and a mechanical engineering degree from the University of Jabalpur, India. He received his master's degree in industrial engineering and management from Oklahoma State University and an MBA from Arizona State. He is a senior member of IIE.

*Reprinted from **Industrial Engineering**, February 1983.*

Integrated Planning Approach Is Needed To Manage Productivity Improvement Efforts

By Kelvin F. Cross
Wang Laboratories Inc.

Although there is universal acceptance of the need to manage productivity, a systematic, comprehensive approach to productivity management remains elusive. Yet through planning, productivity can be managed. It requires a corporate commitment combined with a practical approach. When implemented with the same diligence as financial planning generally is, productivity planning simply becomes good management.

A practical and comprehensive productivity planning process can be outlined as follows:

Preparation:
☐ Document corporate philosophy and strategies.
☐ Establish productivity planning priorities.

Analysis:
☐ Document the present operation.
☐ Visualize the future based on the present operation assuming no improvements.
☐ Document applicable technologies, techniques and future trends.
☐ Develop a scenario for the future, assuming improvements are applied.

Implementation:
☐ Set goals and objectives to realize the scenario developed in the preceding step.

☐ Implement the plan and monitor progress.

Before a corporate productivity planning program is embarked on, care must be taken to place "productivity" in the proper perspective. This can be done by understanding productivity in the context of the entire organization and its relationship to sound business practice. Four principles establish this perspective:

1.) *Improved productivity is the result of any activity which reduces costs in the long run without compromising the corporate philosophy and the quality of the product in service or reducing gross revenue.*

Too often, productivity is viewed only as a means of reducing costs. There are other considerations as well: product quality, management philosophy, marketing strategy, social concerns and environmental concerns.

For example, it does no good to "improve productivity" by reducing product quality, threatening the work force, eliminating advertising, ignoring safety or polluting the environment. While such actions may reduce costs in the short term, they can decimate revenues in the long term.

2.) *Planning for productivity is an action, not a reaction.*

Many productivity improvement programs are implemented to "catch up" in a changing business environment. This "reaction" only perpetuates the already prevalent short-range mentality of management, which usually insists on a short-term payback for productivity improvement projects. Equal emphasis should be placed on planning for the future to avoid having to catch up later.

The lack of productivity programs is especially pronounced in companies experiencing periods of success and rapid growth. At this stage, profitability doesn't depend on productivity, since the company can sell what-

Productivity improvement depends on the successful integration of trends, techniques and technologies in the operation.

ever it makes at a profit.

This is precisely the time when ambitious effort should be devoted to productivity planning. Eventually the product growth curves will level off, whether due to the product life cycle, the economy or the competition.

A good example can be seen in the minicomputer industry. The combination of these factors has led to a situation in which margins are eroding and profitability now does depend very much on productivity.

3.) *Productivity is a universal objective, not an individual function.*

Productivity planning, although it requires a coordinator, is an organizational effort that should be directed at the whole organization. This maximizes the quantity and quality of improvement ideas and paves the way for their implementation.

4.) *Productivity goals must be specific, achievable and readily understood and supported by those charged with their implementation.*

Basing a productivity plan on realistic and detailed objectives will make the successful achievement of those objectives much easier. Specifying projects in detail and setting up comprehensive implementation schedules will minimize ambiguity and enable progress to be measured.

This level of detail will also enable productivity goals to be incorporated as an integral part of the company's financial, strategic and operating plans. Then productivity will be viewed as a serious concern and a basis for determining management performance.

Philosophy and strategy

Productivity planning requires documentation of the corporate philosophy and operating strategies. This document will serve as a guideline for evaluating productivity improvement ideas. Consideration will be given only to those ideas which support the documented policy.

This documentation of corporate philosophy and strategies should be developed by top management and/or the owners of the business. In his book, *Theory Z,* William Ouchi says: "A corporate philosophy must include (1) the objectives of the organization, (2) the operating procedures of the organization and (3) the constraints placed on the organization by its social and economic environment. It specifies not only the ends, but also means."

Although this document may consist of only a few pages, its usefulness in productivity planning can be substantial. For instance, assume a company's stated objective is to "manufacture customer tailored high-quality products."

The term "customer-tailored" implies a commitment to building the product to each customer's specifications. This simple declaration could eliminate from consideration any idea which would inhibit the flexibility of the manufacturing operation.

Similarly, a company with an explicit strategy of providing "an unexcelled level of customer service" will accept the relatively high cost of this effort. Any productivity measure that is implemented solely to reduce costs, and compromises the "big picture," will fail or cause the business to fail.

Successful productivity plans are responsive to the overall objectives of the business as well as the bottom line.

Setting priorities

Obviously, productivity planning for an entire organization is a major undertaking. To facilitate the initial success and effectiveness of the effort, it should be directed first at areas in which the benefits will be most apparent. Unfortunately, the areas that show the highest costs are not necessarily the areas where the most gains can be made.

Uncovering the areas with the greatest potential for improvement is a manageable task. It can be carried out according to the systematic procedure outlined below. A format will be established for displaying where both asset and operational costs are incurred.

In a matrix form this format will also display how much is spent. Productivity potential is then estimated and a percentage entered for each item in the matrix.

Finally, using current costs and productivity potential, the potential dollar savings can be calculated. These savings estimates will both highlight the areas with the greatest improvement potential and point to strategies for achieving that potential.

Designing a cost matrix

One way of identifying areas with the greatest potential for improvement through productivity planning is by looking at the costs which cut across organization lines. Similar functions and expenses may respond to the same productivity improvement techniques.

For example, a variety of functions, such as marketing, finance, research and development and production control, all utilize analysts and clerical personnel. Although each department is independent, the analyst and clerical personnel engage in similar tasks and follow similar procedures.

By identifying the costs and improvement potential of these personnel as a group, rather than by function or department, an area of major potential can become a priority for productivity planning.

Table 1 demonstrates two ways of categorizing a company's costs—in this case, a manufacturer. The vertical categories designate traditional profit centers or departments. The horizontal categories represent assets, functions or other expense areas which cut across organizational lines. In application, this matrix would be broken down to a finer level of detail.

The input for the categories will have to come from upper management, department heads, a financial analyst and a productivity specialist (an industrial engineer, organizational development specialist or internal consultant). The categories and their level of detail must be such that actual costs and the impact of productivity improvements are relatively easy to approximate. The idea is to get a rough estimate of where preliminary efforts should be directed, not to establish specific goals.

Enlisting the cooperation and support of management and staff at this early date may be crucial to the program's success. The manager of each department, utilizing its annual budget history, should be asked to report a cost figure for each of the items designated on the horizontal axis.

Depending on the detail of some of the items and their variance from budgeting categories, some assistance may be required from the finance department or productivity specialist. Assistance from these sources can also help ensure consistency throughout the organization.

In the example used in Table 1, the highest costs can be attributed to the hourly and direct labor groups, at $3.39 million and $3.35 million per year. The bulk of both of these figures is spent on hourly workers doing direct labor at a cost of $2.75 million per year. This is the primary area of expenditure, followed by the cost of analysis and clerical work at $2.05 million and direct material costs at $2.0 million.

Table 1: Annual Costs in $10,000s

Direct Costs:	Direct Material 200	Space	Energy	Hourly Labor	Material Handling	Analysis & Clerical	Engineers	Information Systems	Petty Cash		Total Cost
	Direct Labor	—	—	275	45	15	—	—	—		335
Factory Overhead	Maintenance Department	5	5	40	0	5	5	2	3		65
General Overhead:	R & D Dept.	5	10	4	1	25	50	10	5		110
	Finance Dept.	5	5	5	—	50	—	15	10		90
Sales Costs:	Marketing Department	5	5	5	—	50	—	15	10		90
	Sales Department	5	5	0	—	50	10	15	25		110
	Total Cost:	25	30	339	46	205	65	57	53		1,000

A list of the categories from across the top of the matrix should be presented to each manager and his or her staff. Working with them, a productivity specialist could generate potential improvement ideas for reducing the annual costs in each category. Sketchy ideas will suffice; therefore the process is well suited to a "brainstorming" session.

The improvement ideas and suggestions are likely to fall into two groups: technologies and techniques. The productivity specialist may want to initiate discussion by submitting some examples for discussion, such as:

Technologies:
☐ Robotics.
☐ Microcomputers.
☐ Conveyors.
☐ Software packages.
☐ Teleconferencing.

Techniques:
☐ Management by objectives.
☐ Quality circles.
☐ Profit sharing.
☐ Incentive pay.
☐ Organizational change.

The ideas generated must be evaluated to ensure that they are consistent with corporate philosophy and operating strategy. Some consideration should also be given to the practicality of the expense involved in their implementation. A list of ideas deemed feasible in each category can then be prepared.

A rough estimate of the cost reduction potential of each idea should then be calculated by the department manager and the productivity specialist. For each category, the improvement ideas should be summed and a total theoretical cost reduction percentage calculated. Again, sophisticated analysis is not necessary at this stage; an educated guess is sufficient.

After collecting similar percent-

ages from every department, the productivity specialist can develop a new table containing the best estimate of the productivity improvement potential for each department and category.

As an example, Table 2 illustrates that a major reduction could be achieved in the cost of direct labor for material handling. This cost reduction could come from conveyors, robotics or other mechanical handling systems.

To take another example, the finance department estimates that the money devoted to simple numbers crunching by analysts and clerks could be cut significantly through the use of personal desktop microcomputers. This, combined with a quality circles program, has the potential of reducing analytical and clerical costs by 50% and at the same time eliminating boring tasks.

By multiplying the annual costs in

Table 2: Rough Cut Costs

Direct Costs:		Direct Mat'l 20%	Space	Energy	Hourly Labor	Material Handling	Analysis & Clerical	Engineers	Information Systems	Petty Cash
	Direct Labor	—	—		20%	50%	20%	—	—	—
Factory Overhead:	Maintenance Department		10%	10%	10%	—	20%	0%	0%	0%
General Overhead:	R & D Department		0%	10%	0%	0%	50%	0%	10%	0%
	Finance Department		0%	10%	50%	—	50%		33%	10%
Sales Costs:	Marketing Department		0%	10%	20%	—	50%		33%	10%
	Sales Department		0%	10%	—	—	50%	0%	0%	20%

Table 1 by the estimated cost reduction percentages in Table 2, a cost avoidance matrix can be calculated. This third table would define in dollars the estimated costs which could be avoided if the ideas for improvement were in place. Categories with the most potential for productivity improvement would be highlighted on a completed matrix (see Table 3).

This approach will remove from further study high-cost categories and areas for which no ideas for improvements have been generated. Areas for which both great potential for cost savings and a rough idea of how to achieve that potential are indicated will be prioritized for productivity planning.

Table 3 shows that the greatest potential for cost reduction lies in the costs attributed to "analysis and clerical" work. Hourly labor, although shown to be a primary expense category by Table 1, has a productivity improvement potential that is second to those of both the analysis and clerical category and the direct labor group as a whole. Therefore, the category of analysis and clerical work should receive top priority for productivity planning.

Understanding the present

Once a category of function has been identified as having potential for productivity improvements, the task of detailed planning can begin. The best place to start is with the development of a comprehensive and accurate picture of the current operation. This serves three purposes:
☐ It helps ensure that productivity goals are based on a realistic understanding of the function or category.
☐ It develops a basis for comparison once productivity plans and improvements are implemented.
☐ It helps ensure that goals are specific and achievable.

A process flow chart accompanied by a data sheet is one of the best tools for documenting current operations in a manufacturing environment (see Figure 1). The chart serves to clarify the manufacturing operation and ensures that nothing is overlooked. Each step of the process under study is detailed, including inspections, transport steps, delays and storages.

Values can then be attached to each step denoting production capacity, number of people or machines, square footage requirements, etc. Other current factors such as performance, utilization, attendance, yields, equipment uptimes and rework volumes also should be recorded. Through these charts and figures an accurate picture of present operating conditions is developed.

It should be noted that this exercise is somewhat like taking a picture of a moving target: it's an accurate portrayal of only a single moment. In reality, a company does not stand still; there are always organizational shakeups and new systems and procedures being installed. At some point a limit must be established on what will be included in the picture. Typically, this will require that a deadline be set. The picture will then reflect the operation as of that specific date.

Five years forward

An effective technique for understanding the impact of continuing "as is" and highlighting the potential for improvement is that of envisioning the future operation based on today's operating conditions. The marketing department should be able to provide a five-year forecast for product demand, including new products.

Utilizing the "picture" of the current operation "as is" and the marketing forecast, the data sheets can be revised to reflect the resources necessary to achieve that forecast. This new picture will illustrate how the operation will look in five years assuming no improvement in productivity. It will serve as a "worst-case" scenario for use in establishing and measuring the potential impact of productivity improvements. It will also help ensure that a long-range perspective is maintained.

Although a manufacturing envi-

250

Table 3: Potential Cost Avoidance in $10,000s

Direct Costs:	Direct Mat'l 40.0	Space	Energy	Hourly Labor	Material Handling	Analysis & Clerical	Engineers	Information Systems	Petty Cash		Total Cost
	Direct Labor	—	—	55.0	22.5	3.0	—	—	—		80.5
Factory Overhead	Maintenance Department	.5	.5	4.0	—	1.0	—	—	—		6.0
General Overhead:	R & D Department	—	1.0	—	—	12.5	—	1.0	—		14.5
	Finance Department	—	.5	2.5	—	25.0	—	5.0	1		34.0
Sales Costs:	Marketing Departments	—	.5	1.0	—	25.0	—	5.0	1.0		32.5
	Sales Department	—	.5	—	—	25.0	—	—	5.0		30.5
	Total Cost:	.5	3.0	62.5	22.5	91.5	—	11.0	7.0		198.0

ronment is used in this example, the same technique could be applied to other productivity studies. A flow chart and other data could just as easily be created to illustrate energy usage throughout a building, information flow, customer service routines or even the tasks of analysts and clerical workers.

A five-year projection of resource requirements could then be done based on "no change" in these operations. Later a "gap analysis" comparing this worst case to a "best-case" scenario will enable productivity benefits to be quantified and specific goals established.

This five-year picture will be used to assess the impact of a productivity plan. Emphasis is placed on the far-sighted cost avoidance benefits of medium- to long-term productivity planning rather than on a short-term payback.

Planning the future

Research is the key to planning for the future. Suggested improvement ideas must be investigated. The factors and trends that will have an impact on productivity must be identified and their potential impact evaluated. This is manageable, since most productivity improvements being factored into a five-year productivity plan will incorporate techniques, technologies and trends that are currently documented.

Technology

The primary contributor to great strides in productivity will be technology.

John J. Connell states in a recent article: "The potential of the market-place and the forces of competition will result in the introduction of a bewildering array of machines with constantly expanding capabilities. One can attempt to manage these technologies in an intelligent way or one can let them pour in haphazardly."

To manage intelligently, a list should be made of all currently and soon-to-be-available technologies applicable to the area being studied. For each item on the list, detailed documentation of applicability and productivity potential can be collected from trade magazines, journal articles and recent books.

This list should include facts and figures on capabilities, tolerances, capacity, costs, advantages, disadvantages, etc. Realistic productivity goals can then be established based on the success of previous installations.

Techniques

The same procedure can also be followed for investigating various management techniques and assessing their improvement potential. Quality circles, employee participation and incentive systems require research into similar industries and applications.

Trends

It is necessary to identify not only the potential of various technologies and techniques, but also their compatibility with socioeconomic trends that may impact the enterprise, its products and the ways it conducts its business. Successful planning for the future requires an understanding of trends which may change the nature of business and the structure of the working environment.

Significant demographics and other trends are an important and necessary consideration in any productivity plan. For instance, changes in workers' attitudes, career planning and education level may require a new approach to job design. As career expectations rise, the work force becomes increasingly discontent with mundane, repetitive work. As management consultant Serge Birn put it: "We now have too few idiots to fill idiot-proof jobs."

This trend would indicate that the traditional industrial engineering practice of segmenting jobs may no longer be the most beneficial approach. Techniques which address the human factor would be more productive and compatible with this trend.

Through this analysis of technologies, techniques and trends a usable data base has been gathered. This information can be used to manage the future.

Managing the future

The future can be managed and controlled with an effective productivity plan which is aggressively incorporated into a business's operating and strategic plans. This requires specific, achievable, readily understood productivity goals that are supported by those charged with their implementation. Such goals, based on technologies and techniques that are compatible with future trends and the company's operating philosophy, should become an integral part of a company's operation.

Defining potential

Potential can be illustrated by developing a detailed scenario which illustrates in identifiable, concrete form the results of productivity planning. This new scenario, illustrating the improved operation in five years, also can be used to quantify the impact of productivity planning by comparing it with the picture of the operation assuming no improvement.

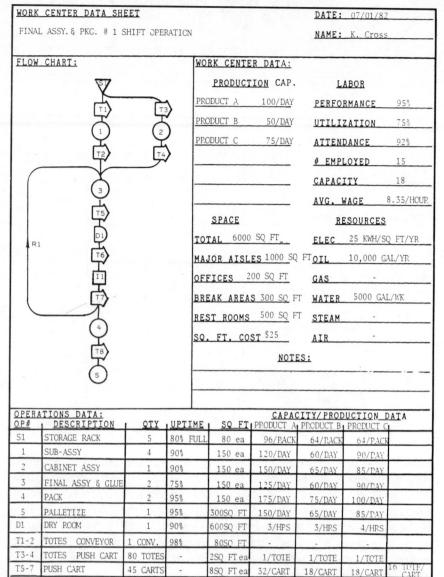

OP#	DESCRIPTION	QTY	UPTIME	SQ FT	PRODUCT A	PRODUCT B	PRODUCT C
S1	STORAGE RACK	5	80% FULL	80 ea	96/PACK	64/PACK	64/PACK
1	SUB-ASSY	4	90%	150 ea	120/DAY	60/DAY	90/DAY
2	CABINET ASSY	1	90%	150 ea	150/DAY	65/DAY	85/DAY
3	FINAL ASSY & GLUE	2	75%	150 ea	125/DAY	60/DAY	90/DAY
4	PACK	2	95%	150 ea	175/DAY	75/DAY	100/DAY
5	PALLETIZE	1	95%	300 SQ FT	150/DAY	65/DAY	85/DAY
D1	DRY ROOM	1	90%	600 SQ FT	3/HRS	3/HRS	4/HRS
T1-2	TOTES CONVEYOR	1 CONV.	98%	80 SQ FT	-	-	-
T3-4	TOTES PUSH CART	80 TOTES	-	2 SQ FT ea	1/TOTE	1/TOTE	1/TOTE
T5-7	PUSH CART	45 CARTS	-	8 SQ FT ea	32/CART	18/CART	18/CART 16 TOTE/CART
T8	CONVEYOR	1	98%	80 SQ FT	-	-	-
I1	INSPECTION	2	90%	150 ea	110/DAY	55/DAY	82/DAY
R1	REWORK	-	-	-	5%	8%	5%

Measuring productivity against the future, rather than the present, also serves to ensure a commitment to long-range thinking.

Two sets of data developed earlier will be used to construct this new scenario: the flow charts and other data that illustrate the operation five years ahead with no productivity improvement and the documentation of technologies, techniques and trends. When the two are combined, a new scenario will emerge that shows the potential of the operation in five years.

This new scenario must be developed in detail. Each step and aspect of the operation needs to be analyzed and factored to reflect the successful

application of the technologies and techniques to be employed.

As was stated earlier, practical productivity planning should specify achievable goals and document the means of achieving them. Once a detailed scenario for the future is prepared, an implementation schedule or project plan must be developed.

This schedule will establish benchmarks and target dates for the completion of tasks throughout the five-year planning period. Benchmarks will serve to promote a sense of accomplishment and can also be used to measure progress.

Annual objectives can then be set based on these benchmarks. Assuming the objectives are realistic and supported, commitment to their achievement should be total.

The company's financial, strategic and operating plans should also incorporate these objectives. Productivity would then attain the status it deserves with management. The more productivity is incorporated into business planning, the more seriously it will be taken, and the more likely the scenario for the future is to be achieved.

Like any business planning, productivity planning cannot be effective if it is done once every five years. It requires constant modification and reevaluation. In addition, adjustments will have to be made to adapt to changing business conditions.

Productivity management

In addition to management commitment, productivity planning requires coordination by a central group of productivity specialists and the participation and cooperation of the employees. It is not a simple function which can be delegated to or made the mandate of one individual or group.

The approach to productivity planning presented here is a beginning, a method for integrating productivity into an organization's planning process. After a time, productivity planning should become less rigid and more routine.

As with the budgeting process, it should be middle management's responsibility to pull the plan together. This includes handling the details and documentation related to how and when productivity improvements will take place. But productivity is not a solitary effort.

The employee

Employee participation in the productivity planning process can go a long way toward ensuring its success. Planning for productivity requires creative, fresh ideas and a long-range orientation. To generate these fresh ideas and a commitment to the future, emphasis should be placed on employee involvement.

Howard Tuttle points out in a recent article that enlisting employee participation requires a non-threatening approach, and suggests that, as one such approach, managers ask employees, "What can management do to help you get your job done more easily?"

Productivity specialists

Productivity specialists are needed to coordinate, control and support the effort. Depending on the size of the company, productivity specialists may range from a full-time staff under a vice president of productivity to an industrial engineer who devotes part time to the task. In any case, the productivity specialist should have a strong people orientation balanced by a sound understanding of business practices. He or she should command the respect of others and have the explicit support of top management.

A productivity specialist should not be over-specialized, but rather should be a generalist able to interface with any aspect of the business.

Controlling the planning process means not only setting up the procedures, but also monitoring the progress of the plan. Supporting management by bringing out ideas, evaluating their potential and providing recommendations would be a key function of the productivity special-

ist as well.

A corporate commitment to productivity planning, combined with a practical approach, can ensure future profitability. Productivity planning improves decision making, increases future operational effectiveness and uncovers current inefficiencies.

When this planning process is given the same emphasis as budgeting normally receives, it can be an effective tool for a serious attack on the productivity problem. As the process is implemented and goals are established, they can be incorporated into the company's strategic, financial and operating plans.. Productivity planning successfully integrated into a company's operation simply becomes good management.

For further reading:

Birn, Sirge A., "Productivity Slump: Broader IEs Needed," *Industrial Engineering,* February, 1979.

Connell, John J., "Managing Human Factors in the Automated Office," *Modern Office Procedures,* March, 1982.

Cross, Kelvin F., "Manufacturing Planning: Key to Improving Productivity," *Industrial Engineering,* May, 1981.

Drucker, Peter, *Managing in Turbulent Times,* Harper and Row, New York, NY, 1980.

Mundel, Marvin E. (editor), "Productivity: A Series from *Industrial Engineering,"* AIIE, 1978.

Ouchi, William G., *Theory Z,* Addison-Wesley, Reading, MA, 1981.

Scott, Sir Walter, "Profit Improvement Procedures," Section 13, Chapter 5, in H. B. Maynard (editor), *Industrial Engineering Handbook,* McGraw-Hill Book Co., New York, NY, 1971.

Toffler, Alvin, *The Third Wave,* Morrow, New York, NY, 1980. **IE**

Kelvin F. Cross is a senior (facilities) planning analyst for Wang Laboratories Inc. He is primarily responsible for developing and implementing a systems approach to manufacturing planning. Cross holds a master's degree in industrial engineering and an undergraduate degree in psychology from the University of Massachusetts, Amherst. He is a member of the Boston chapter of the Institute of Industrial Engineers and of Alpha Pi Mu.

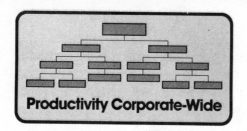

Productivity Corporate-Wide

*Reprinted from **Industrial Engineering,** February 1983.*

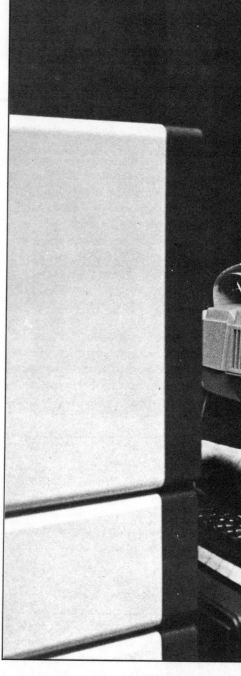

Program Enables Workers To Present Ideas On Investments For Productivity

By Ronald L. Bible
Hospital Corp. of America

When we think of capital investments to improve productivity, what usually come to mind are large expenditures to rebuild manufacturing equipment or purchase new facilities. Often, smaller projects that would be valuable methods of improving performance are ignored or never brought to the surface for implementation.

This usually happens because working level employees have no vehicle for presenting their ideas to management in a logical, analytical format.

The Department of Defense has developed a productivity enhancing capital investment program designed to take advantage of such projects. In the U.S. Air Force, it is referred to as the Fast Payback Capital Investment Program or FASCAP. The program is intended to put to use ideas that save on labor, energy, materials, facilities and other cost factors that can have a significant impact on the production of required output.

The FASCAP program basically involves money set aside at headquarters level. Units submit requests for these funds for use in purchasing equipment that is readily available "off-the-shelf" and will reduce operating and support costs.

The request itself is in the form of a simple economic analysis that is formatted so that employees at all levels of the organization can fill it out and submit a good idea for possible funding.

The proposal must meet certain criteria to be funded. The equipment cost per unit of equipment cannot be less than $3,000 or greater than $100,000. It must produce enough tangible savings in the manpower, material/energy, supplies, maintenance and/or "other" category over a two-year period to amortize the acquisition cost. Two examples of savings in the "other" category are documented overtime savings and cancellation of an outside contract to accomplish the work.

Leased equipment can be purchased if it was leased with the purpose of verifying its potential for productivity gains or if productivity benefits are realized in addition to reduced lease costs. Leased equipment must have 80% or more of the economic life of similar new equipment. This prevents purchase of equipment that won't last long enough to show substantial productivity gains.

Other rules preclude the purchase of equipment that has already been funded through normal budgeting channels and prohibit the establishment of an in-house capability that could be obtained more readily and economically through a commercial contract.

Projects are usually given processing priority based on the date they are received. If a backlog occurs or funds set aside are nearly depleted, the projects are prioritized in the following order: shortest payback period, number of whole manpower positions saved, amount of overtime costs saved, amount of consumable materials saved and other tangible savings.

Management analyst Dave Albers operates one of 15 minicomputers purchased under two different FASCAP projects.

Funds are requested by filling out a form specifically designed for this purpose. The form identifies the project and gives specific information on the unit that is requesting funds. It includes a clear, concise description of the project in nontechnical terms. The type of equipment to be acquired is stated as well as how it will be used and how it will improve productivity.

A detailed justification summarizes the expected costs and benefits, any significant changes from present processes or operations and the anticipated results. Table 1 is an excerpt of the request form showing how the economic analysis is derived.

A published regulation governs completion of this form and gives details on how it should be accomplished. By following the guidelines step by step, anyone, at any level, can submit a request for funds to improve productivity.

Acquisition costs as listed on the form include the cost of the equipment and applicable shipping and installation costs charged by the vendor. Any sales credits offered by the vendor should be deducted.

Installation costs are charges for installation not performed by the vendor and not included in acquisition costs, including any modification or restoration costs for existing facilities.

Transportation costs are also identified if they aren't included under acquisition costs. Other costs include those for spare parts, start-up kits, training, etc., needed to make the equipment operational.

Cost savings derived according to Table 1 quantify the project's two-year savings in dollars. These are "hard" tangible savings that can be quantified and documented, with an audit trail, over the period of time allotted for payback. These requirements can be waived on a by-exception basis.

Manpower costs saved include basic salary, bonuses, holiday pay, fringe benefits, retirement, shift differentials and the like. Material/energy savings include all expendable equipment, fuels, supplies, utilities and materials. If any material or item of equipment is turned in as a result of the project, credit is given as a one-time savings.

Maintenance savings are the expected value of labor and parts, not already accounted for, that would otherwise be expended to maintain equipment. Any warranty is included under one-time savings. For example, if the warranty is for nine months and the contract maintenance agreement cost is $100 per month, the warranty savings is $900.

Other costs include overtime and other relevant costs not included in the previous categories. These numbers represent the result of a lot of data gathering and detailed calculations. The data must be included as an attachment to show the source of all calculations and to serve as back-up for audit purposes.

To compute the payback period, the costs and savings are categorized as one-time, annual old or annual new. Annual costs are used to find the recurring cost difference between the present method and the proposed method using the new equipment.

The result is the annual savings, which is multiplied by two and added to the one-time savings for the total savings over the two-year payback period.

Life cycle savings are also computed to give a better idea of the total value of a project. Economic life in years is used, and is based on the expected physical, technological or "mission" life of the equipment, whichever is shortest.

Since the Defense Department is a non-profit organization, payback period is used as the primary means of evaluating projects, and the forms are constructed specifically for this type of analysis. The format could easily be modified to show deprecia-

Table 1: Request for Fast Payback Capital Investment (FASCAP) Funds

Investment Costs:

Acquisition A	Installation B	Transportation C	Other D	Total E

Cost Savings:

Category of Savings A	Onetime Savings B	Annual Cost of Old System C	Annual Cost of New System D	Annual Savings (C-D) E	Two-year Total Savings ((2 × E) +B) F
(1) Manpower					
(2) Material/Energy					
(3) Supply Credit for Equipment					
(4) Maintenance					
(5) Other (Specify)					
(6) Totals					

Payback Period—To qualify for FASCAP, payback period must be no more than 24 months computed as follows:

$$\frac{\text{Total investment costs}}{\text{Two year total savings}} = \underline{\hspace{1cm}} \times 24 = \underline{\hspace{1cm}} \text{ Payback period (Months)}$$

Life Cycle Savings:

$$\frac{\text{Economic Life}}{\text{of Equipment}} \times \frac{\text{Net Annual}}{\text{Savings}} + \frac{\text{One-Time}}{\text{Savings}} = (\underline{\hspace{1cm}} \text{ Life Cycle Savings})$$

Table 2: Report of Fast Payback Capital Investment (FASCAP) Expenditures and Benefits

Investment Expenditures Summary:

Item of Expense A	Authorized Funding B	Total Expenditures C	Difference (B-C) D
(1) Acquisition			
(2) Installation			
(3) Transportation			
(4) Other			
(5) Totals			

A. Original Estimate of 2-Year Savings: _____

B. Actual Savings to Date:

Item of Cost Savings A	Total of one-time Savings (First Report Only) B	Old System Cost C	New System Cost D	Savings This Report Period (C)-(D) + (B) E	Previous Report's Total Savings F	Cumulative Savings to Date (E) + (F) G
(1) Manpower						
(2) Material/Energy						
(3) Supply Credit for Equipment						
(4) Maintenance						
(5) Other						
(6) Totals						

C. Revised Estimate of 2-Year Savings:

$$\frac{\text{Cumulative Savings—One-Time Savings}^*}{\text{No. Months Reported to Date}} \times 24 = \text{_____} + \frac{\text{One Time}}{\text{Savings}^*} = \text{_____}$$

*Obtain from first 180-Day Report

D. Difference between original and revised estimates: _____

Amortization:

$$\frac{\text{Total Actual Savings}}{\text{To Date}} \text{_____} - \frac{\text{Total Investment Expenditures}}{\text{To Date}} \text{_____} = \frac{\text{Net Savings}}{\text{To Date}^*} \text{_____}^*$$

*If "Net Savings to Date" is "zero" or a "positive" number, project has been amortized, and no further reporting is required. If not, submit another report in 180 days.

Payback Period (in months):
  A. Original Estimate: _____
  B. Revised Estimate: _____ Compute as follows:

$$\frac{\text{Total Investment Expenditures}}{\text{Revised Estimate of 2-Year Savings}} \times 24 = \text{(Revised Payback Period)}$$

Life Cycle Savings:
  A. Original Estimate: _____
  B. Revised Estimate: _____ Compute as follows:

$$\frac{\text{Revised Estimate of 2-Year Savings—One Time Savings}^*}{2} \times \text{(Economic Life)} + \text{(One Time Savings)} = \text{(Revised Estimate)}$$

*Obtain from first 180-day report.

Figure 1: Program for Analyzing FASCAP Requests

```
PROGRAM NAME - FASCAPROJ
10 REM      FASCAP REQUEST
20 ? CHR$(12)
30 ? : ? "ENTER"
40 INPUT "ACQUISITION, INSTALLATION, TRANSPORTATION, OTHER";A,I,TR,O
50 ? : ? "ENTER"
60 INPUT "MANPOWER SAVINGS (i.e., ONE TIME, ANNUAL OLD, and ANNUAL NEW)";M1,M2,M3
70 ? : ? "ENTER"
80 INPUT "MAT/ENERGY SAVINGS (i.e., ONE TIME, ANNUAL OLD, and ANNUAL NEW)";ME(1),ME(2),ME(3)
90 ? : ? "ENTER"
100 INPUT "SUPPLY SAVINGS (i.e., ONE TIME, ANNUAL OLD, and ANNUAL NEW)";S1,S2,S3
110 ? : ? "ENTER"
120 INPUT "MAINT. SAVINGS (i.e., ONE TIME, ANNUAL OLD, and ANNUAL NEW)";MA(1),MA(2),MA(3)
130 ? : ? "ENTER"
140 INPUT "OTHER SAVINGS (i.e., ONE TIME, ANNUAL OLD, and ANNUAL NEW)";OS(1),OS(2),OS(3)
150 ? : ? "ENTER"
160 INPUT "ECONOMIC LIFE (YEARS)";EL
170 T=A+I+TR+O
180 AS(1)=M2-M3
190 AS(2)=ME(2)-ME(3)
200 AS(3)=S2-S3
210 AS(4)=MA(2)-MA(3)
220 AS(5)=OS(2)-OS(3)
230 MS(2)=2*AS(1)+M1
240 ME(4)=2*AS(2)+ME(1)
250 SS(2)=2*AS(3)+S1
260 MA(4)=2*AS(4)+MA(1)
270 OS(4)=2*AS(5)+OS(1)
280 TiT=M1+ME(1)+S1+MA(1)+OS(1)
290 TWO=M2+ME(2)+S2+MA(2)+OS(2)
300 TNN=M3+ME(3)+S3+MA(3)+OS(3)
310 TA=AS(1)+AS(2)+AS(3)+AS(4)+AS(5)
320 TS(2)=MS(2)+ME(4)+SS(2)+MA(4)+OS(4)
330 TS(1)=2*(TWO-TNN)+TiT
340 PP=TiT/TS(1)*24
350 LC=EL*TA+TiT
360 ? : INPUT "DO YOU NEED A PAPER COPY ?(Y or N)";A$
370 IF A$="Y" THEN 380 ELSE 400
380 ASSIGN #CONOUT TO "$lo"
390 ? : ? TAB(35),"FASCAP REQUEST" : ? : ?
400 ? " ACQUISITION  INSTALL  TRANSPORT  OTHER  TOTAL" : ? : ? A,I,TR,O,T
410 ? : ?
420 ? " CATEGORY  ONETIME  ANNUAL OLD  ANNUAL NEW  ANNUAL SAV  TWO YEAR SAV"
430 ?
440 ? " MANPOWER   ";M1,M2,M3,AS(1),MS(2)
450 ?
460 ? " MAT/ENERGY ";ME(1),ME(2),ME(3),AS(2),ME(4)
470 ?
480 ? " SUPPLY CREDIT";S1,S2,S3,AS(3),SS(2)
490 ?
500 ? " MAINTENANCE ";MA(1),MA(2),MA(3),AS(4),MA(4)
510 ?
520 ? " OTHER      ";OS(1),OS(2),OS(3),AS(5),OS(4)
530 ?
540 ? " TOTALS     ";TiT,TWO,TNN,TA,TS(2)
550 ?
560 ? " PAYBACK PERIOD    ";PP;"MONTHS"
570 ?
580 ? " ECONOMIC LIFE    ";EL;"YEARS"
590 ? : ? " LIFE CYCLE SAVINGS  $";LC
600 RELEASE #CONOUT
610 ? : INPUT "DO YOU WANT TO RUN AGAIN(Y OR N)";QS
620 IF QS="Y" THEN 10
630 IF QS="N" THEN CHAIN "FAS*PIF"
640 ? CHR$(12) : ? TAB(30);"INVALID RESPONSE" : ? : ? : ? : ? : ? : ? : ? : ? : ?
650 FOR I=1 TO 1000
660 REM      PAUSE
670 NEXT I
680 ? CHR$(12) : GOTO 610
690 END


PROGRAM NAME - FASCAP180
10 REM      FASCAP REPORT OF EXPENDITURES AND BENEFITS
20 ? CHR$(12)
30 ?
40 ? "ENTER" : INPUT "Acquisition Expense (i.e., authorized, total to date)";A,T
50 ? : ? "ENTER"
60 INPUT "Installation Expense (i.e., authorized, total to date)";AI,T4
70 ? : ? "ENTER"
80 INPUT "Transportation Expense (i.e., authorized, total to date)";A2,T2
90 ? : ? "ENTER"
100 INPUT "Other Expense (i.e., authorized, total to date)";A3,T3
110 ? : ? "ENTER"
120 INPUT "Original Estimate of 2 yr Savings";OE
130 ?
140 INPUT "Is this the first report? (Y or N)";B$
150 IF B$="Y" THEN 170 ELSE 160
160 M=0 : ME=0 : S=0 : HA=0 : O=0 : GOTO 190
170 ? : ? "ENTER"
180 INPUT "One Time Savings (i.e., manpower, material/energy, supply credit, maintenance, other)";M,ME,S,MA,O
190 ? : ? "ENTER"
200 INPUT "Old System Savings (i.e., manpower, material/energy, supply credit, maintenance, other)";M1,ME(1),S1,MA(1),O1
210 ? : ? "ENTER"
220 INPUT "New System Savings (i.e., manpower, material/energy, supply credit, maintenance, other)";M2,ME(2),S2,MA(2),O2
230 IF B$="Y" THEN 260 ELSE 240
240 ? : ? "ENTER"
250 INPUT "Previous Report's Total Savings (i.e., manpower, material/energy, supply credit, maintenance, other)";M3,ME(3),S3,MA(3),O3
260 ? : ? "ENTER"
270 INPUT "No. Months Reported to date";MN
280 ? : ? "ENTER"
290 INPUT "Economic Life (Years)";EL
300 ? : ? "ENTER"
310 INPUT "One Time Savings from First Report";TT
320 ?
330 AD=A-T
340 TD=A1-T4
350 TD=A2-T2
360 OD=A3-T3
370 TA=A+A1+A2+A3
380 TE=T+T4+T2+T3
390 DT=TA-TE
400 SA=(M1-M2)+M
410 SA(1)=(ME(1)-ME(2))+ME
420 SA(2)=(S1-S2)+S
430 SA(3)=(MA(1)-MA(2))+MA
440 SA(4)=(O1-O2)+O
450 SA=SA+M3
460 CS(1)=SA(1)+ME(3)
470 CS(2)=SA(2)+S3
480 CS(3)=SA(3)+MA(3)
490 CS(4)=SA(4)+O3
500 TL=M+ME+S+MA+O
510 TL=M1+ME(1)+S1+MA(1)+O1
520 TN=M2+ME(2)+S2+MA(2)+O2
530 TS=(TL-TN)+TT
540 TP=M3+ME(3)+S3+MA(3)+O3
550 CT=TS+TP
560 RP=((CT-TT)/MN)*24
570 RE=AB+TT
580 DI=RE-OE
590 AM=CT-TE
600 RP=(TE/RE)*24
610 LC=((RE-TE)/2)*EL+TT
620 INPUT "Do you need a paper copy? (Y or N)";A$
630 ?
640 IF A$="Y" THEN 650 ELSE 680
650 ASSIGN #CONOUT TO "$lo"
660 ? : ? TAB(35);"FASCAP 180 DAY REPORT"
670 ? : ?
680 ? "Item of Expense","Auth Fund","Total Exp","Dif"
690 ?
700 ? "Acquisition    ",A,T,AD
710 ?
720 ? "Installation   ",AI,T4,ID
730 ?
740 ? "Transportation ",A2,T2,TD
750 ?
760 ? "Other          ",A3,T3,OD
770 ?
780 ? "Totals         ",TA,TE,DT
790 ?
800 ? "Original Estimate of 2 yr Savings   $";OE
810 ? : ? : ?
820 ? "Item","1 Time","Old Sys","New Sys","This Period","Prev Sav","Cum Sav"
830 ?
840 ? "Manpower  ",M,M1,M2,SA,M3,CS
850 ?
860 ? "Mat/Energy  ",ME,ME(1),ME(2),SA(1),ME(3),CS(1)
870 ?
880 ? "Supply Credit",S,S1,S2,SA(2),S3,CS(2)
890 ?
900 ? "Maintenance  ",MA,MA(1),MA(2),SA(3),MA(3),CS(3)
910 ?
920 ? "Other   ",O,O1,O2,SA(4),O3,CS(4)
930 ?
940 ? "Totals    ",TL,TL,TN,TS,TP,CT
950 ?
960 ? "Revised 2 Yr Savings  $";RE;"    Difference (Revised - Original)   $";DI
970 ?
980 ? "No. of Months Reported  ";MN;"Months"
990 ?
1000 ? "Net Savings    $";AM
1010 ?
1020 ? "Revised Payback   ";RP;"Months"
1030 ?
1040 ? "Economic Life   ";EL;" Yrs"
1050 ?
1060 ? "Revised Life Savings  $";LC
1070 ?
1080 RELEASE #CONOUT
1090 INPUT "Do you want to run again? (Y or N)";A$
1100 IF A$="Y" THEN 10
1110 IF A$="N" THEN CHAIN "FAS*PIF"
1120 ? CHR$(12) : ? TAB(30);"INVALID RESPONSE" : ? : ? : ? : ? : ? : ? : ? : ? : ?
1130 FOR I=1 TO 1000
1140 REM      PAUSE
1150 NEXT I
1160 ? CHR$(12) : GOTO 1090
1170 END
```

tion, taxes, terminal value and other factors that affect profitability.

The modified procedure could then be used to compute payback, average rate of return, net present value, return on investment or any other desired qualifying criteria based on the company's objectives.

Although these additional computations may be a little more complex, they should not prevent the nonfinancial worker from submitting a good project for funding consideration. Help in completing the calculations could be obtained from the financial staff.

No program is complete without an accounting system to track implementation of the project and ensure that it does what it was supposed to do. This is accomplished by means of semiannual reports, the first of which is due six months from the operational date of the project. The regulation dealing with the project details how to complete a form for this report. Table 2 gives the details of the numerical analysis. Again, it is based on the costs and benefits derived from the project.

As an incentive to participate in productivity enhancement, workers are given the opportunity to share in the savings. They can submit projects through the suggestion program and recommend that they be funded by FASCAP. If they are adopted and produce tangible savings, the suggestors may be awarded a percentage of the savings.

FASCAP has worked very well within the existing organizational structure. No new office or function was created to administer the program. It is a simple, straightforward procedure that is easy to use at all levels.

The program has been successful since its inception. Since it was started in 1977, more than 450 projects have been funded. By the end of fiscal year 1981, total outlays of slightly more than $20 million had produced two-year savings greater than $36 million and life-cycle savings in excess of $143 million.

Among the items of equipment that have been purchased under FASCAP are some earth movers equipped with telescopic arms with buckets on the ends of them. These machines, which were proposed for purchase by civil engineers involved in earth-moving projects, have produced considerable labor savings by allowing mechanized earth removal in areas formerly inaccessible to machines where crews of workers had to be used in a very labor-intensive type operation.

Another FASCAP project involved the procurement of aerospace ground electronic equipment for powering planes sitting on the ground. In contrast to the portable units in use previously, the new units are set into the concrete ramp on which the plane is parked. Wear and tear has been reduced, and equipment life lengthened, with the stationary units, and energy is saved as well.

Some 300 word processors and 15 minicomputers have also been purchased under the FASCAP program following studies undertaken to determine the feasibility and economic advantages (or disadvantages) of their use.

It is important to point out that these are projects that, in all probability, would never have been implemented without some means of presenting them to management for consideration. Thanks to FASCAP, they are opportunities that were not lost.

The program used by Military Airlift Command to analyze FASCAP requests and reports is shown in Figure 1. It is written in C-Basic as used on the COMPUCORP 675 minicomputer. **IE**

Ronald L. Bible is a consulting management engineer with the Hospital Corp. of America's western region headquarters in Dallas. His duties include implementing productivity reporting systems and quality control systems. When the article was written, Bible was a productivity principal for military airlift command in the U.S. Air Force. In that position he was responsible for the conduct of a productivity enhancement program in an 88,000-person organization. Bible received his BA from Colorado State University and an MBA from Southern Illinois University. He is a member of IIE and the Society for the Advancement of Management.

About the editor. *Jerry L. Hamlin's career spans two decades. He served with the U.S. Air Force, and has held a variety of positions with General Motors, Eastman Kodak, and Cities Service Company. His career has encompassed the positions of chief industrial engineer, senior technical advisor, project manager, productivity consulting manager, vice president of supply, and related corporate staff appointments. He holds a Ph.D. in business management from Columbia Pacific University and an M.S. in industrial engineering from the University of Tennessee.*

Dr. Hamlin teaches productivity management as an adjunct professor at Oral Roberts University in Tulsa. For two years he was an associate at the American Productivity Center in Houston, where he did productivity research. A past division director and a senior member of IIE, he is presently chairman of IIE's Special Productivity Projects Committee.

DATE DUE			